UNION INTERNATIONALE DES SCIENCES PRÉHISTORIQUES ET PROTOHISTORIQUES
INTERNATIONAL UNION FOR PREHISTORIC AND PROTOHISTORIC SCIENCES

PROCEEDINGS OF THE XV WORLD CONGRESS
(LISBON, 4-9 SEPTEMBER 2006)
ACTES DU XV CONGRÈS MONDIAL (LISBONNE, 4-9 SEPTEMBRE 2006)

Series Editor: Luiz Oosterbeek

VOL. 14

Session C54

Prés du bord d'un abri:
Les histories, théories et méthodes
de recherches sur les abris sous roche

On Shelter's Ledge:
Histories, Theories and Methods
of Rockshelter Research

Edited by

Marcel Kornfeld, Sergey Vasil'ev, and Laura Miotti

BAR International Series
2007

This title published by

Archaeopress
Publishers of British Archaeological Reports
Gordon House
276 Banbury Road
Oxford OX2 7ED
England
bar@archaeopress.com
www.archaeopress.com

BAR S

Proceedings of the XV world Congress of the International Union for Prehistoric and Protohistoric Sciences
Actes du XV Congrès mondial de l'Union Internationale des Sciences Préhistoriques et Protohistoriques

Outgoing President : Vítor Oliveira Jorge
Outgoing Secretary General: Jean Bourgeois
Congress Secretary General : Luiz Oosterbeek (Series Editor)
Incoming President: Pedro Ignacio Shmitz
Incoming Secretary General: Luiz Oosterbeek

*Prés du bord d'un abri: Les histories, théories et méthodes de recherches sur les abris sous roche /
On Shelter's Ledge: Histories, Theories and Methods Of Rockshelter Research*

© UISPP / IUPPS and authors 2007

ISBN 978 1 4073

Signed papers are the responsibility of their authors alone.
Les texts signés sont de la seule responsabilité de ses auteurs.

Contacts :
Secretary of U.I.S.P.P. – International Union for Prehistoric and Protohistoric Sciences
Instituto Politécnico de Tomar, Av. Dr. Cândido Madureira 13, 2300 TOMAR
Email: uispp@ipt.pt
www.uispp.ipt.pt

Printed in England by Chalvington Digital

All BAR titles are available from:

Hadrian Books Ltd
122 Banbury Road
Oxford
OX2 7BP
England
bar@hadrianbooks.co.uk

The current BAR catalogue with details of all titles in print, prices and means of payment is available free from Hadrian Books or may be downloaded from www.archaeopress.com

Note of the series editor

The present volume is part of a series of proceedings of the XV world congress of the International Union for Prehistoric and Protohistoric Sciences (UISPP / IUPPS), held in September 2006, in Lisbon.

The Union is the international organization that represents the prehistoric and protohistoric research, involving thousands of archaeologists from all over the world. It holds a major congress every five years, to present a "state of the art" in its various domains. It also includes a series of scientific commissions that pursue the Union's goals in the various specialities, in between congresses. Aiming at promoting a multidisciplinary approach to prehistory, it has several regional or thematic associations as affiliates, and on its turn it is a member of the International Council for Philosophy and Human Sciences (an organism supported by UNESCO).

Over 2500 authors have contributed to c. 1500 papers presented in 101 sessions during the XVth world Congress of UISPP, held under the organisation of the Polytechnic Institute of Tomar. 25% of these papers dealt with Palaeolithic societies, and an extra 5% were related to Human evolution and environmental adaptations. The sessions on the origins and spread of hominids, on the origins of modern humans in Europe and on the middle / upper Palaeolithic transition, attracted the largest number of contributions. The papers on Post-Palaeolithic contexts were 22% of the total, with those focusing in the early farmers and metallurgists corresponding to 12,5%. Among these, the largest session was focused on prehistoric mounds across the world. The remaining sessions crossed these chronological boundaries, and within them were most represented the regional studies (14%), the prehistoric art papers (12%) and the technological studies (mostly on lithics – 10%).

The Congress staged the participation of many other international organisations (such as IFRAO, INQUA, WAC, CAA or HERITY) stressing the value of IUPPS as the common ground representative of prehistoric and protohistoric research. It also served for a relevant renewal of the Union: the fact that more than 50% of the sessions were organised by younger scholars, and the support of 150 volunteers (with the support of the European Forum of Heritage Organisations) were in line with the renewal of the Permanent Council (40 new members) and of the Executive Committee (5 new members). Several Scientific Commissions were also established.

Finally, the Congress decided to hold its next world congress in Brazil, in 2011. It elected Pe. Ignácio Shmitz as new President, Luiz Oosterbeek as Secretary General and Rossano Lopes Bastos as Congress secretary.

L.O.

Table of Contents

Introduction

Variation, Continuity and Change in Rockshelters and Rockshelter Studies: A Global Perspective 1
by *The Coordinators*

Regional Overviews

Rockshelters of the Périgord: 25 Ans Après 9
by *Jean-Philippe Rigaud*

Approaches to the Middle Paleolithic Rockshelter and Cave Research in Croatia 13
by *Ivor Karavanić, Nikola Vukosavljević, Rajna Šošić, and Sanjin Mihelić*

Archaeological Research in Rockshelters and Caves in Slovenia 23
by *Martina Knavs*

125 Years of the Rockshelter Studies in Russia 31
by *Sergey A. Vasil'ev*

Desert Caves and Rockshelters in the Great Basin of North America 37
by *C. Melvin Aikens*

Rockshelter Archaeology in the Middle Tennessee Valley of North America 43
by *Boyce Driskell*

Rockshelter of the Middle Rocky Mountains: 70 years of research 51
by *Marcel Kornfeld*

A Century of Basketmaker II Rockshelter Research in the American Southwest:
the Archaeology of Transition to Farming Across the Colorado Plateau 61
by *Francis E. Smiley and Susan Gregg Smiley*

Subterranean Caves, Their Morphology and Archaeological Content:
The Mortuary Caves in Coahuila, Mexico 69
by *Leticia González Arratia*

Current Research in Europe and Africa

Rockshelter Studies in Southwest Iberia: The Case of Vale Boi (Algarve, Southern Portugal) 75
by *Nuno Bicho, Mary C. Stiner, Delminda Moura, and Armando Lucena*

Early Tardiglacial Human Uses of el Mirón Cave (Cantabria, Spain) 83
by *Lawrence Guy Straus and Manuel González Morales*

Answer to the Problem of the Diachronic and Synchronic Relationship of Arqueopaleontological
Elements in Sites with Homogeneous Sediments in the Middle-Pleistocene:
The Example of Gran Dolina, Sierra de Atapuerca 95
by *Rosana Obregón and Antoni Canals*

Stratigraphie et Chronologie an Archéologie Préhistorique 101
by *Françoise Delpeche*

Caves and Rockshelters of the Trieste Karst (Northeastern Italy) in Late Prehistory 109
by *Manuela Montagnari Kokelj*

The Secret Cave City Hidden in the Cliffs (Lovranska Draga Canyon, Istria, Croatia) 119
by *Darko Komšo and Martina Blečić*

Reflections on the Takarkori Rockshelter (Fezzan, Libyan Sahara) 125
by *Stefano Biagetti and Savino di Lernia*

Current Research in Americas

Collapsed Rockshelters in Patagonia 135
by *Louis Alberto Borrero, R. Barbereña, F.M. Martin, and K. Borrazzo*

Chorrillo Malo 2 (Upper Santa Cruz Basin, Patagonia, Argentina):
New Data on its Stratigraphic Sequence 141
by *Nora Franco, Adriana Mehl, and Clara Otaola*

The Paleoindian Occupations at Bonneville Estates Rockshelter, Danger Cave, and Smith Creek Cave
(Eastern Great Basin, U.S.A.): Interpreting Their Radiocarbon Chronologies 147
by *Ted Goebel, Kelly Graf, Bryan Hocket, and David Rhode*

A GIS Perspective on Rockshelter Landscapes in Wyoming 163
by *Mary Lou Larson*

The Geologic and Geomorphic Context of Rockshelters in the Bighorn Mountains, Wyoming 173
by *Judson Finley*

Closed Site Investigation in the American Northeast: The View from Meadowcroft 181
by *James Adovasio*

The Madness Behind the Method: Interdisciplinary Rockshelter Research in the Northeastern United States 191
by Jonathan *A. Burns, John S. Wah, and Robert E. Kruchoski*

List of Figures

1.1 Location of study areas included in the following papers. 3

2.1. Résume les étapes d'une démarche tenant compte de ce qui vient d'être évoqué. 11

3.1 Map showing Middle Paleolithic sites in Croatia discussed in the paper. 14
3.2 Excavations of the site at Krapina 14
3.3 Stratigraphic profile at Krapina 15
3.4 Vindija cave entrance 16
3.5 Vindija stratigraphic profile 17
3.6 View from the area in front of Mujina Pećina. 18
3.7 Excavating in Mujina Pećina. 19
3.8 Stratigraphic profile of Mujina Pećina 19

4.1 Position of Slovenia 24
4.2 Slovenian karst systems 24
4.3 Slovene informal provincial divisions 25
4.4 Locations of the some of most known archaeological sites in caves and rock shelters in Slovenia 26
4.5 Ground plan of Viktorjev spodmol with the test trench outlines. 28

5.1 Main cave site concentrations under study at Russia and adjacent countries. 32
5.2 Spatial distribution of cultural debris in the lower stratum (VI) of the Kiik-Koba grotto, Crimea 33

7.1 Physiographic subdivisions and the locations of important sheltered sites within the Middle Tennessee Valley. 44
7.2 Suggested chronology for the Paleoindian and Early Archaic periods in the Middle Tennessee Valley 46

8.1 Location of the Middle Rocky Mountains (left) and the Bighorn Region (right). 52
8.2 Medicine Lodge Creek Site 54
8.3 Rock images at Medicine Lodge Creek Site. 54
8.4 Temporal distribution of site densities in the Bighorn region (right), compared to the western Black Hills (left). 56
8.5 Backplot (top) and stratigraphic section (bottom) of Two Moon shelter. 57

9.1 Photorealistic simulation of the northern American Southwest. 62
9.2 Photorealistic simulation showing the Black Mesa - Marsh Pass region. 62
9.3 Photorealistic simulation showing the Grand Gulch - Cedar Mesa - Comb Ridge region. 63
9.4 White Dog Cave near Kayenta, AZ. 64
9.5 Interior of Three Fir Shelter on northern Black Mesa. 64
9.6 Tseahatso (Big Cave) in Canyon del Muerto in eastern Arizona. 65
9.7 Fish Mouth Cave in the Comb Ridge. 65
9.8 Boomerang Shelter in Comb Ridge. 66
9.9 Interior of Boomerang Shelter showing the main prehistoric habitation area. 66

11.1 Map of Portugal with the location of Vale Boi. 75
11.2 General view of the site of Vale Boi. 76
11.3 Schematic section of the site. 77
11.4 Group of bone tools from the Gravettian and Solutrean middens. 78
11.5 Solutrean points found in the Midden. 78

12.1 Features at El Miron, 2000-2005 86
12.2 Features at El Miron, 2004-2006 88

13.1 Stratigraphic section of Gran Dolina 98
13.2 Backplot showing a high concentration of material and lateral change in density 99
13.3 Backplot showing a possible paleochannel. 99

14.1 Situation géographique des gisements cités dans le texte 102
14.2 La Ferrassie. Evolution schématique des associations d'ongulés prenant en compte. 103

iv

14.3 Le Pech de l'Azé II. Evolution des associations d'ongulés et interprétations biostratigraphiques 105
14.4 Association d'ongulés des gisements de La Ferrassie (Fer), Le Flageolet I (Flag) et Roc de Combe (RdeC).. 105

15.1 Location map of the Trieste Karst 110
15.2 The C.R.I.G.A. project: example of database forms, showing topographic (left) and archaeological data (right) 112
15.3 The C.R.I.G.A. project: simplification of the database structure 112
15.4 The C.R.I.G.A. project, GIS analyses: values of the walkable surface. 113
15.5 The C.R.I.G.A. project, GIS analyses: values of the sunlight surface... 113
15.6 The C.R.I.G.A. project, GIS analyses: map of the site-resource Euclidean distance. 114
15.7 The C.R.I.G.A. project, GIS analyses, maps of accessibility to primary resources. 115

16.1 The eastern cliffs of the Lovranska Draga Canyon 120
16.2 Oporovina, remains of the grooves in the cave wall 120
16.3 Oporovina, human burials in the second trench 121
16.4 Abri Uho, carved and hewn crosses 122

17.1 The area under concession to 'The Italian-Libyan Archaeological Mission.' 126
17.2 The Holocene cultural sequence in the Acacus Mountains and its surroundings. 126
17.3 Painting in the Middle Pastoral style at Uan Amil (central Acacus). 127
17.4 The Takarkori rockshelter during excavation. 127
17.5 Open-air sites in the vicinity of Takarkori. 128
17.6 Late Acacus layers in the Takarkori rockshelter. 129
17.7 Naturally mummified burial of the Middle Pastoral period at the Takarkori rockshelter. 130
17.8 Uses of rockshelters in the Acacus Mountains throughout the Holocene. 130
17.9 Burial tumuli in the Acacus Mountains. 131

18.1 Location of areas mentioned in the text. 136

19.1 Location of Chorrillo Malo 2 site. 142
19.2 Chorrillo Malo 2 rockshelter at the beginning of the excavation. 142

20.1 Map showing location of archaeological sites. 148
20.2 A view of the Smith Creek Cave excavation. 150
20.3 Hearth feature 03.15.. 150
20.4 Calibrated ^{14}C ages that accurately measure the ages of the early cultural occupations. 157

21.1 Location of rockshelter research in northern Wyoming. 164
21.2 The Bighorn shelter and Paint Rock Archaeological Landscape District (PRCALD) project areas. 164
21.3 Artifacts from Southsider Shelter and Spring Creek Caves. 166
21.4 Percentage of archaeological components per shelter for Bighorn rock shelter data set............................... 166
21.5 Paint Rock Canyon looking west towards the Bighorn Basin.. 167
21.6 Paint Rock Archaeological Landscape district overlaid on Allen Draw 7.5' quadrangle. 167
21.7 Paint Rock Canyon Archaeological Landscape district showing ArcGIS 9.1 map and data table (top) and rock shelters with those containing archaeological material on the surface of the shelter highlighted (bottom). 168
21.8 Aspect of All Paint Rock Canyon Rock Shelters. 169
21.9 Slope in degrees of Paint Rock Canyon rock shelters. 170

22.1 Location map of the Bighorn Mountains and Basin in north-central Wyoming. 174
22.2 Examples of rockshelter stratigraphy 178

23.1 Location of Meadowcroft Rockshelter in southwestern Pennsylvania.. 183
23.2 Plan map of excavations at Meadowcroft Rockshelter. 185

24.1 Geographic location of the State of Pennsylvania in Northeastern North America. 192
24.2 Locations in the Ridge and Valley Province of the two rockshelter case studies. 192
24.3 Excavations at Sheep Rock Shelter... 193
24.4 Excavations at Mykut Rockshelter.. 193
24.5 Sample distributional contour map. 194
24.6 Camelback Rockshelter (2004). 194
24.7 Surface contour map of Camelback Rockshelter and immediate vicinity... 195

24.8 Schematic particle-size distribution with depth for Camelback Rockshelter. 196
24.9 Excavation units measuring 50cm^2. 196
24.10 Combined sample plan view plots from Camelback Rockshelter referred to in the text. 197
24.11 Cache of lithic bifaces from Camelback rockshelter. 197

List of Tables

4.1 Viktorjev spodmol: all stone artifacts 28

8.1 Summary data on excavated shelters in the Bighorn region. 55

10.1 Reported inhumations in subterranean caves. 69
10.2 Cave measurements and characteristics. 70
10.3 Artifacts found in subterranean burial caves. 71

11.1 AMS dates from Vale Boi 78
11.2 Faunal List from G25 and Z27 (slope area). 81

14.1 Laugerie-Haute-Ouest, niveaux Solutréens. 104

20.1 Radiocarbon Dates for the Terminal Pleistocene and Early Holocene Deposits at Danger Cave. 152
20.2 Radiocarbon Dates for the Terminal Pleistocene and Early Holocene Deposits at Smith Creek Cave............. 154
20.3 Radiocarbon Dates for the Terminal Pleistocene and Early Holocene Deposits at Bonneville Estates
    Rockshelter. 155
20.4 Reliable ^{14}C ages dating cultural occupation events. 157

VARIATION, CONTINUITY, AND CHANGE IN ROCKSHELTERS AND ROCKSHELTER STUDIES: A GLOBAL PERSPECTIVE

THE COORDINATORS*

*Marcel Kornfeld
George C. Frison Institute, 1000 East University Avenue, University of Wyoming, Laramie, WY 82071, U.S.A.;
anpro1@uwyo.edu
*Sergey A. Vasil'ev
Institute for the Material Culture History, 18 Dvortsovaya emb. 191186 St. Petersburg, Russia;
sergevas@av2791.spb.edu
*Laura Miotti
Department of Archaeology, Facultad de Ciencias Naturales y Museo, Universidad de La Plata, Paseo del Bosque s/n, 1900 La Plata, Buenos Aires, Argentina, 1900; lmiotti@fcnym.unlp.edu.ar

Abstract. Rockshelters have always held a special place in archaeology. Early in the history of the discipline rockshelters provided much information on early hominids and hominid behavior. Later and in many parts of the world some shelters were found to have exceptional preservation, thus providing a glimpse of perishable items of prehistoric peoples. Because of their importance to unraveling the mysteries of prehistory, special field, analytical and interpretive techniques were developed to deal with shelters. Rockshelters are unique features of the landscape in that they offer naturally produced shelter lasting innumerable generations. However, not all rockshelters are the same and the differences between them are as large as the difference between shelters and open air sites. Nevertheless, the variation in formation, morphology, deposition, occupation, and other characteristics of shelters is not well understood. In this paper we introduce the special place of rockshelters and rockshelter research in archaeology and prehistory.
Keywords: Rockshelters, shelter variation, methods, interpretive theory

Résumé. Les abris sous roche occupent une place spécial dans l'archéologie. Dès le commencement du développement de la discipline les abris ont fourni beaucoup d'informations sur les hominidés et leur comportement. Plus tard et dans plusieurs régions du monde quelques abris se sont avérés pour avoir la conservation exceptionnelle, en fournissant un aperçu des artefactes périssables préhistoriques. En raison de leur importance pour eclaircir les mystères de la préhistoire, des techniques de travaux de terrain, les méthodes analytiques et interprétatives spéciales ont été mis au point pour la recherche sur les abris. Les abris sous roche sont les éléments uniques du paysage parce qu'elles offrent l'abri naturel pour les générations innombrables. Cependant, les abris sont loin d'être identiques et les différences entre eux sont parfois aussi grandes que la différence entre les abris et les sites de plein air. Néanmoins, nos connassances sur la variation de la formation, de la morphologie, du sédimentation, de l'occupation, et d'autres caractéristiques des abris sont maigres. En cet article nous présentons une vue d'ensemble sur des abris et la recherche des abris dans l'archéologie et la préhistoire.
Mots clés: Rockshelters, variation d'abri, méthodes, théorie interprétative.

Caves and rockshelters are unique features of the landscape in that they offer naturally produced shelter lasting innumerable generations. However, closed sites (to borrow a concept introduced in this session by Jim Adovasio) vary among themselves nearly as much as they vary from other (notably open air) archaeological manifestations. It is widely accepted that cave site have a special and controversial place in prehistoric studies. The main idea of our meeting is to combine the historical perspective and regional research experience with pivotal contemporary problems provoking hot debates. In spite of apparent difference in the national research traditions and differences in prehistoric studies, the main trends are similar. Cave site exploration is among the oldest in archaeological traditions, its development being shaped by the complex interplay of field discoveries with the theoretical concepts prevailing during various historic periods. Caves began to be explored systematically in the first half of the 19th century (in Belgium, England, France, and Germany), followed by the new discoveries in Central and Eastern Europe. The general line of the development in cave site studies during the 19th to early 20th centuries was the sophistication of methods oriented toward the detailed stratigraphic resolution. The identification of long-term culture successions based on comparative study of superimposed assemblages served as the base for the establishment of cultural evolution, central to the culture-historical archaeology of the time (Trigger 1990). Among structural features, mostly fireplaces were identified by the early investigators.

Later, from the 1920s onwards a new dimension, namely spatial analysis, was added to cave research, and scholars begun to study the human adjustment to rockshelters as habitations. The foundations of modern approach to the cave site excavations, with careful individual plotting of artifacts and bones, combined with detailed stratigraphic planimetric studies is associated with two remarkable exploration campaigns carried out at the late 1940s to the early 1950s. These were directed by François Bordes at

Pech-del-Azé and André Leroi-Gourhan at Arcy-sur-Sure (e.g., a brief history of cave exploration in Europe, Groenen 1994).

In the Americas closed site investigations probably began with the work of Lund in the 1830s in Minas Gerais region of Brazil and his discovery of the Lagoa Santa Man (Lavallee 1995:22-24). However, as Mercer sometimes lamented in the late 1800s, very few in the Americas followed Lund's early cave explorations.

In North America archaeological cave and rockshelter research is at least 150 years old, or roughly comparable to that on other continents, as archaeology generally did not exist in the field until about that time (Willey and Sabloff 1993). Among early mentions of closed site investigation is an 1861 recommendation for excavation procedures of "caverns" offered by George Gibbs (1862) in the Annual Report of the Board of Regents of the Smithsonian Institution. For its age, 150 years ago, this is a remarkable recommendation for several reasons. First, it clearly recognizes the significance of stratigraphy. The importance of stratigraphy was only recognized a few years earlier with the works of Lyell and its various applications to early archaeology, generally by European scholars (Lyell 1833). American prehistorians were clearly aware of these important developments and understood their significance, although the heyday of stratigraphic excavations was to come more than a half century later (Willey and Sabloff 1993). Second, Gibbs's statement employs the **principal of association** ("position relative to other remains"). Again, the principal was known at least since Worsaae (Willey and Sabloff 1993) associated grave goods from individual internments, but was not widely used in the Americas until after Max Uhle's work in Peru (Willey and Sabloff 1993). Third, the determination of the "actual circumstances" of the remains is a clear call for recording and interpreting the context of the finds and a precursor for the investigation of site formation processes and taphonomy a very modern endeavor. Gibbs' paper is followed by other closed site reports throughout the late 1800s (e.g., Whitney 1872; Meigs 1872; Mercer 1894a, 1894b, 1894c, 1896).

The overarching question during this part of American Archaeology was the origin and time depth of American Indians (Willey and Sabloff 1993). Specifically it was believed that modern groups were not related to earlier groups, but were rather relatively modern immigrants into the area. Leaving aside the political implications of this point of view, the above investigations of caves had these questions in mind. After the solution of the "Moundbuilder" myth, in 1894 (Thomas 1894), the question turned to the age of the American Indians with the well known American Paleolithic controversies (Wilmsen 1965; Willey and Sabloff 1993). Henry Chapman Mercer excavated several caves at the end of the last century with these questions clearly in mind. His work ranges from Pennsylvania to the Yucatan.

Mercer saw caves as stratigraphic panaceas, certainly a common view of the time and for a long time after Mercer's life. He (Mercer 1896:10) states: "Continued investigation has established the fact that of all searching-grounds known to archaeology caves best answer the question which lies at the bottom of the science, namely, the question of sequence, which came first and which next? When and where were the beginning, middle, and end of the story?" In their investigations of American closed sites, Mercer and others drew heavily on the immediate preceding and European contemporaries such as Lyell, Schmerling, MacEnry, Pengelly, Lartet, and Christy.

In spite of great achievements of multidisciplinary studies of cave sites, focused on detailed paleoenvironmental reconstructions, the archaeological methods remain underdeveloped. There are several lines of archaeological inquiry only slightly touched in literature and we hope to concentrate on these points in our discussions. Among those the following topics can be mentioned:

1. Accessibility of natural shelters in different regions through prehistory, due to the processes of cave formation, erosional destruction and possible burial;
2. Problems of stratigraphic correlation between different portions of a site that inform on the identification of living floors which could be interpreted as reflections of short-term occupation episodes;
3. Problems of the identification of constructed features, which is the most difficult in cave site context. Many stone distributions, such as pavements, rounded or oval shaped concentrations, ramparts, alignments, etc. can be interpreted as natural or as constructed features. In many cases it is far from clear if the visible concentrations of cultural debris are the reflection of activity areas or simply erosional remnants of culture-bearing strata;
4. Taphonomic problems related to the differentiation of human from natural agencies in bone accumulation, the illumination of the role of carnivores (cave bears, cave hyenas, etc.) in bone modification

These are only some of the more promising lines of inquiry based on the confrontation of newest theoretical and methodological advances and rich factual data accumulated in different regions of the world that we hope will lead to further information exchange and future international co-operation on closed sites.

The purpose of the papers in this volume is to provide a global assessment of today rockshelter studies, in particular:

1. The histories and synthesis of rock shelters research in various region;
2. The theoretical perspectives of the role of rockshelters in prehistory of different regions; and

1.1. Location of study areas included in the following papers. 1) Iberian peninsula (Bicho et al.; Straus and Morales; Obregon); 2) Perigord (Rigaud; Delpech) 3) Karst region (Slovenia, Knavs; Italy Montagnari) and Croatia (Karavanić et al.; Komšo and Blečić); 4) Russia and the former Soviet republics (Vasil'ev); 5) North Africa (Biagetti and di Lernia); 6) Southern Cone-Argentina (Borrero et al.; Franco et al.); 7) Mexico (Gonzales Arratia); 8) North American Southwest (Smiley and Smiley); 9) North American Great Basin (Aikens; Goebel et al.); 10) North American Middle Rocky Mountains (Finley; Kornfeld; Larson); 11) East central North America (Adovasio; Burns et al.); 12) Southeastern North America-Tennessee Valley (Driskell). Areas presented at the session, but not included: a) Ukraine (Stepanchuk and Kovalyukh); b) Uzbekistan (Krivoshapkin; Wrinn et al.; Anoikin); c) Altai (Shunkov; Zenin). Also missing from the papers presented at the congress are some papers from the same regions as those presented here, including: Mentzer et al. Martinez-Moreno et al., Tixier., Collins, Postnov as well as those submitted but not presented at the congress. (Modified from www. Paxromana.org/files/iages/world-map.preview.png)

3. The field, analytical, and interpretive methods specific to rock shelter investigations utilized by the participants or their colleagues; and
4. To get a sense of global variation in character and use of closed sites

This monograph provides a global view on variability of rock shelter formation, deposition, and evolution. This is a particularly significant aspect of the colloquium as it is currently unclear how rock shelters vary globally and how this variability affected their prehistoric use, as well as past and present investigative approaches.

In spite of our efforts to have a global representation of rockshelter research, the geographical focus of the papers is biased (**Figure 1.1**). The majority of the contributions at the conference dealt with the Americas and Eurasian evidence, while only one paper dealt with Africa, a profile that is even narrower in the present volume. In our session we had a series of presentations on central and northern Asia, unfortunately, the authors chose not to submit manuscripts for this monograph. East and Southeast Asian, and Oceanian papers were not included in the session, although we strenuously tried to contact colleagues in those areas that specialized or have excavated and reported on closed sites. The one Oceanian submission to the colloquium was, unfortunately not presented. The lack of representation from these continents is exceedingly unfortunate, as many caves and shelters have been reported and excavated and their lack of representation severely curtailed the success of the session as well as this monograph. On the other hand this monograph represents one of the rare instances where shelters from multiple continents, excavated over the entire range of archaeology as a science, are synthesized in a series of articles. But even more significantly, we see global variation in rockshelters, approaches to studying and interpreting these types of archaeological manifestations, and cutting edge research approaches.

The chapters included in this volume vary in scope and size. There are several methodological papers, emphasizing such issues as mapping of the regional networks of closed sites (Komšo and Blečić, Larson, and Montagnari Kokelj) or stratigraphic resolution of shelter sediments (Delpech, Obregon and Canals). Many of the papers are more or less detailed regional reviews, summarizing the developmental history and current state of the art in rockshelter studies in different parts of the world. As such we divide the papers into three groups. First, the introduction and regional overviews, second the Current Research in European and North African including methodological and substantive contributions, followed by Current Research in the American.

Following this introduction we begin the monograph with Rigaud's examination of the 'classical' region of southwestern France, where archaeology began. He is followed by Knavs and Karavanić et al. overviews of Slovenian and Croatian closed sites, the latter a location of one of the earliest as well as continuous hotbeds of Neandertal research. Moving further east and completing the European overviews is Vasil'ev presentation of the history of research in the vast territory of Russia that has as rich a tradition in closed sites studies as the other areas just covered. The remaining overviews are of several regions of North America and Mexico. Aikens reports the results of investigations of dry caves of the Great Basin, where the excellent preservation conditions provided a unique opportunity to study in detail the adaptations of Pleistocene/Holocene boundary occupants. Although known for their preservation of a variety of perishable materials, these were not the first closed site investigations in North America, rather such cases were in the eastern part of the continent. Driskell takes the southeastern part of North American and presents an overview of cave site exploration, focusing on the middle Tennessee Valley. The following three chapters (Kornfeld, Smiley and Gregg Smiley, and Gonzáles Arratia) cover western North America and Mexico. Especially of interest are the intensive and sometimes complex occupations of the southwestern caves containing at least Basketmaker II villages if not complete cities and the mortuary caves of Mexico.

In the second part of the monograph are the European papers covering current investigations. The Iberian Peninsula is one of the main concentrations of closed sites in the Old World and the research traditions are particularly strong in the area. Bicho and colleagues focus on the Upper Paleolithic occupations of Vale Boi (southern Portugal), describing the methods for artifact plotting and analysis. Advanced computer-based methods for maximizing the stratigraphic resolution within essentially homogeneous cave sediments are presented by Obregon and Canals at the Lower Paleolithic site of Gran Dolina. Straus and Morales report on their meticulous analyses of Magdalenian occupations at El Miron, a cave site located in northern Spain. While Delpech challenges the traditional views on site chronology, providing a careful stratigraphic analysis of several key sites located in southwestern France and offering several implications from this re-examination. A relatively new type of approach is presented by Montagnari Kokelj, emphasizing the main results of an ambitious GIS-based project considering cave site distributions in the Trieste Karst area of northeastern Italy. In a nearby area Komšo and Blečić analyze the rich evidence of a cluster of caves in the Lovranska Draga of Croatia. The final paper in this section, by Biagetti and di Lernia, provides preliminary results of recent excavation campaign at Takarkori Rockshelter in Libya and their implications for the Holocene prehistory of the region.

In the third part of the monograph are the current investigations in the Americas beginning with the South American papers of Borrero and colleagues and Franco and colleagues. Borrero et al. explore two cases where the inner space of rockshelters were reduced by collapse of roofs or refilled by sediments and rocks. These changes results in the abandonment by human groups and challenge our ability to find, investigate, and interpret these sites, since they are not longer shelters. These events occurred many times during Late Pleistocene and Holocene. The authors review several cases in Southern Patagonia and two are explored in some detail. Franco and colleagues present the results of excavations of an early Holocene component at Chorrilo Malo 2, in which they emphasize the complexity of geoarchaeological context. In North America, Goebel and colleagues provide an overview of earliest Great Basin shelters including recent reinvestigations of several classic localities, but focus their attention on ongoing investigations and discoveries at Bonneville Estates Rockshelter of Nevada. Larson along with Finley, and Kornfeld's overview (the latter in the first section of the monograph) are part of an ongoing multidisciplinary study of rockshelters in the Bighorns, a part of the Middle Rocky Mountain physiographic province. Larson institutes a first GIS based analysis of rockshelters in the region and perhaps in North America, while Finley devotes his time to geoarchaeological investigation of previously investigated and newly discovered closed sites. His analysis provides for the first time in this region and perhaps in North America for a direct link between paleoenvironmental reconstruction and the cultural assemblages (e.g., Bryan 2006). The final two papers in this section are by Adovasio and Burns and colleagues. Adovasio presents the sophisticated methodology of an unprecedently long-term multidisciplinary research project at Meadowcroft Rockshelter in Pennsylvania and the need for such studies to resolve important questions of site formation. Burns paper takes advantage of such previous studies as Adovasio's to develop an even more sophisticated set of field techniques, and discusses the problems of relationship between natural and human factors in rockshelter studies and reconstruction of hunter-gatherer adaptations in Pennsylvania.

One final note is necessary regarding dates reported and discussed in this monograph. Unless specifically stated otherwise in individual chapters, all dates are reported in uncorrected radiocarbon years before present (RCYBP).

CONCLUSION

Although geographically more restricted than we had hoped, the papers in this volume touch on a great many of the issues that we identified at the beginning of this introduction. As such they represent the history of rockshelter research on several continents and in different regions of these continents. In the second and third parts of the monograph the papers represent the state of the art of rockshelter research and provide a valuable reference to ongoing closed site studies and investigations throughout the world. We hope our colleagues will find the papers useful in developing their own research plans and agendas for rockshelter and cave, that is, closed site research.

Acknowledgments. We would like to thank Dena Dincauze and Richard Woodbury for pointing us in several pertinent directions in our search of early North American cave and rockshelter excavations. The French translation of the monograph title benefited greatly from James Enloe's multilingual abilities. We further acknowledge the editorial assistance of Caroline Charles and Joe Gingerich. Without them this monograph would not have been completed. Finally, we thank all the contributors to this volume for sending their manuscripts and associated materials, sometimes multiple times.

REFERENCES

GIBBS, GEORGE (1862). Instructions for archaeological investigations in the U. States. *Annual Report of the Board of Regents of the Smithsonian Institution.* Government Printing Office, Washington.

GROENEN, M. (1994). *Pour une histoire de la préhistoire.* Ed. Jérôme Millon, Grenoble.

LYELL, CHARLES (1833) *Principles of Geology.*

MEIGS, J. AITKEN (1872). Description of a human skull in the collection of the Smithsonian Institution. Annual *Report of the Board of Regents of the Smithsonian Institution, Showing, The Operations, Expenditures, and Conditions of the Institution for the Year 1867*, pp. 412-415. Government Printing Office, Washington.

MERCER, HENRY C. (1894a). Progress of field work of the department of American and Prehistoric archaeology of the University of Pennsylvania. *The American Naturalist*, July, pp. 626-628.

MERCER, HENRY C. (1994b). Progress of field work of the department of American and Prehistoric archaeology of the University of Pennsylvania. *The American Naturalist*, April, pp. 355-357.

MERCER, HENRY C. (1994c). Geilenreuth Cave 1894. *The American Naturalist*, September, pp. 821-824. Re-exploration of Hartman Cave, near Stroudsburg, Pennsylvania. Proceedings of the Academy of Natural Sciences of Philadelphia.

MERCER, HENRY C. (1896). *The hill-caves of Yucatan.* Lippincott, Philadelphia.

THOMAS, CYRUS (1894). *Report of the Mound Explorations of the bureau of Ethnology.* Washington D.C.

TRIGGER, BRUCE G. (1990). *A History of Archaeological Thought.* Cambridge University Press, Cambridge.

WHITNEY, J.D. (1872). Cave in Calveras County, California. *Annual Report of the Board of Regents of the Smithsonian Institution, Showing, The Operations, Expenditures, and Conditions of the Institution for the Year 1867*, pp. 406-407. Government Printing Office, Washington.

WILLEY, GORDON R. AND JEREMY A. SABLOFF. (1993). *A History of American Archaeology*, 3rd edition. W.H. Freeman and Company, New York.

WILMSEN, EDWIN, N. (1965). An outline of Early Man studies in the United States. *American Antiquity* 31(2):172-192.

www.paxromana.org/files/images/world-ap.preview.png (accessed December 22, 2006).

PART I – REGIONAL OVERVIEWS

ROCK SHELTERS OF THE PÉRIGORD: 25 ANS APRÈS

Jean-Philippe RIGAUD

Institut de Préhistoire et de Géologie du Quaternaire, Université Bordeaux 1, Av. des Facultés, 33405 Talence, Français;
jp.rigaud@ipgq.u-bordeaux1.fr

Abstract. Rockshelters of the Perigord contributed significantly to establishing the chronological and cultural framework of the European middle and late Pleistocene. A large number of rockshelter sites, long cultural sequences, and good preservation of cultural material have been responsible for a long standing reputation of the Perigord as a prehistorian's paradise. In what way has this situation marked the theoretical and methodological orientation of prehistoric research since the middle of the 19th century? The Perigordian model has been exported, imported, duplicated, misinterpreted, plagiarized all over Europe, Africa and Middle East with no consideration for its specificity and for its limits of use, generating criticisms, inappropriate sometime, clear-sightedness occasionally.

Now, 25 years after Rockshelters of the Perigord has been published, what has changed in our methodology, in our interpretation of the raw data? What still makes rockshelter archaeology an unrivaled source of information?

Keywords: Rock shelters, Périgord, methodology, archéo-stratigraphy, data validation

Résumé. L'archéologie des abris-sous-roche du Périgord a contribué largement à l'établissement du cadre chronologique et culturel du Pléistocène moyen et supérieur européen. Grâce à ses nombreux abris, de longues séquences culturelles et une bonne conservation ders vestiges, le Périgord a justifié sa réputation d'éden des préhistoriens. De quelle façon cette reconnaissance a marqué l'orientation théorique et méthodologique des recherches depuis le milieu du XIX ème siècle ? Le modèle périgourdin a été largement exporté, importé, reproduit tant en Europe, qu'en Afrique ou au Moyen-Orient, mais il a aussi parfois été mal interprété sans tenir compte de sa spécificité et de ses limites générant ainsi des critiques parfois judicieuses mais aussi parfois inappropriées. 25 ans après la publication de « Rock Shelters of the Périgord » quels ont été les changements dans notre méthodologie et dans l'interprétation de nos données ? Pour quelles raisons l'archéologie des abris-sous-roche est-elle toujours une remarquable source d'information ?

Mots-clés: Abris-sous-roche, Périgord, méthodologie, archéo-stratigraphie, validation des données

Dès le début du XIXe siècle, les premiers amateurs qui furent à l'origine de l'archéologie préhistorique avaient remarqué la présence de vestiges très anciens dans des dépôts quaternaires fluviatiles et éoliens exploités en carrière dans le nord de la France. Cependant, très rapidement, la curiosité de ces pionniers fut attirée par le contenu préhistorique des sites sous abri que les caprices de la géologie localisaient très majoritairement dans le sud de la France et dans lesquels eurent lieu par la suite de très nombreuses fouilles.

Ces sites privilégiés suscitaient la faveur des premiers préhistoriens pour plusieurs raisons. En premier lieu, ces abris-sous-roche n'étant que rarement enfouis totalement sous des dépôts quaternaires et pour les prospecteurs, il était facile de les remarquer dans le paysage. D'autre part, ces sites abrités favorisaient un état de conservation des vestiges bien supérieur à celui de la majorité des sites de plein air où, de plus, les restes fauniques étaient fréquemment absents, rares ou partiellement conservés. L'abondance et la qualité des vestiges contenus dans ces espaces clos et aux dimensions modestes attiraient les chercheurs et les collectionneurs. Enfin, certains abris contenaient des séquences archéologiques exceptionnellement longues, plus particulièrement dans le nord de l'Aquitaine où l'on pouvait soupçonner une relative continuité dans l'occupation pléistocène de ces sites abrités.

Très fortement inspirée par les théories de la Paléontologie humaine de l'époque et par la curiosité naturaliste des précurseurs, plusieurs paradigmes étaient à la base des premières recherches en archéologie préhistorique:

1. Les évènements culturels ont un caractère universel et synchrone et les produits techniques (outillages, industries) sont spécifiques d'une culture (culture matérielle) ou d'une phase chronologique de celle-ci. Un « outillage » différent signifie une « culture » différente ou une phase différente de celle-ci.

2. Une « culture » est le fait d'un type humain ou d'une population. Un changement culturel implique un changement anthropologique et réciproquement.

3. Une « culture matérielle » est caractérisée par le contenu archéologique d'un stratotype qui représente sa « définition légale », son étalon de référence.

Ce dernier point comme le premier, directement issus de la géologie, sont à la base des travaux des contemporains de D. Peyrony, mais aussi de F. Bordes et D. de Sonneville-Bordes, et même plus récemment bien d'autres encore (Demars et Laurent, 1989, Djindjian, 2000, Bosselin, 1996).

C'est sur ces bases théoriques qu'à partir de la fin du XIXe siècle commença l'exploitation des riches gisements nord aquitains qui s'est poursuivie jusqu'au milieu du XXe siècle sans que les méthodes de fouille ne se soient beaucoup améliorées si ce n'est parfois, un plus grand respect pour ce qui était alors appelé la « stratigraphie », dans les limites de l'aptitude du fouilleur à la déchiffrer. Au cours de cette période, à partir des séquences des grands sites périgourdins et de leur contenu typologique, que D. Peyrony, a dessiné les grandes lignes du « modèle périgourdin » qui, pendant des décennies, a servi de référence, a été importé, interprété et transposé, parfois sans discernement, dans des régions éloignées du Périgord originel.[1]

Poursuivant l'œuvre de Peyrony, F. Bordes prit une part active dans l'avancée méthodologique qui a marqué la seconde moitié du XXe siècle par la mise en oeuvre de disciplines naturalistes (géologie, palynologie, paléontologie) et physico-chimiques (datations numériques) et par l'usage de méthodes quantitatives élémentaires pour l'analyse et la description des techno-complexes. Si les bases théoriques évoquées précédemment étaient resté globalement inchangées, ces perfectionnements méthodologiques ont eu pour conséquences d'induire de nouvelles méthodes de terrain consistant à déchiffrer dans les dépôts de remplissage des abris les « entités stratigraphiques » les plus fines possibles, considérées comme les plus « brèves » et donc (sic) les plus homogènes. Ce fut le fondement de la « méthode stratigraphique » critiquée par les tenants de la «méthode par décapages » mise au point dans les sites de plein air à stratigraphie considérée comme simple.

Deux remarques s'imposent cependant. En premier lieu, les moyens mis en œuvre pour définir les « entités stratigraphiques les plus fines possibles » étaient en fait un mélange empirique de techniques inspirées de la géologie, de la pédologie et de l'archéologie. L'enregistrement des variations de couleur, de texture et de structure n'ayant qu'un objectif, celui de distinguer des « niveaux» dont le contenu était supposé représenter des « moments » aussi brefs que possible dans l'occupation préhistorique de l'abri. La prise en compte de ces entités à des fins strictement sédimentologiques, paléoclimatiques ou biostratigraphiques n'était pas nécessairement judicieuse ou justifiée et leur intérêt dépendait surtout des objectifs du sédimentologue, du paléontologue ou du palynologue.

D'autre part, la technique de fouille par décapages de grandes surfaces développées par A. Leroi-Gourhan et ses collaborateurs pour l'étude des habitats de plein-air magdaléniens, nécessitait une adaptation pragmatique à la topographie et à la dynamique sédimentaire sous abri qui étaient fort différentes de celle qui prévalaient dans le Bassin Parisien.

« **Rock Shelters of the Périgord** » que J. Sackett, H. Laville et moi-même avons publié en 1980, était destiné à présenter à nos collègues anglophones, au-delà des données relatives au Périgord préhistorique, les bases théoriques et méthodologiques qui étaient les nôtres dans les années 70 pour l'étude des abris sous roche du Périgord. Nous avions eu toutefois la prudence de conclure en précisant « *In the meanwhile, we offer the Perigord scheme as an enterprise to be emulated rater than as a model to be reproduced.* (Laville *et al*, 1980 : 355).

Un quart de siècle plus tard, nous sommes naturellement tentés de faire un bilan et de mesurer le chemin parcouru dans ce domaine. Les techniques de laboratoire ont connu un essor dans les domaines de la sédimentologie, de la géochimie et de la biochimie, des datations numériques, les techniques et méthodes informatiques ont multiplié à l'extrême les performances de l'outil analytique et le développement de l'archéologie préventive nous a apporté avec ses méthodes spécifiques une masse documentaire considérable. Mais au-delà de ces améliorations technologiques majeures, ce quart de siècle a surtout été marqué par une réflexion méthodologique concomitante à une révision de nos concepts et de nos données de base.

Les méthodes des recherches dans les abris sous roche étaient bien souvent un mélange de techniques empruntées à des domaines scientifiques variés (géologie, paléontologie, physique, chimie, mathématiques, anthropologie et archéologie) qui participèrent à l'élaboration de modèles fragiles et instables, véritables « châteaux de cartes » fondés sur des raisonnements circulaires et débouchant sur de graves contradictions (Djindjian 2000) maintes fois dénoncées.

En 1995, avec J.-P. Texier et F. Delpech nous avons entrepris une révision critique des interprétations chronologiques, paléoenvironnementales et culturelles des principaux « sites de référence » périgourdins. Au cours de ces travaux, auxquels nous renvoyons pour plus de détails Delpech F. *ce volume*, Texier J.-P. *ce volume*), il nous est apparu nécessaire de distinguer :

- la *lithostratigraphie* qui est établie par le géologue sur la base de critères strictement sédimentaires. Celle-ci prend en compte les origines des sédiments, la dynamique sédimentaire, les altérations biologique, physico-chimique et éventuellement anthropique post-dépositionnelles. Elle permet une analyse taphonomique du site et contribue à l'établissement du processus de formation du site.

- la *biostratigraphie* qui est établie sur la base des méthodes de la paléontologie. Elle permet de reconstruire les traits essentiels de l'environnement animal et végétal à partir des vestiges osseux

---

[1] Cette fréquente référence au modèle périgourdin lui a conféré parfois des allures de dogme que certains ont stigmatisé en évoquant un véritable "impérialisme périgourdin" !

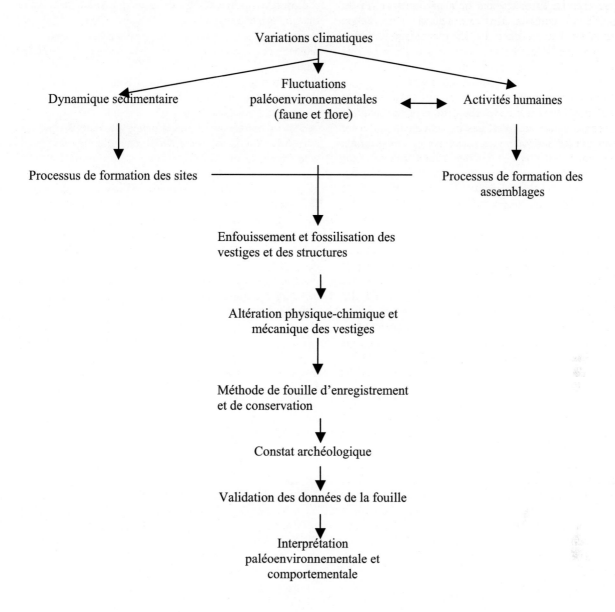

Figure 2.1. Résume les étapes d'une démarche tenant compte de ce qui vient d'être évoqué.

recueillis dans les sites archéologiques compte tenu du biais qu'ont introduit les activités humaines passées.

- l'*archéostratigraphie* qui est établie sur la base des données de la fouille (*constat archéologique*[2]) et de l'application des techniques de validation des ensembles archéologiques : évaluation de leur synchronicité par la répartition spatiale (diagrammes de projection, analyse de fabric), les remontages, raccords et appariements, la présence éventuelles de *réelles* structures évidentes (foyers structurés, dallages intentionnels, nappes de pigment…)

S'agissant de démarches ayant leur propre méthode, la lithostratigraphie, la biostratigraphie et l'archéostratigraphie doivent être établies indépendamment jusqu'à leur confrontation. Le débat auquel nous assistons actuellement sur la prétendue interstratification Castelperronien/Aurignacien à la Grotte des Fées à Châtelperron montre tout l'intérêt de cette démarche.

D'autre part, l'interprétation « pompéienne » des nappes de vestiges restera dans le domaine de l'utopie et des « belles histoires » tant que la contemporanéité des vestiges et leur localisation spatiale initiale n'auront pas été établies objectivement. L'interprétation spontanée et univoque de la répartition des vestiges doit préalablement laisser la place à la prise en compte de **tous** les facteurs potentiellement responsables de la disposition actuelle observée, le « constat archéologique », dont l'apparence

---

[2] Le "constat archéologique" (*archaeological statement*) est ce que l'archéologue peut constater après avoir mis au jour les vestiges et structures. C'est une stricte constatation objective sans interprétation.

doit être soumise à une critique taphonomique prenant en compte la dynamique sédimentaire, la conservation différentielle des vestiges et les perturbations post-dépositionnelles (**Figure 2.1**).

Ces considérations méthodologiques ne sont pas spécifiques de l'archéologie des grottes et des abris, elles concernent également les sites de plein air. Cependant, la bonne conservation des vestiges généralement observée dans les dépôts sous abri a souvent donné l'impression d'une intégrité presque parfaite, peu ou pas critiquée dans la publication des résultats (Bordes, 1972). Nos travaux en Afrique du sud dans l'abri-sous-roche de DIEPKLOOF, nous ont confronté à cette situation (Rigaud et al, 2006). Dans cet abri, plus de 2,5 m de sédiments variés contiennent des industries du MSA dont l'âge est compris entre 80 et 50 000 ans. Dans certains niveaux, l'état de conservation des vestiges organiques est exceptionnellement bon et nous avons observé fréquemment des accumulations de branches, de feuilles et d'herbe sur des surfaces de plusieurs décimètres carrés. De telles accumulations permettent de considérer qu'il s'agit de vestiges *in situ* qui n'ont pas été affectés par la moindre érosion. Cependant, ces zones d'accumulation sont séparées par de petits chenaux correspondant à un ruissellement diffus qui a provoqué une légère érosion et des dépôts d'accumulation lenticulaires de couleur noire (microcharbons), grise (cendres) ou brun rouge (sable ruisselé) dans lequel les artefacts et autres vestiges sont en position secondaire. Cette mosaïque de surfaces en positions primaires et de zones perturbées – même si les déplacements ont été faibles – ne peut être considérée comme un authentique « sol d'habitat » homogène (Bordes *et al*, 1972). L'empilement de ces formations lenticulaires ne permet pas de suivre lors de la fouille une surface isochrone et la contemporanéité des zones *in situ* ne pourra être établie par les méthodes habituelles: projections, raccords, remontages, appariements, etc.

Il est évident que dans des sites ne présentant pas d'aussi bonnes conditions de conservation des vestiges organiques notamment végétaux, et ils sont très largement majoritaires sous nos latitudes, la distinction entre les zones perturbées et celle qui ne le sont pas est beaucoup plus délicate et le risque de produire de pseudo-sols d'habitat n'est pas négligeable.

CONCLUSIONS

Les contraintes que nous venons d'évoquer ne sont pas spécifiques à l'archéologie des abris sous roche, elles concernent tous les types de site qu'ils soient de plein air, ou sous abri. Ce qui est particulier à ces derniers est qu'ils présentent souvent des lithostratigraphies plus longues et plus complexes, mises en place par une dynamique sédimentaire et une diagénèse fortement marquées par la morphologie de l'abri. Cette complexité augmente les possibilités de contamination entre les nappes de vestiges par de causes naturelles (ruissellement, cryoturbation, solifluxion, bioturbation) ou anthropiques (aménagements de l'habitat, creusement de sépultures).

A la question « l'archéologie des abris est-elle spécifique ? » il est tentant de répondre négativement. Elle implique une étroite collaboration pluridisciplinaire, des méthodes de fouille et d'enregistrement rigoureuses et pragmatiques, une certaine expérience et un savoir-faire acquis sur le terrain. Tout le reste n'étant qu'une question de détails techniques (matériel de fouille, topographique et autre équipement électronique). Rien que de très banal en somme si ce n'est que la fouille d'un abri-sous-roche s'inscrit dans la durée et qu'une dizaine de campagnes suivies de quelques années d'analyses critiques des données seront nécessaires pour atteindre les objectifs attendus.

BIBLIOGRAPHIE

BORDES F., 1972. Compte rendu critique de H. de Lumley « La Grotte de l'Hortus », *Quaternaria*, 16, 1972 :299-305.

BORDES F., RIGAUD J.-Ph. et Sonneville-Bordes (de) D. 1972. Des buts problèmes et limites de l'archéologie paléolithique, *Quaternaria,* 16, 1972 : 15-34.

DEMARS P.-Y. et LAURENT P., 1989. Types d'outils lithiques du Paléolithique supérieur en Europe. *Cahiers du Quaternaire* 14, CNRS, Paris.

DJINDJIAN F, 2000. Human adaptation to the climatic deterioration of the last Pleniglacial in southwestern France (30,000 to 20,000 BP). In *The Mid Upper Pallaeolithic (30 000 to 20 000 BP). In France in Hunters of the golden age* edited by WILL ROEBROEKS, MARGHERITA MUSSI, et JIRI SVOBODA (eds.), pp. 313-324. University of Leiden.

LAVILLE H., RIGAUD J.-Ph., SACKETT J. Rock Shelters of the Périgord; Geological stratigraphy and archaeological succession. Studies in archaeology, Academic Press, 1980.

RIGAUD J.-Ph. 1994. L'évaluation contextuelle préalable à l'analyse de la répartition spatiale des vestiges. *Préhistoire, Anthropologie Méditerranéennes*, 1994, t. 3 : 39-41.

RIGAUD J.-Ph., TEXIER P.-J., PARKINGTON J. et POGGENPOEL C., 2006. Le mobilier Stillbay et Howiesons Poort de l'abri Diepkloof. La chronologie du Middle Stone Age sud-africain et ses implications. *C.R. Palévol* 5 (2006) 839-849.

APPROACHES TO THE MIDDLE PALEOLITHIC ROCKSHELTER AND CAVE RESEARCH IN CROATIA

ASPECTS DE LA RECHERCHE DES ABRIS ET DES GROTTES DU PALÉOLITHIQUE MOYEN EN CROATIE

Ivor KARAVANIĆ[*], Nikola VUKOSAVLJEVIĆ[*], Rajna ŠOŠIĆ[*], and Sanjin MIHELIĆ[**]

[*]Department of Archaeology, Faculty of Humanities and Social Sciences, University of Zagreb Ivana Lučića 3, HR-10000 Zagreb, Croatia; ikaravan@ffzg.hr
[**]Archaeological Museum in Zagreb, Zrinjevac 19, HR-10000 Zagreb, Croatia

Abstract. This paper presents the research history of the Middle Paleolithic cave sites in Croatia traced on the basis of investigations of three sites: Krapina, Vindija (both in northwestern Croatia) and Mujina Pećina (southern Croatia). The first Middle Paleolithic site to be discovered was Krapina. It was excavated by Dragutin Gorjanović-Kramberger. His excavation methodology was very sophisticated at the time, and included stratigraphic excavation and notes on the horizontal distribution of finds in certain parts of the site. The Vindija cave yielded important finds of late Neanderthals associated with Mousterian industry (level G3) and possibly with Upper Paleolithic bone industry in level G1. In southern Croatia, the only systematically excavated Middle Paleolithic site is Mujina Pećina. The Mousterian people in both regions of Croatia (northwestern and southern) successfully adjusted the production of their tools to various types of the most easily accessible raw materials, and they were successful predators.

Keywords: Middle Paleolithic, Krapina, Vindija, Mujina Pećina, Croatia

Résumé. Cette communication présente l'historique de la recherche des sites en grottes du Paléolithique Moyen en Croatie, en s'appuyant plus particulièrement sur l'historique des fouilles de trois sites: Krapina, Vindija (en Croatie du Nord-ouest) et Mujina Pećina (en Croatie méridionale). Le premier site datant du Paléolithique Moyen découvert en Croatie était Krapina. Ce site a été fouillé par Dragutin Gorjanović-Kramberger. Ses méthodes de fouilles étaient en avance sur son temps, car il suivait la stratigraphie et notait la distribution horizontale des trouvailles à plusieurs endroits du site. Des trouvailles importantes de Néanderthaliens tardifs en association avec de l'industrie moustérienne (niveau G3) et peut-être aussi en association avec des outils en os datant du Paléolithique Supérieur (niveau G1) furent découvertes dans la grotte de Vindija. En Croatie méridionale, le seul site du Paléolithique Moyen à avoir été fouillé systématiquement est la grotte de Mujina Pećina. Les hommes du Moustérien dans les deux régions de la Croatie (Nord-ouest et Sud) avaient su adapter leur production d'outils aux sources de matières premières disponibles et étaient des prédateurs efficaces.

Mots-clé: Paléolithique Moyen, Krapina, Vindija, Mujina Pećina, Croatie

There is a long tradition of cave and rockshelter research in Croatia. Even though systematic excavations of Paleolithic sites in Croatia have not been numerous, several sites are of great importance for research on the adaptation of the Paleolithic populations and their material cultures. These sites are located in two regions within different environmental zones. The most famous sites, such as Krapina and Vindija, are situated in the continental zone, which distinguishes them geographically and environmentally from the Mediterranean sites situated farther south, on the eastern Adriatic coast and its hinterland. Besides Krapina and Vindija important sites from continental (northwestern) Croatia include Velika Pećina and Veternica, while in southern Croatia the only systematically excavated Middle Paleolithic site is Mujina Pećina. This paper presents a brief history and different approaches to the research of the Middle Paleolithic rockshelters and caves in Croatia, traced on the basis of investigations of three sites: Krapina, Vindija and Mujina Pećina (**Figure 3.1**).

SITES

Krapina

The first Middle Paleolithic site discovered was Krapina, located in northwestern Croatia. The story of the Krapina Neanderthals started on 23rd August 1899, when Dragutin Gorjanović-Kramberger arrived in Krapina to visit the site on *Hušnjakov Brijeg* ("Hušnjak Hill"), from where as early as 1895 the inquisitive school-teacher Josip Rehorić and a gentleman named Kazimir Semenić, sent him a couple of "curious" bones and teeth (Barić 1978; Radovčić 1988). Due to his intensive work on the geological mapping of Croatia he was not able to visit the Krapina site before. Upon his arrival at the site, he immediately observed the sequence of deep cultural layers with hearths and ash; further, the lithic industry, pieces of animal bones etc., and on the very day the prehistoric site was discovered he also found a human tooth. Realizing that the site was of exceptionally great importance, with the help of eminent local citizens he put

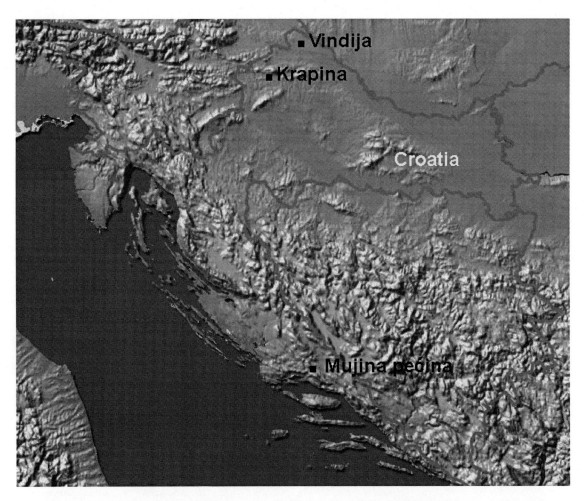

3.1 Map showing Middle Paleolithic sites in Croatia discussed in the paper.

3.2 Excavations of the site at Krapina: man in the black suit in the centre is most likely
D. Gorjanović-Kramberger and near him is his assistant S. Osterman (after Gorjanović-Kramberger 1913).

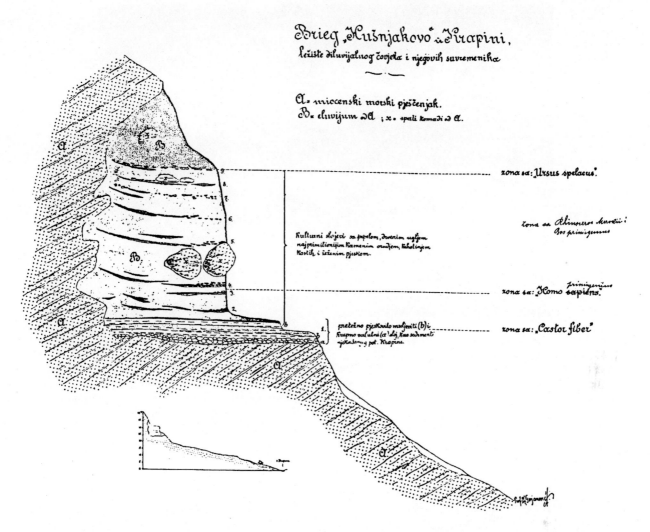

3.3 Stratigraphic profile at Krapina. Drawing by D. Gorjanović-Kramberger (after Gorjanović-Kramberger 1906)

a stop to the further devastation of the site, often quarried for sand by the local population. A little later he arrived at the site with his assistant, Stjepan Osterman, and, having first made a plan, started to excavate the uppermost, ninth layer (Figure 3.2).

In the course of the excavation he renumbered the layers and separated four zones within them based on the frequency of faunal finds: the 1st zone with *Castor fiber* (beaver), the 2nd zone with *Homo sapiens* (man), the 3rd zone with *Rhinoceros merckii* (rhino) and a zone with *Ursus spelaeus* (cave bear) (Figure 3.3). He marked the finds with the number of the layer they belonged to. He continued the excavation of the rockshelter at Hušnjak Hill until 1905. In 1901 he fell ill of tuberculosis, so that year the excavation was supervised by S. Osterman, who himself participated in the excavation as Gorjanović's assistant from the beginning. Already in 1903 Gorjanović once again led the excavations in Krapina himself, and he carried out very extensive final excavations in 1905. The documentation of the Geological and Paleontological Department of the Croatian Museum of Natural History in Zagreb, as well as published works (Malez 1978a),

reveal that Gorjanović left a witness section of the site, which disappeared after his death. During the entire excavation more than five thousand objects were collected (bones of prehistoric man and animals, artifacts), of which 874 were human remains (Radovčić et al. 1988) and 1191 were lithic artifacts (Simek and Smith 1997). He carried out his excavations with precision, following the horizontal distribution of the finds, so he made a note that most of the finds were adjacent to the rockshelter wall. He also excavated according to the natural sequence of layers, which was a methodological feature more advanced than the usual excavation methods of the time. He was the first in the world to apply the fluor-test for relative age determination, and to make X-ray images of bones.

From the very start Gorjanović believed that the fossil remains of human bones belonged to the prehistoric man of the *Homo sapiens* species. He later adopted the standard term of that time for prehistoric man, labeling the finds with the name *Homo primigenius* (Gorjanović-Kramberger 1906). This was a term used by contemporary German prehistorians for almost all

3.4 Vindija cave entrance. Photo taken by I. Karavanić

remains of Pleistocene humans, and it wasn't before the beginning of the 1920s that he accepted the term *Homo neanderthalensis*.

He correctly attributed the lithic industry to the Mousterian period and demonstrated the exploitation of local sources of raw materials from the nearby Krapinica stream (Gorjanović-Kramberger 1913), for tool production, as was later substantiated in several papers (Zupanič 1970; Simek and Smith 1997).

From the time of discovery until today the Krapina finds have been essential for the research on the relationship between morphology, culture and the way of life of the Neanderthals. The human remains from Krapina have been analyzed several times by different scholars (Gorjanović-Kramberger 1899, 1906; Smith 1976; Wolpoff 1979; Russel 1987; Radovčić et al. 1988; Russel et al. 1988; Turk and Dirjec 1991; Laluze Fox and Frayer 1997). Furthermore, Krapina is today considered the site with the most numerous finds of Neanderthal dental remains, and the most numerous Neanderthal individual remains, which according to Wolpoff (1979) belong to as much as seventy or so individuals. Some of the levels were dated by ESR to about 130 ka BP, and the entire stratigraphic sequence is to be attributed to the late Riss glacial or to the Riss-Würm interglacial (the end of oxygen isotope stage 6 or beginning of 5e) (Rink et al. 1995; Simek and Smith 1997).

A technological and typological analysis of lithic assemblages from Krapina indicates that the site was occupied only by the Mousterian people at least twice and that local river cobbles were used for tool production (Zupanič 1970; Simek 1991; Simek and Smith 1997). The raw material inventory is dominated by tuffs and silicified tuffs (65%) while chert amounts to 23% (Simek and Smith 1997:566). Various scrapers dominate the Charentian Mousterian assemblages of the site. Only a small proportion of the tools were brought into the site in finished form while most were produced *in situ* on diverse blanks. Though examples of various raw materials were uncovered from the earlier levels, high quality cherts and silicified tuffs were found only in the later levels (Simek and Smith 1997). Both the Levallois method and the cobble wedge exploitation method were used. The small number of lithics in all levels and the presence of some non-local materials suggest a brief occupation by mobile groups of people (Simek 1991), although long-term occupation at some occupation levels is also possible (Patou-Mathis 1997). In addition to being efficient tool makers, the Krapina Neanderthals appear to have been successful hunters of large game, including Merck's rhinoceros (Miracle 1999). Other identified species in the Krapina faunal assemblage include *Bos/Bison, Ursus spelaeus, Castor fiber,* etc. (Patou-Mathis 1997).

Several authors have suggested that Neanderthals from Krapina were cannibals (e.g. Gorjanović-Kramberger 1906; Smith 1976; White 2001). However, this hypothesis is controversial because it is also possible that human bodies were defleshed with stone tools in preparation for secondary burial (Russell 1987), or buried by natural or human processes after death (Trinkaus 1985).

Vindija

Another important site in northwestern Croatia is the Vindija cave, situated 20 km west of Varaždin. Its entrance lies in a narrow gorge 275 m a.s.l. The cave is more than 50 m deep, up to 28 m wide and more than 10 m high (**Figure 3.4**).

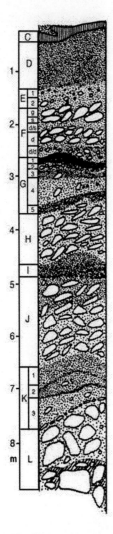

3.5 Vindija stratigraphic profile
(modified after Ahern et al. 2004 : Figure 1)

Stone artifacts and fauna on the site were discovered and first published by Stjepan Vuković (1953), who in the autumn of 1928 came to the site for the first time. For more than thirty years, with some interruptions, he carried out excavations of the cave and the area in front of it, mostly in the uppermost layers. Due to insufficient financial means, Vuković's fieldwork was limited to small trenches, generally between 3-4 m, and often even smaller. Under the circumstances, Vuković achieved relatively good results. He excavated with respect to the stratigraphy of the site and he determined the material cultures of the site.

Systematic excavations of the Vindija cave were started by Mirko Malez in 1974. The finds of fossil humans and their material cultures make Vindija a site of particular scientific value. The excavations started in the area in front of the cave with the aim of opening longitudinal and transversal sections along the central part of the cave by means of a spacious trench, in order to establish the precise stratigraphic relations of the quaternary sediments (Malez 1975, 1978b). The fieldwork continued every year until 1986. It was during this period that most of the lithic and faunal material, as well as all of the human fossil remains, were recovered. Attention was paid to following the stratigraphic relations but unfortunately there is very little data regarding the horizontal distribution of finds, and only a small part of the sediment was sieved.

The stratigraphic profile, which is about 9 m high, comprises about twenty strata which according to Malez and Rukavina (1979) covered the period from the onset of the Riss glaciation (oxygen isotope stage 6 or earlier) through the Holocene (**Figure 3.5**). One U-Th date on a bone from Level K of 114 ka suggests an age from the last interglacial for these deposits, while U-Th dates for older levels (L & M) are inconsistent and unreliable (Wild et al. 1988). Level G3, which contains Neanderthal remains associated with the late Mousterian industry was dated to 42.4 ka BP by amino-acid racemization (Smith et al. 1985). Neanderthal remains from the same level were directly dated over to 42 ka BP and to approximately 38 ka BP by AMS (Krings et al. 2000; Serre et al. 2004). There is the possibility that these dates overlap if two standard deviations for the later date are taken into account (Serre et al. 2004). There are several radiocarbon dates from level G1 (see Wild et al. 2001; Ahern et al. 2004: Table1) but the most important are direct dates from the samples of Neanderthal bones from the same level. Neanderthal remains from level G1 have been directly dated by AMS to 29 and 28 ka BP (Smith et al. 1999) and recently redated to approximately 33 ka BP (Higham et al. 2006).

Neanderthal remains from level G3 show distinct changes in facial morphology when compared to earlier Neanderthals (see Wolpoff et al. 1981; Smith 1984; Wolpoff 1999). Among such changes are supraorbital tori with a shape somewhere between the Krapina Neanderthals and early modern Europeans (Karavanić and Smith 1998:239).

In the lower Mousterian levels tools were produced on local raw materials (Kurtanjek and Marci 1990; Blaser et al. 2002) using the Levallois technique. In contrast, the Levallois technique was not employed in level G3, where local raw materials (chert, quartz, tuff, etc.) were also used. Level G3 yielded 350 lithic finds of which 60 (17.1%) are tools. The Late Mousterian industry from the same level is dominated by sidescrapers, notched pieces and denticulates, but also includes some Upper Paleolithic types (e.g., endscrapers). In addition to flake technology, level G3 also includes evidence of bifacial technology and blade technology (Karavanić and Smith 1998).

As in level G3, a mixture of the Middle and Upper Paleolithic typological characteristics is also present in the stone tool assemblage from level G1, where bone points and Neanderthal remains were found. While the poor lithic industry of this level suggests a continuation of the Mousterian technological and typological tradition (with the lack of the Levallois method), bone tools from the same level are typical of the Upper Paleolithic. Although an unusual association of Neanderthal remains and Upper Paleolithic bone points in level G1 can be explained as a result of the mixing of these remains through different strata of origin (Zilhão and d'Errico 1999), it may well also represent contemporaneously used tools, that is, an assemblage (Karavanić and Smith 1998, 2000). However, the small archaeological assemblage from the level might suggest very brief occupation by mobile groups of Neanderthal people.

According to the evidence from stable isotopes, these Neanderthals behaved as highly efficient carnivores, fulfilling almost all of their dietary needs from animal sources (Richards et al. 2000). They were efficient predators and possibly cannibals like Krapina Neanderthals, who occupied the same area some 100 ka earlier (White 2001).

Mujina Pećina

Although there is evidence of many Middle Paleolithic open air sites from southern Croatia (Batović 1988), and sediments from several caves were attributed to the Middle Paleolithic period, Mujina Pećina is the only systematically excavated site with homogenous Mousterian deposits in clear stratigraphic context to have been well-dated chronometrically (Rink et al. 2002). The cave, which is about 10 m deep and 8 m wide, is located in Dalmatia, north of Trogir and west of Split, in typical karstic terrain **(Figure 3.6)**. Lithic material was initially collected from the surface inside and in front of the cave entrance (Malez 1979). The results from the first excavation, undertaken in 1978, were briefly reported by Petrić (1979). From 1995 to 2003 systematic excavations were carried out as part of a joint project of the Department of Archaeology, University of Zagreb and the Kaštela City Museum. Excavation levels followed the natural stratigraphy, all sediments were sieved, and all artifacts and ecofacts over 2 cm in size were entered in three dimensions on site plans **(Figure 3.7)**. The stratigraphic profile in the cave is only about 1.5 m deep. The deposits comprise poorly sorted Quaternary sediments composed of large fragments of carbonate rock (debris), gravel and sand grains, rarely silt, and some clay **(Figure 3.8)**.

Electron spin resonance (ESR) dating was conducted on two teeth from Mousterian level E1 at Mujina Pećina, while five bone samples and one charcoal sample coming from five different strata were dated by AMS ^{14}C (Rink et al. 2002). For level E1, mean ESR age estimates of 40±7 ka (EU) and 44±5 ka (LU) have been obtained (assuming 30% moisture for both the gamma and beta dose rate calculations). The interface between Level E2 and E1 has been dated by AMS to 45,170 ± 2780/2060 BP (GrA-9635) while the AMS age of overlying levels (D2, D1, C and B) calculated as the mean of the five dates from these levels is 39,222 ± 2956 BP (for individual dates and lab numbers see Rink et al. 2002, Table 3), and the true (calibrated) age of this mean age is about 42 kya. These

3.6 View from the area in front of Mujina Pećina. The karstic landscape is marked by hilly terrain and the Kaštela Bay, which was

3.7 Excavating in Mujina Pećina. Bedrock is visible in excavated squares in the middle of the cave. Photo taken by S. Burić.

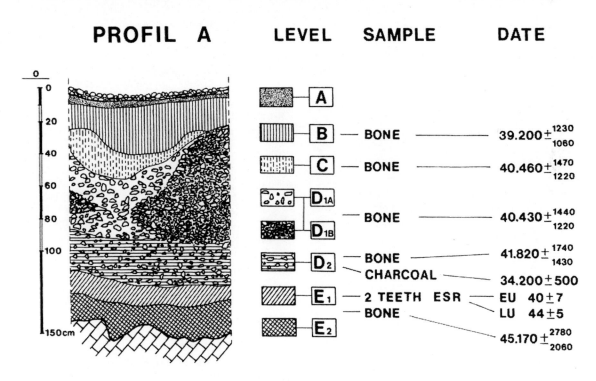

3.8 Stratigraphic profile of Mujina Pećina (modified after Rink et al. 2002: Figure 3)

dates suggest that all the Mujina Pećina Mousterian materials were deposited very rapidly during oxygen isotope stage 3 and they indicate that the radiocarbon results are in close agreement with the ESR ages calculated at 30% moisture (Rink et al. 2002).

Levels D1 and D2 are characterized by a significant presence (about 20% of all debitage) of Levallois debitage. In contrast, there is a modest representation of Levallois debitage in overlying level B, including one small Levallois core which suggests that the "micro-Levallois" technique may have been used, but regular flakes, flakes ≤ 2 cm in size, and debris are all more frequent. Over 50% of all primary, secondary and regular flakes in level B were transformed into tools, including small tools similar to those of the Micromousterian type.

Such small tools were found in different levels of Mujina Pećina along with "typical" Mousterian tools, the majority of which are retouched flakes, notched pieces and denticulates. Sidescrapers are also present in all levels, but we only note a significant presence of "Upper Paleolithic" types in level B.

The small size of tools at Mujina Pećina can probably best be explained by the size of available local raw material, which consists of small chert pebbles and nodules whose diminutive size impeded the removal of much cortex on the tools and cores from level B, for instance. The low flaking quality of local cherts present in larger packages also limits tool size, since knapping experiments have demonstrated that it was rarely possible to detach large and regular flakes from the larger blocks of such materials and, moreover, that many of the large flakes produced experimentally would break during retouching.

Miracle (2005) has been able to demonstrate the human action as a factor on the remains of numerous animals including chamois, ibex, deer, auroch and bison. On the other hand the accumulation of bones of animals such as equids and hare can most likely be attributed to carnivore activity. Various carnivores were using the cave when people left, often feasting on the remains of human meals.

Based on the fetal and neonatal faunal remains, Miracle (2005) was also able to establish the exact seasons during which some of these episodes took place. During the period that level B was accumulated, most of these visits took place during fall, and possibly in spring. Spring episodes are also likely during the formation of level D1. People were not present at the site during summer or during winter, when the cave served as a bear den.

Starting at the bottom of the Mujina Pećina sequence, the oldest levels (E3, E2, E1) are the richest in archaeological material, which indicates much more intensive use of the site by humans than in more recent levels (D2, D1, B and C) where artifacts are less dense. The richness of levels E3, E2, and E1 may represent long-term occupations of the cave but may also have been deposited as the result of the site being occupied for many brief occupations (Conard 1996). Levels D2, D1 and B probably represent short hunting episodes while the level C is only present in a very limited part of the cave.

Two fireplaces were found in level D2. They have no organized structure and are basically the remains of fires. Near one of the fireplaces, in a niche on the right side of the cave, a part of a deer antler and several stone artifacts and bones were found. An analysis of the charcoal from the fireplaces, done by Meta Culiberg (personal communication), showed that the most commonly used firewood was juniper (*Juniperus* sp.), probably from the vicinity of the cave and dried before use. The finds of bones in level D2 are concentrated within the niche on the right side of the cave, where one of the fireplaces was found. This is not a surprise, as this area of the cave is sheltered enough and provides protection from the wind and lower temperatures. In contrast to the bones, stone artifacts show no such concentration and they are much more evenly distributed.

CONCLUSION

The Middle Paleolithic rockshelters and caves in Croatia are analyzed today on two basic levels. One comprises the use of contemporary analytic methods for the material from previously excavated sites, while the other encompasses reconnaissance surveys and excavation of sites with state-of-the-art methods. It is also important to compare the results obtained for north-western Croatia with those for Dalmatia, as these regions belonged to two different climatic and paleo-ecological zones. The results show that in both zones the Neanderthals successfully adjusted the production of their tools to the types of most easily accessible raw materials, and that they were successful predators. The success of their adaptation was therefore not strictly conditioned by the environment, climate and available raw materials, as it was already suggested for some other European sites (Patou-Mathis 2000). It is expected that, along with reviews of old collections, current fieldwork will provide new evidence important for the interpretation of the Middle Paleolithic of Croatia.

Acknowledgements. We thank Marcel Kornfeld for the invitation to participate in the session "On shelter's ledge. Histories, theories and methods of rockshelter research". Our research was sponsored by the Ministry of Science, Education and Sports of the Republic of Croatia.

REFERENCES

AHERN, J. C. M.; KARAVINIĆ, I.; PAUNOVIĆ, M.; JANKOVIĆ, I. and SMITH, F.H. (2004). New discoveries and interpretations of hominid fossils and artifacts from Vindija Cave, Croatia. *Journal of Human Evolution* 46, p. 25 – 65.

BARIĆ, Lj. (1978). Dragutin Gorjanović-Kramberger i otkriće krapinskog pračovjeka. In MALEZ, M., ed. *Krapinski pračovjek i evolucija hominida*. Zagreb: Izdavački zavod Jugoslavenske akademije znanosti i umjetnosti, p. 23 – 46.

BATOVIĆ, Š. (1988). Paleolitički i mezolitički ostaci s Dugog otoka. *Poročilo o raziskovanju paleolita, neolita in eneolita v Sloveniji*. Ljubljana. 16, p. 7 – 54.

BLASER, F. ; KURTANJEK, D. ; and PAUNOVIĆ, M. (2002). L'industrie du site néandertalien de la grotte de Vindija (Croatie): une révision des matières premières lithiques. *L'Anthropologie* 106, p. 87 – 398.

CONARD, N. J. (1996). Middle Paleolithic Settlement in Rhineland. In CONARD, N. J., ed. *Middle Paleolithic and Middle Atone Age Settlement System*. U.I.S.P.P. - XIII Congrés. Forlì: ABACO, p. 255 – 268.

GORJANOVIĆ-KRAMBERGER, D. (1899). Paleolitički ostaci čovjeka i njegovih suvremenika iz diluvija u

Krapini. *Ljetopis Jugoslavenske akademije znanosti i umjetnosti*. Zagreb. 14, p. 90 – 98.

GORJANOVIĆ-KRAMBERGER, D. (1906). *Der diluviale Mensch von Krapina in Kroatien. Ein Beitrag zur Paläoanthropologie*. Wiesbaden: Kreidel.

GORJANOVIĆ-KRAMBERGER, D. (1913). *Život i kultura diluvijalnog čovjeka iz Krapine u Hrvatskoj*. Djela Jugoslavenske akademije znanosti i umjetnosti 23. Zagreb: Jugoslavenska akademija znanosti i umjetnosti.

HIGHAM, T.; BRONK RAMSEY C.; KARAVANIĆ, I,; SMITH, F. H. and TRINKAUS, E. (2006). Revised direct radiocarbon dating of the Vindija G1 Upper Paleolithic Neandertals. *Proceedings of the National Academy of Sciences*, USA 103, p. 553 – 557.

KARAVANIĆ, I. and SMITH F. H. (1998). The Middle/Upper Paleolithic interface and the relationship of Neanderthals and early modern humans in the Hrvatsko Zagorje, Croatia. *Journal of Human Evolution* 34, p. 223 – 248.

KARAVANIĆ, I. and SMITH F. H. (2000). More on the Neanderthal problem: The Vindija case. *Current Anthropology* 41, p. 838 – 840.

KRINGS, M.; CAPELLI, C.; TSCHENTSCHER, F.; GEISERT, H.; MEYER, S.; VON HAESELER, A.; GROSSCHMIDT, K.; POSSNERT, G.; PAUNOVIĆ, M. and PÄÄBO, S. (2000). A view of Neandertal genetic diversity. *Nature Genetics* 26, p. 144 – 146.

KURTANJEK, D. and MARCI, V. (1990). Petrografska istraživanja paleolitskih artefakata spilje Vindije. *Rad Jugoslavenske akademije znanosti i umjetnosti*. Zagreb. 449(24), p. 227 – 238.

LALUZE FOX, C. and FRAYER D. W. (1997). Non-dietary marks in the anterior dentition of the Krapina Neanderthals. *International Journal of Osteoarchaeology* 7, p. 133 – 149.

MALEZ, M. (1975). Die Höhle Vindija – eine neue Fundstelle fossiler Hominiden in Kroatien. *Bulletin scientifique Conseil des Academies des Sciences et des Arts de la RSF de Yougoslavie*. Zagreb. 20/5 – 6, p. 139 – 141.

MALEZ, M. (1978a). Stratigrafski, paleofaunski i paleolitski odnosi krapinskog nalazišta. In MALEZ, M., ed. *Krapinski pračovjek i evolucija hominida*. Zagreb: Izdavački zavod Jugoslavenske akademije znanosti i umjetnosti, p. 61 – 91.

MALEZ, M. (1978b). Novija istraživanja paleolitika u Hrvatskom zagorju. In RAPANIĆ, Ž., ed. *Arheološka istraživanja u sjeverozapadnoj Hrvatskoj*. Zagreb: Hrvatsko arheološko društvo, p. 6 – 69. (Izdanja Hrvatskoga arheološkog društva; 2).

MALEZ, M. (1979). Nalazišta paleolitskog i mezolitskog doba u Hrvatskoj. In BENAC, A., ed. *Praistorija jugoslavenskih zemalja*, vol. I. Sarajevo: Svjetlost, p. 227 – 276.

MALEZ, M. and RUKAVINA, D. (1979). Položaj naslaga spilje Vindije u sustavu članjenja kvartara šireg područja Alpa. *Rad Jugoslavenske akademije znanosti i umjetnosti*. Zagreb. 383, p. 187 – 218.

MIRACLE, P. T. (1999). Rhinos and beavers and bears, oh my! Zooarchaeological perspectives on the Krapina fauna 100 years after Gorjanović. *International Conference „The Krapina Neandertals and Human Evolution in Central Europe"*. Program and Book of Abstracts, Zagreb – Krapina, p. 34.

MIRACLE, P. T. (2005). Late Mousterian subsistence and cave-use in Dalmatia: the zooarchaeology of Mujina Pećina, Croatia. *International Journal of Osteoarchaeology* 15, p. 84 – 105.

PATOU-MATHIS, M. (1997). Analyses taphonomique et palethnographique du matériel osseux de Krapina (Croatie): nouvelles données sur la faune et les restes humains. *Préhistoire Européenne*. Liège. 10, p. 63-90.

PATOU-MATHIS, M. (2000). Neanderthal subsistence behaviour in Europe. *International Journal of Osteoarchaeology* 10, p. 379 – 395.

PETRIĆ, N. (1979). Mujina pećina, Trogir – paleolitičko nalazište. *Arheološki pregled*. Beograd. 20 (1978), p. 9.

RADOVČIĆ, J. (1988). *Dragutin Gorjanović Kramberger i krapinski pračovjek: počeci suvremene paleoantropologije*. Zagreb: Hrvatski prirodoslovni muzej and Školska knjiga.

RADOVČIĆ, J.; SMITH, F. H.; TRINKAUS, E. and WOLPOFF, M. H. (1988). *The Krapina Hominids: An Illustrated Catalog of Skeletal Collection*. Zagreb: Mladost and Croatian Natural History Museum.

RICHARDS, M. P.; PETTITT, P. B.; TRINKAUS, E.; SMITH F.H.; PAUNOVIĆ, M. and KARAVANIĆ, I. (2000) - Neanderthal diet at Vindija and Neanderthal predation: The evidence from stable isotopes. *Proceedings of the National Academy of Sciences*, USA 97, p. 7663 – 7666.

RINK, W. J.; SCHWARCZ, H. P.; SMITH, F. H. and RADOVČIĆ, J. (1995), ESR ages for Krapina hominids. *Nature* 378, p. 24.

RINK W. J.; KARAVANIĆ, I.; PETTITT, P. B.; VAN DER PLICHT, J.; SMITH, F. H. and BARTOLL, J. (2002), ESR and AMS based ^{14}C dating of Mousterian levels at Mujina Pećina, Dalmatia, Croatia. *Journal of Archaeological Science* 29, p. 943 – 952.

RUSSELL, M. D. (1987). Bone breakage in the Krapina hominid collection. *American Journal of Physical Anthropology* 72, p. 373 – 379.

RUSSELL, M. D.; VILLA, P. and COURTIN, J. (1988). A reconsideration of the Krapina cutmarks. *Collegium Antropologicum*. Zagreb. 12, Suppl., Abstracts, p. 348.

SERRE, D.; LANGANEY, A.; CHECH, M.; TESCHLER-NICOLA, M.; PAUNOVIC, M.; MENNECIER, P.; HOFREITER, M.; POSSNERT, G. and PÄÄBO, S. (2004). No evidence of Neanderthal mtDNA contribution to early modern humans. *Public Library of Science Biology* 2, p. 313 – 317.

SIMEK, J. F. (1991). Stone Tool Assemblages from Krapina (Croatia, Yugoslavia). In MONTET-WHITE, A.; HOLEN, S., eds. *Raw Material Economies Among Prehistoric Hunter-Gatherers*. Lawrence: University of Kansas, p. 59 – 71. (Publications in Anthropology; 19).

SIMEK J. F. and SMITH, F. H. (1997). Chronological changes in stone tool assemblages from Krapina (Croatia). *Journal of Human Evolution* 32, p. 561 – 75.

SMITH, F. H. (1976). *The Neandertal Remains from Krapina: A Descriptive and Comparative Study.* Reports of Investigation 15. Knoxville: University of Tennessee, Department of Anthropology.

SMITH, F. H. (1984). Fossil hominids from the Upper Pleistocene of Central Europe and the origin of modern Europeans. In SMITH, F. H.; SPENCER, F., eds. *The Origins of Modern Humans: A World Survey of the Fossil Evidence.* New York: Alan R Liss, p. 137 – 209.

SMITH, F. H.; BOYD, C. D. and MALEZ, M. (1985). Additional Upper Pleistocene Human Remains from Vindija Cave, Croatia, Yugoslavia. *American Journal of Physical Anthropology* 68, p. 375 – 388.

SMITH, F. H.; TRINKAUS, E.; PETTITT, P. B.; KARAVANIĆ, I. and PAUNOVIĆ, M. (1999). Direct radiocarbon dates for Vindija G1 and Velika Pećina Late Pleistocene hominid remains. *Proceedings of the National Academy of Sciences,* USA 96, p. 12281 – 12286.

TRINKAUS, E. (1985). Cannibalism and burial at Krapina. *Journal of Human Evolution* 14, p. 203 – 216.

TURK, I. and DIRJEC, J. (1991). Krapinski kanibalizem, kult lobanj in pokopi: primerjalna tafonomska analiza fosilnih ostankov *Homo sapiens neanderthalensis* in Krapine (Hrvaška). *Poročilo o raziskovanju paleolita, neolita in eneolita v Sloveniji.* Ljubljana. 19, p. 131 – 144.

VUKOVIĆ, S. (1953). Pećina Vindija kao prethistorijska stanica. *Speleolog.* Zagreb. 1/1, p. 14 – 23.

WILD, E. M.; STEFFAN, I. and RABEDER, G. (1988). Uranium-series dating of fossil bones, progress report. *Institut für Radiumforschung und Kernphysik* Wien 53, p. 53 – 56.

WILD, E. M.; PAUNOVIĆ, M.; RABEDER, G.; STEFFAN, I. and STEIER, P. (2001). Age determination of fossil bones from the Vindija Neanderthal site in Croatia. *Radiocarbon* 43, p. 1021 – 1028.

WHITE, T. D. (2001). Once were Cannibals. *Scientific American* 265(2), p. 47 – 55.

WOLPOFF, M. H. (1979). The Krapina dental remains. *American Journal of Physical Anthropology* 50, p. 67 – 114.

WOLPOFF, M. H. (1999). *Paleoanthropology.* 2nd ed. Boston: McGraw-Hill.

WOLPOFF, M. H.; SMITH, F. H.; MALEZ, M.; RADOVČIĆ, J. and RUKAVINA, D. (1981). Upper Pleistocene Human Remains from Vindija Cave, Croatia, Yugoslavia. *American Journal of Physical Anthropology* 54, p. 499 – 545.

ZILHÃO, J. and D'ERRICO, F. (1999). The Neanderthal problem continued: Reply. *Current Anthropology* 40, p. 355 – 364.

ZUPANIČ, J. (1970). Petrografska istraživanja paleolitskih artefakata krapinskog nalazišta. In MALEZ, M., ed. *Krapina 1899 – 1969.* Zagreb: Jugoslavenska akademija znanosti i umjetnosti, p. 131 – 140.

ARCHAEOLOGICAL RESEARCH IN ROCKSHELTERS AND CAVES IN SLOVENIA

RECHERCHES ARCHÉOLOGIQUES DANS DES ABRIS-SOUS-ROCHE ET GROTTES EN SLOVÉNIE

Martina KNAVS

Department of Archaeology, University of Ljubljana, Slovenia; Martina_Knavs@hotmail.com

Abstract. Western and southern Slovenia are predominantly karstic regions. Therefore these regions are rich in caves, overhangs and other karst manifestations. In this study, the author deals with the historiography of rock shelter and cave research in Slovenia and, more generally, with the human use of such places, from prehistory to present time. As specific sedimentary processes operating in rockshelters and caves demand a different methodological approach than those at classical open-air sites, this essay also presents a critical analysis of the methods and techniques developed and used by Slovene researchers, studying especially archaeological items and other artefacts.

Keywords: caves, rockshelters, Slovenia, archaeological research

Résumé. La région de la Slovénie, sur ses parties les plus occidentales et méridionales, est recouverte de karst. C'est pourquoi ces régions s'avèrent riches en grottes, surplombs et autres manifestations karstiques. Dans cette étude, l'auteur traite de l'historiographie des recherches menées en Slovénie dans des abris rocheux et grottes et, plus généralement, de l'usage humain fait de tels lieux, de la Préhistoire à l'époque actuelle. Comme les règles de sédimentation, s'appliquant à l'étude des abris-sous-roche et grottes, demande une approche méthodologique différente comparée aux fouilles classiques sur des sites à ciel ouvert, cet essai présente aussi une analyse critique des méthodes de travail et techniques développées et utilisées par les chercheurs slovènes, étudiants en particulier des items archéologiques et autres artefacts.

Mots-clés: grottes, abris, Slovenie, recherches archéologiques.

Four major European geographic regions meet in Slovenia: the Alps, the Dinaric area, the Pannonian plain and the Mediterrannean (**Figure 4.1**). The western and southern parts of Slovenia are covered with karst. In the past two centuries numerous caves and other karst forms were explored on the Kras region (southwestern part of Slovenia). Today they present the properties of Classical Karst from where the name of karstology derives and where the roots of speleology are found.

Some 9000 km² or 44% of the territory of Republic of Slovenia can be classified as karstic. Over two thirds of this territory (6300 km²) consist of limestones, mainly Mesozoic, whereas karst areas on other rocks (dolomite, conglomerate, calcarenite and breccia) occupy some 30% of the entire karst areas of Slovenia. The karst in Slovenia is commonly divided in relation to geological, hydrological and geomorphological characteristics into three major units: the alpine karst, dinaric karst, and isolated karst of the intermediate area (Habič 1992:33) (**Figure 4.2**).

Slovene archaeology closely links caves with Paleolithic or/and prehistoric research. This is partly due to the fact that the majority of the excavations in caves were undertaken by (early) Stone Age researchers (Srečko and Mitja Brodar, Franc Osole, France Leben, Vida Pohar, Ivan Turk). Unfortunately, often the only goal of the archaeological research in caves or rockshelters was to determine the Paleolithic layer. For the most part the finds from younger periods discovered in cave environment were acquired coincidentally and without controlled excavation (Czoeringerjeva Jama, Jama Nad Jezerom, Svetinova Dvorana, Pečine v Valah, Jama Rokavc, Gorenja Jama, Sveta Jama, Luknja v Lazu and others).

A close link of caves and Paleolithic studies is also a consequence of the fact that most Paleolithic sites in Slovenia were in fact discovered in caves or rockshelters (till today we only know a few open Paleolithic sites in Slovenia: Kostanjevica na Krki, Podrisovec, Solkan, Vrhnika, Nevlje, Ruperč vrh, Zemono; see also Karavanić et al. in this volume, and Komšo and Blečić in this volume). Younger artifacts and archaeological layers were also discovered in many caves and rockshelters, but due to the lack of complete publication about excavations we have incomplete knowledge about post-Paleolithic occupations of some sites. Because of the close link of caves and rockshelters with the Paleolithic period the review of cave and rockshelter research in Slovenia is at the same time a review of Paleolithic research. In the last 80 years a great many of cave and rockshelter sites were discovered but sadly the majority of them were

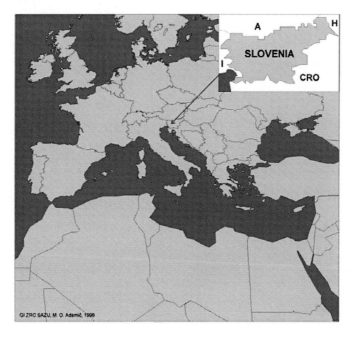

4.1 Position of Slovenia (Anton Melik Geographical Institut, ZRC SAZU, Slovenia).

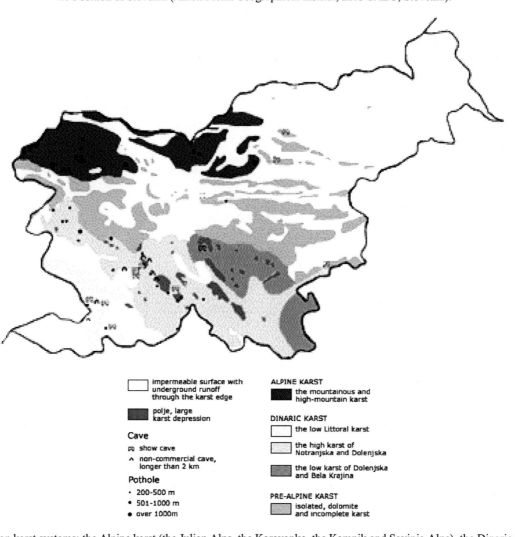

4.2 Slovenian karst systems: the Alpine karst (the Julian Alps, the Karavanke, the Kamnik and Savinja Alps), the Dinaric karst (the Littoral Karst, the Karst Of Notranjsko and the Karst of Dolenjsko), isolated karst in sub-Alpine and sub-Dinaric Slovenia (subdivided into several homogenous isolated units) (Karst in Slovenia, Authors: Dr. Peter Habič and eng. Vili Kos, The Encyclopedia of Slovenia, Vol. 5, Reference KRAS. Mladinska knjiga Publishing, Ljubljana 1991).

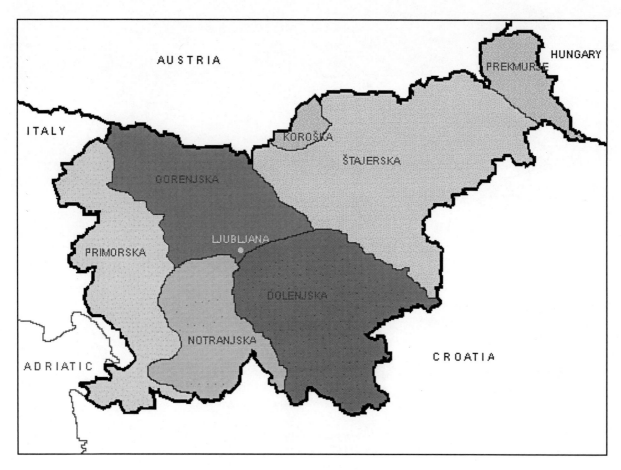

4.3 Slovene informal provincial division.

apparently not published. At the moment cave and rockshelter archaeology in Slovenia is not very active. Since the past archaeological work in caves and rockshelters in Slovenia in my opinion is not well known outside the country or region I begin with a short history of research. I will also present some recent work in Viktorjev spodmol rockshelter because two different field methods were used on the site and because the processing of material gained with different methods are instructive.

HISTORY OF RESEARCH

The earliest notes about the discovery of fossil animal bones in a cave in Primorsko-notranjski karst date to the year 1818. The discovery was the result of road construction in Postojna cave system. H. Volpi, F. Hochenwart, H. Frayer and A. Schmidl were among the first explorers. The first major excavations were conducted in year 1878 by Hochenwart in Križna Jama near Lož. The first researchers were paleonthologists and only at the end of the 19th century the idea about exploring archaeological evidence in caves emerged. From the end of the 19th century till the beginning of the 1st World War most of the cave sites in Trieste and rest of Mediterrannean karst were discovered. The main researchers in this period were K. Moser, J. Müllner, F. Müller, B. Wolf, E. Neumann, C. Marchesetti and G.A. Perko. Although they began their studies independently, they soon received support from local museums and institutions from Trieste and from the Central Commision and Anthropological Society from Vienna. Excavation reports from first excavators are deficient and do not provide information about chronological and cultural aspects of the finds. Most of the graphic material is useless as it lacks stratigraphic position (Leben 1967:44; Osole 1979:19-25).

After the 1st World War the cave sites on Notranjska and Primorska karst, territory occupied by Italians, were subjected to more systematic research under the supervision of R. Battaglia and E. Boegen **(Figure 4.3)**. R. Fabiani worked on Pleistocene faunal assemblage and Battaglia on chronological interpretation of Paleolithic finds from shelter Pod Kalom. The main research center was still in Trieste, excavation reports were published in scientific publications in Trieste and Vienna (Leben 1967:44). Also in 1932 F. Anelli excavated several test pits in Betalov spodmol near Postojna. The excavations in this cave site contionuoed till the beginning of the 2nd World War (1941). The results of his work were never published. When the speleo – archaeological center moved from Trieste to Padova, the intensity of the research in Trieste karst rapidly diminished. At about the same time the Italian Speleological Institute was founded

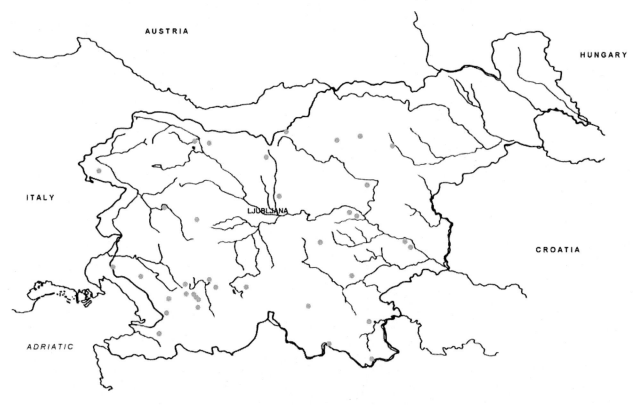

4.4 Locations of the some of most known archaeological sites in caves and rock shelters in Slovenia.

in Postojna. Their research focused on Slovene classical karst in Primorska and Notranjska region. Unfortunately the main goal of this research was finding new locations for military strategic instalations (Leben 1967:44).

The most important moment for Slovene archaeology between the wars was the discovery of the Potočka Zijalka, an alpine cave discovered by Brodar in 1928 (Bayer and Brodar 1928; Ložar 1941:130). The discovery, the first Paleolithic cave in the Slovene teritory is also considered a foundation for Paleolithic archaeology in Slovenia (Novaković 2002:330). In following years S. Brodar extended his paleolithic cave and rockshelter research to other regions of Slovenia and Yugoslavia (Novaković 2002:330). Ten years after the discovery of Potočka Zijalka, S. Brodar published a synthetic review of Yougoslav Paleolithic which was also the review of all cave archaeological research until then (Brodar 1938;140-172). After the 2nd World War, with the incorporation of Paleolithic archaeology at the University of Ljubljana in 1946, cave and rockshelter archaeology research blossomed. The annexation of the Primorska and Istria to the Yougoslav Republics of Slovenia and Croatia made it possible for S. Brodar and his co-workers to explore caves on Notranjska and Primorska karst (Leben 1974/75:262).

In the first decade after the 2nd World War the archaeological cave studies in Trieste karst gradually revived. Members of the speleo section of Trieste mountaineering society and the Institute for the Protection of Cultural Heritage of Slovenia explored new archaeological sites in caves and made test pits in already known cave sites. Discovered finds were chronologically, culturally and stratigraphically defined since in every cave undisturbed layers were discovered (Leben 1967:44).

Until 1979 24 cave or rock shelter sites were confirmed or excavated in Slovenia (Brodar and Osole 1979:135-159; Osole 1979:19). After the death of S. Brodar (1893 - 1987), France Osole (1920 – 2000) continued with archaeological excavations in caves and rockshelters. His scientific and research work was devoted to discovering and research of Pleistocene cave sediments and their palentological and archaeological contents (Pohar 2000:257). Excavations and topographic reconnaissance of caves and rockshelters, first at the Karst Research Institute and later at Institute of Archaeology (both Scientific Research Centre of the Slovenian Academy of Science and Arts) was the professional interest of France Leben (1928 – 2002) (Dular 2003:449).

Mitja Brodar was also involved in developing methodological work in investigating Early Stone Age in southeastern Alps. He conducted several archaeological excavations in caves and rockshelters (Dular 1986:13-15). M. Brodar is still active in field of Stone Age study which is closely linked with cave and rockshelter sites. Among Slovene archaeologists at the present only Ivan Turk is, through Paleolithic research, intensely involved in archaeological research in caves and rockshelters (Scientist at the Institute of Archaeology at Scientific Research Center of the Slovenian Academy of Science and Arts). His research is directed to Upper Pleistocene study with emphasis on Divje Babe I site.

Department of Archaeology (Faculty of Arts at the University of Ljubljana) conducted two major subterranean archaeological excavations, in Ajdovska Jama near Nemška vas (Brodar 1953:7-40; Korošec 1975:170-209; Horvat 1989; Josipovič 1991:145-150) and on-going project in Mala Triglavca rockshelter (Northern Adriatic Project, University of Edinburgh).

In 1997 a joint project began (Department of Geology, Faculty of Natural Sciences and Engineering, University of Ljubljana and Institut für Paläontologie, Universität Wien) in form of multiple short terms excavations in Potočka Zijalka. The goal was to acquire new data, especially regarding paleontological finds (Pacher et al. 2004). Within the same project short term excavations in Križna Jama, Ajdovska Jama near Nemška vas and Herkove Peči near Radlje ob Dravi were conducted (Rabeder 2003/2004:114-116; Kralj 2003/2004:130-133) **(Figure 4.4)**.

HUMAN USE OF CAVES AND ROCK SHELTERS

On the basis of moderate finds in Risovec cave, Betalov spodmol and Jama v Lozi we can track the oldest trail of human presence in Upper Paleolithic. Betalov spodmol near Postojna and Divje babe I in Idrijca river valley are two sites where the presence of the temporary residence can be tracked back to the Neanderthals (Turk 1999).

Karstic caves and rock shelters were periodically used by prehistoric and by much later herdsmen (Velušček 1999:142; Kregar 1985). Human use of caves and rockshelters is shown for all periods of human history. Luknja v Lazu yielded remains of two dry-stone walls which were closing the entrance in prehistory. Despite a new settlement type, kaštelir/castellier, there are still known layers in caves and rockshelters with finds that demonstrate human occupation also in the Bronze age. Caves and rockshelters were used as refuges also in the restless Roman Period. Late medieval and modern finds allow conclusions that even in younger archaeological and historical periods caves and rockshelters were used as refuges and storage places.

First documented examples of cave as a place of burial is Ajdovska jama near Nemška vas (S. Brodar 1953:7-40; Korošec 1975: 170-209) and Tominčeva jama (Leben 1974:246). Later also Jama 1 na Prevalu (also known as Okostna jama) and Pecova jama by Merče (Velušček 1999:143) were used as burial places. Iron Age burials in caves and rock shelters on Karst are especially interesting because they represent an exception regarding the prevailling ritual of this period. In such cases one could anticipate that special place of burial (in caves and rockshelters) may indicate special status of the deceased (Slapšak 1999:155).

Individual archaeological finds (perhaps washed into caves), above all from Iron age, are known from several caves and rockshelters in Slovenia as well (Leben 1974:248). We first see caves as a ritual places in Slovene territory with the discovery of Jama 2 at Preval (also known as Mušja jama). Bronze artefacts, charcoal and burned animal bones discovered in this site permit the interpretation of the site as a ritual and place of sacrifice at the end of the Bronze and beginning of the Iron Age (P. Turk 1996:102). The region of Skocjan Caves Park, which also include Mušja jama, is interpreted to be one of the most significant pilgrimage sites in Europe. In some other caves and rockshelters Roman inscriptions and artefacts with Christian motifs from later periods were found (Slapšak 1999:153).

During the 1st World War several caves and rockshelters known as archaelogical sites, were used as military bases (caverns, refuges). One construction company used Trhlovca cave near Divača as an explosive storehouse during road construction. In Jama pri Korincovih (Korinceva jama) prehistorical cemetery was excavated; today the cave serves as garbage dump. Similarly, the Paleolithic cave site Marovška zijalka near Šentlovrenc is today used as a goat house.

Review of human use of caves in Slovenia shows the need for better legal regulation and enforcement for the protection of caves and rockshelters.

METHODOLOGY

Archaeological methodology in caves and rockshelters in Slovenia is not uniform. Above all older publications on excavations in subterranean environments lack good descriptions. Recently a research on quality and quantity of published data about methodological work in caves and rockshelters was completed (Knavs 2006). The research showed several deficiencies in published data concerning organization of sites, aerial extend and size of the excavation, establishment of system for spatial documentation, determination of excavation method, site recording system and sampling technique. The lack of the excavators self criticism regarding excavation methods was detected in all publications examined.

Improvement in every field mentioned above is apparent in the recently published monograph on Viktorjev spodmol and Mala Triglavca rockshelters (Turk 2004). High quality published data is also evident with Divje Babe I cave site (Turk et al. 1988; Turk et al. 1989; Turk et al. 1995; Turk 1997; Turk et al. 2001; Turk et al. 2002; Turk et al. 2003; Turk 2003), which I attribute to the fact that Divje Babe I has been given the greatest archeological attention in the last several years in Slovenia.

Below I summarize some aspects concerning the archaeological methodology in Viktorjev spodmol

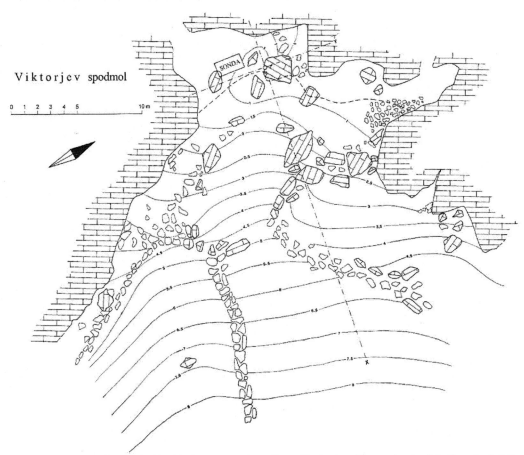

4.5 Ground plan of Viktorjev spodmol with the test trench outlines. The broken line marks the drip line, the full lines are contour lines of relative heights. Measured and executed by J.Dirjec jun., drawn by J.Dirjec sen. (Turk ed. 2004, Fig. 3.1.)

Excavation phases	Tools	Debris (total)	Debris (< 3 or 5 mm)	TOTAL
Viktor	9	164	0	173
Viktor IzA	98	2386	1097	2484
IzA	69	10 158	8958	10 227
TOTAL	176	10 708	10 055	12 884

Table 4.1: Viktorjev spodmol: all stone artefacts (adopted from Turk 2004, Table 6.4.1).

rockshelter. I chose to present Viktorjev because two different field methods were used during its excavation and the analysis of material gained by the different methods is informative. In this way I want to draw attention to the relativity of the results and explanations, and how these depend on the method and its execution. The results and the explanations depend among other things on the scale of error that occurs in establishing a specific situation (Turk 2004a, 24-26).

In the archaeological sense, the site was discovered by Viktor Saksida, the founder of the archaeological section of the Caving Society in Sežana. Together with a member of the same society, he excavated a 1 x 2 meters wide and approximately one meter deep test pit in the lower part of the rock shelter. The depth was restricted by a large fallen rock over the entire area of the test trench. He used the method of removing the sediment by means of 20 – 30 cm thick horizontal spits and concurrent examination of the sediments during excavation and afterwards without sieving (Turk 2004a:24).

The second method consisted of removing the sediment by means of 5 cm deep horizontal spits on an area of 200 x 20 cm and examing the entire sediment after excavations with the aid of wet sieving with 3 mm and 1 mm (or 0.5 mm) mesh and with the use of a magnifying glass. This method was carried out during the excavations of the Institute of Archaeology (Turk 2004a:24).

The method described had been first used in Palaeolithic excavations and test excavations, with few improvements, until 1986 (Turk 2003). Investigations in the field at Viktorjev spodmol took place in three phases, which I. Turk named Viktor Phase (the results and interpretations of Viktor's test trench), Viktor and IzA Phase (the results and interpretations of a re-examination of the sediments from Viktor's test trench, when all the sediments were washed and sieved) and IzA Phase (the results and interpretations of the stratigraphic excavations of the section of Viktor's test trench) (Turk 2004b:32).

In the first phase the excavators did not record the depth at which individual finds lay. They divided all the excavated sediments and finds (in the test trench 1 x 2 x 1 m) into two stratigraphic units (Turk 2004b:32). The second phase was the re-examination of the already examined sediments. Since the finds were not stratified Turk treated them as a whole, and so he lost important information (Turk 2004b:36). The main purpose of the stratigraphic excavations (third phase) was to establish the position of individual finds in the section of the site to the accuracy of 5 cm. In this manner a block of sediments of 0.2 x 2 x 1 m was excavated along the section. All finds were collected with the same thickness of horizontal spits which Turk later coordinated with layers, visually determined in the section. Since the section was 2 m long, the finds were collected separately by spits for its left and right parts, in order to establish all possible concentrations of finds. Since the finds were equally distributed in all spits in both parts of the profile, in the end he treated both parts as a single spatial unit and the finds stratigraphically by spits. Since the majority of objects were found during wet sieving they are without individual coordinates (Turk 2004b:41).

Here I present a comparison of all stone artefacts gained in individual phases (Table 4.1). Comparison of the various phases of excavation in which different techniques and methods were used clearly shows that an interpretation of sites which is based on finds is very dependent on the methods and accuracy of work in the field and later in the laboratory (Turk 2004b:51). The comparison shows that the classical method of concurrent examination of sediments in the field without sieving (Viktor phase) gave the worst results. All the information critical for a robust "culturological" definition of the site were overlooked. On the other hand the stratigraphic excavation (IzA phase) gave the best results and allowed for the same interpretation in a small area as excavations in an area five times bigger with lesser accuracy (Turk 2004b:52). The author of the comparison, I.Turk, concluded that the reduction of the accuracy of the excavations dictates investigation of extremely large area in order to obtain the results equal to investigations in the IzA phase (Turk 2004b:52).

And since such a deficient method of work was until recently a rule in investigating cave and rockshelters sites in Slovenia (Turk 2003; 2004a:24; 2004b:52), one can question the results and interpretations deriving from such methods. It should be in archaeologists best interest that archaeological methodology be standardized to the highest quality of excavation and recording and that any deviations to the standard methodology be carefully described and theoretically grounded. Also a reinvestigation of previously excavated caves and rockshelters is justified, as seen from the results presented for the case of Viktorjev spodmol (**Figure 4.5**).

Acknowledgements. Participation at UISPP XV Congress was enabled by means of European Union within Socrates II programme, Gruntvig 3 action – mobility of individuals.

REFERENCES

BAYER, J. and S. BRODAR (1928). *Die Potočka Höhle, eine Hochstation der Aurigacschwankung in den Ostalpen.* Prähistorica 1.

BRODAR, S. (1938). Das Paläolithikum in Jugoslawien. *Quartär* 1, p. 140-172.

BRODAR, S. (1953). Ajdovska jama. *Razprave SAZU* 3, p. 7-40.

BRODAR, M. and F. OSOLE (1979). Paleolitik na ozemlju Slovenije. In *Praistorija jugoslavenskih zemalja I. Paleolitsko i mezolitsko doba*, p. 135 - 175, Sarajevo.

DULAR, J. (1986). Ob petinšestdesetletnici Mitje Brodarja. *Arheološki vestnik/Acta archaeologica* 37, p. 13 - 15.

DULAR, J. (2003). France Leben (1928 – 2002). *Arheološki vestnik/Acta archaeologica* 54, p. 449 - 452.

HABIČ, P. (1992). Kras and Karst in Slovenia. In PAK, M. and M. ORAŽEN ADAMIČ, eds., *Slovenia. Geographic aspects of a new independent European nation. Published on the occasion of the 27th International Geographical Congress at Washington 1992*, p. 31 - 39. Ljubljana, The Association of the Geographical Societies of Slovenia.

HORVAT, M., (1989). *Ajdovska jama pri Nemški vasi* (with German Zusammenfassung). Razprave Filozofske fakultete Ljubljana.

JOSIPOVIČ, D. (1991). Spuren eisenzeitlicher besiedlung in der höhle Ajdovska jama bei Nemška vas. *Poročila o raziskovanju paleolita, neolita in eneolita v Sloveniji* 19, p. 145 - 150.

KNAVS, M. (2006). *Archaeological excavations in caves and under rock shelters. Study cases: Potočka zijalka, Crvena Stijena, Divje babe I, Spila Nakovana, Podmol pri Kastelcu, Ajdovska jama pri Nemški vasi* (with English Abstract). Undergraduate Thesis. Ljubljana.

KOROŠEC, P. (1975). Bericht über die Forschungen in der Ajdovska – Höhle in J. 1967. *Poročila o raziskovanju paleolita, neolita in eneolita v Sloveniji* 4, p. 170 - 209.

KRALJ, P. (2003/2004). Sedimentološke značilnosti jamskih usedlin na nekaterih paleolitskih postajah

Slovenije. *Glasnik slovenske matice* 27/28, 1 - 2, p. 130 - 133.

KREGAR, V. (1985). Pastirske kulture v jamah. *Naše jame* 25, p. 47 - 48.

LEBEN, F. (1967). Stratigraphie und zeitliche Einreihung der Höhlenfundstätten auf dem Triester Karst (with German Zusammenfasung). *Arheološki vestnik/Acta archaeologica* 18, p. 84 - 86.

LEBEN, F. (1974). Höhlenarchäologie des Klassischen Karstes. Vortrag am 6. Kongress der jugoslawischen Höhlenforscher (Sežana – Lipica, 10.-15. Oktober 1972) (with German Zusammenfasung). *Acta Carsologica* 6, p. 252 – 253.

LEBEN, F. (1975). Analiza speleo-arheoloških raziskovanj v Sloveniji. Proteus, 37/6-7, p. 261- 264.

LOŽAR, R. (1941). Razvoj in problemi slovenske arheološke vede. *Zbornik za umetnostno zgodovino* 17, p. 107 - 148.

NOVAKOVIĆ, P. (2002). Archaeology in five states – A peculiarity or just another story at the crossroads of »Mitteleuropa« and the Balkans: A case study of Slovene archaeology. In BIEHL P.F.; GRANSCH A.; MARCINIAK A. hrsg. *Archäologien Europas/ Archaeologies of Europe,* 2002, p. 323 - 352.

OSOLE, F. (1979). Ledenodobne kulture Slovenije. In BRODAR, S. ed. *Ledenodobne kulture v Sloveniji.* Katalog, p. 19 – 25, Ljubljana.

PACHER, M.; POHAR V. and G. RABEDER (2004). *Potočka Zijalka : palaeontological and archaeological results of the campaigns 1997-2000.* Mitteilungen der Kommission für Quartärforschung der Österreichischen Akademie der Wissenschaften, Bd.13.

POHAR, V. (2000). Franc Osole (1920-2000). *Arheološki vestnik/ Acta archaeologica* 51, p. 257.

RABEDER, G. (2003/2004). Brlogi jamskega medveda v Sloveniji. *Glasnik slovenske matice* 27/28, 1-2, p. 112 - 116.

SLAPŠAK, B. (1999). Slovenski Kras v poznejši prazgodovini in v rimski dobi. In CULIBERG M. et al. *Kras, pokrajina, življenje, ljudje,* p. 145 - 163. Karst research Institute at ZRC SAZU, ZRC Publishing, Ljubljana.

TURK, I. (1999). Ledena doba – čas velikih naravnih sprememb. In *Zakladi tisočletji. Zgodovina Slovenije od neandertalcev do Slovanov,* p. 24 - 27. Ljubljana. Modrijan Publishing, Ljubljana.

TURK, I. (2003). How to make better use of archaeological methods of excavation in post – excavation analysis and interpretation of the results. Experience of excavation at Divje babe I, Slovenija. *Arheološki vestnik/Acta archaeologica* 54, p. 9 - 30.

TURK I. (2004a). Methodology of the Archaeological Work. In TURK I. ed. *Viktorjev spodmol and Mala Triglavca. Contributions to understanding the Mesolithic period in Slovenia.* Opera Instituti Archaeologici Sloveniae 9 (2004), p. 24 – 31, Institute of archaeology at ZRC SAZU, ZRC Publishing, Ljubljana.

TURK, I. (2004b). Different Archaeological Methods – Different Results in Investigations of Viktorjev spodmol. In TURK I. ed. - Viktorjev spodmol and Mala Triglavca. Contributions to understanding the Mesolithic period in Slovenia. *Opera Instituti Archaeologici Sloveniae* 9 (2004), p. 32 – 52, Institute of archaeology at ZRC SAZU, ZRC Publishing, Ljubljana.

TURK, I. (ed.) (1997). Moustérienska koščena piščal in druge najdbe iz Divjih bab I v Sloveniji. *Opera Instituti Archaeologici Sloveniae* 2 (1997), Institute of archaeology at ZRC SAZU, ZRC Publishing, Ljubljana.

TURK I. (ed.) (2004). Viktorjev spodmol and Mala Triglavca. Contributions to understanding the Mesolithic period in Slovenia. *Opera Instituti Archaeologici Sloveniae* 9:24 - 31. Institute of archaeology at ZRC SAZU, ZRC Publishing, Ljubljana.

TURK, I.; KOGOVŠEK J.; KRANJC A. and DIRJEC J., (1988). Fosfati in tanatomasa v sedimentih iz jame Divje babe I. *Acta carsologica* 17, p. 107 - 127.

TURK, I.; DIRJEC J.; STRMOLE D.; KRANJC A. and ČAR J. (1989). Stratigraphy of Divje babe I. Results of excavations 1980-1986. *Razprave 4. razreda SAZU* 30, 5, p. 161 - 192.

TURK, I.; CIMERMAN F.; DIRJEC J.; POLAK S. and MAJDIČ J. (1995). Fossilised cave bear hairs from 45,000 years ago found at Divje babe I in Slovenia. *Arheološki vestnik/Acta archaeologica* 46, p. 49 - 51.

TURK, I.; D. SKABERNE B; A. B. BLACKWELL and J. DIRJEC (2001). Morphometric and hronostratigrafic sedimentary analysis and paleoclimatic interpretation for the profile at Divje babe I, Slovenia (English Summary). *Arheološki vestnik/Acta archaeologica* 52, p. 221 - 247.

TURK, I.; D. SKABERNE; B. A. B. BLACKWELL and J. DIRJEC (2002). Assessing humidity in an Upper Pleistocene Karst environment palaeoclimates and palaeomicroenvironments at the cave Divje Babe I, Slovenija. *Acta carsologica* 31/2, p. 166 - 175.

TURK, I.; SKABERNE D. and ŠMIT Ž. (2003). Reliability of Uranium Series Dating in Divje babe I. Effect of sedimentation gaps on uranium concentrations in sediments and on uranium series dating. *Arheološki vestnik/ Acta archaeologica* 54, p. 41 - 44.

TURK, P. (1996) The Dating of Late Bronze Age Hoards. In *Hoards and Individual metal finds from the Eneolithic and Bronze Ages in Slovenia 2.* Catalogi et monographiae 30, p. 89 - 124. National museum of Slovenia.

VELUŠČEK, A. (1999). Prazgodovinska in zgodovinska jamska najdišča na Krasu. In Culiberg M. et al. *Kras, pokrajina, življenje, ljudje,* p. 142 - 145. Karst research Institute at ZRC SAZU, ZRC Publishing, Ljubljana.

www.rrr.de/.../pic/dscn2619.jpg, ONLINE 25.08.2006

125 YEARS OF THE ROCKSHELTER STUDIES IN RUSSIA

Sergey A. VASIL'EV

Institute for the Material Culture History, 18 Dvortsovaya emb. 191186 St. Petersburg Russia;
sergevas@av2791.spb.edu

Abstract. The paper deals with the developmental history and contemporary state of art in the rockshelter studies in Russia. In spite of the fact that the majority of advances in the Russian Paleolithic archaeology are associated with the exploration of the open-air occurrences, the territory of Russia and adjacent countries, especially those areas as Caucasus, Crimea, the Ural Mountains, Southern Siberia, and Central Asia, are abound in cave sites. The history of fieldwork began with the discoveries of Merejkowsky at Crimea in 1879. First attempts to develop the methodology for the stratigraphic subdivision of cave sediments were put forward by Schurovsky. The 1920s saw the important developments in field methods. Bonch-Osmolovsky pioneered the study of the stratigraphy linked with the spatial distribution of artifacts at Kiik-Koba (Crimea). Combined with new approaches to the lithic (technological studies, statistical classification) and bone analysis this study was of great originality and importance. It is worth noting that this intriguing line of inquiry foreshadowed the methodology of François Bordes which were to appear in the late 1940s during the exploration campaign at Pech de l'Azé. Further developments in cave site studies in the 1950-1980s in Russia were mostly associated with the works of Liubin at the Caucasus. Today an important cluster of the rockshelters and caves located at Altai and extensively excavated under direction of Derevyanko serves as a focal center for improvement of field methods.

Keywords: Russia, Paleolithic, caves

Résumé. L'article vise à présenter une vue d'ensemble d'historique et le caractère contemporaine de la recherche sur les abris sous roche en Russie. Malgré le fait que les progrès considérables dans les études paléolithiques en Russie sont associés plutôt avec des travaux sur les gisements de plein air, le territoire de la Russie et les pays voisins est très riche en abris sous roche, particulièrement le Caucase, la Crimée, les montagnes de l'Oural, la Sibérie du Sud et l'Asie Centrale. Les premiers vestiges du Paléolithique en abris sous roche ont été découverts en Crimée par Merejkowsky en 1879. Des tentatives pour faire l'analyse stratigraphique des sediments en grottes ont été avancés par Shchurovskiï. Les années 1920 sont le témoin des progrès remarquables des méthodes des fouilles. Bontch-Osmolovskiï a mis au point des méthodes combinantes des études stratigraphiques et planimetriques pendant les fouilles de la grotte de Kiik-Koba en Crimée. Il est à signaler l'orinalité de cette approche aux etude d'industrie lithique et osseuse y compris l'analyse technologique et le traitement statistique des données. Cette nouvelle orientation de recherche anticipait les méthodes utilisées par François Bordes en Pech de l'Azé dès la fin des années 1940. Les années 1950-1980 ont été marquées de travaux de Lioubine en Caucase. Les années récentes sont le témoin de la réalisation d'un programme de fouilles de la grande concentration des grottes de l'Altai dirigée par Derevïanko.

Mots-clés: Russie, Paléolithique, grottes

In spite of the fact that the majority of advances in the Russian Paleolithic archaeology are associated with the exploration of the open-air occurrences, the territory of Russia and adjacent countries, especially those areas as Caucasus, Crimea, the Ural Mountains, Southern Siberia, and Central Asia, are abound in cave sites (**Figure 5.1**). In spite of being 'at the shadow' of the impressive open-air habitation sites with dwelling structures, the cave sites played an important role at different phases of the developmental history of prehistoric archaeology in Russia. The aim of my paper is twofold. First, I would like to illuminate the main historical stages of the cave sites exploration and analysis pointing out important methodological advances pioneered by the Russian scholars. Second, I would like to present a brief overview of the current state of art in the rockshelter studies in our country.

Two preliminary remarks need to be made. First, the historical part of the overview dwells not only on the territory of contemporary Russia, but also on the territories belonging to the Russian Empire before the Revolution (Poland) as well as the former Soviet republics, now the independent states. Second, I confine myself by the history of the research of the Pleistocene archaeology, thus omitting the Mesolithic, Neolithic and late prehistoric habitations, burial structures, sacred places and rock art connected with caves and rockshelters.

125 YEARS OF EXPLORATION: THE HISTORICAL BACKGROUND

Historically the discovery of the Paleolithic remains in caves in Russia coincided with the discovery of the first traces of Early Man in general. If the first open-air site was discovered in our country by Chersky and Chekanovsky at Irkutsk (Siberia) in 1871, roughly the same time evidenced the discovery of the Paleolithic remains in the western most point of the Empire, at Poland, where Zawicza excavated the Mamutowa cave near Krakow. Early attempts to develop the methodology for the stratigraphic subdivision of cave sediments were put forward by Schurovsky (1878), a Russian geologist with strong interest in cave studies. It is worth mentioning that, as it was the case many times in the history of our discipline, the principles of the excavation strategy were put forward before the start of the actual

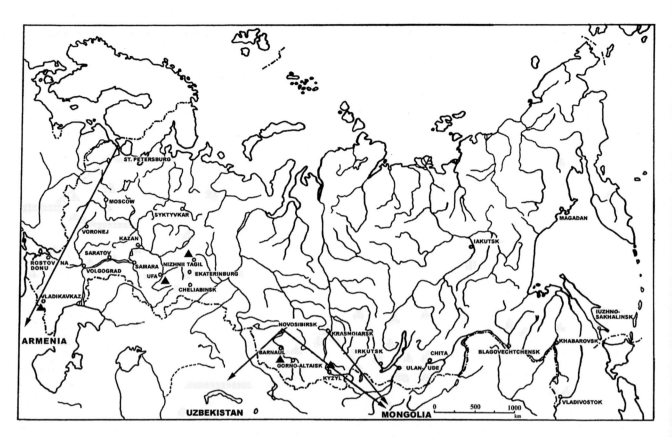

5.1 Main cave site concentrations under study at Russia and adjacent countries.

fieldwork, not as a result of long-term field experience. Schurovsky mentioned several 'scientific methods of cave excavation,' plotting of floor plans, tracing of baselines, stratigraphic excavation by 'cube feet' perpendicular to the central line, enumeration of all findings, etc. He cited the excavations of Pengelly at Kent's Cavern as an example of such kind of investigation.

Famous Russian geographer and archaeologist Anuchin (1879) was a real 'ambassador' of the Russian science throughout Europe, extensively traveling and publishing. During the preparation of the Anthropological Exhibition held in Moscow in 1879, he visited the main Paleolithic localities of southern France and even took part in cave excavations in France. He participated in the excavations of the grotto of Bize with Cartailhac and Laugerie-Basse with Massénat.

But the real history of the cave site exploration in Russia started in 1879, when Konstantin Merejkowsky (1884) after the first reconnaissance trip to Crimea discovered both Upper Paleolithic (Siuren, Kachinskiy Rockshelter) and Mousterian remains. Thus the Volchiy Grotto became the first Middle Paleolithic site discovered in Russia. Later, from 1879 to 1881, Ossowski (1895) excavated the Maszycka cave at Poland, presenting in his final publication the carefully made complete plan and profile of the cave under study.

After a break in exploration, the 1910s witnessed an increase of information on cave sites, mainly contributed by the European scholars. At Okiennik (Poland) the first Middle Paleolithic remains in this country were investigated. Later the Austrian scholar Leon Kozlowski in collaboration with the German archaeologist Schmidt began to explore the numerous caves at Imetetia (western Georgia), thus discovering the first clear evidence of the Upper Paleolithic cave use in the Caucasus. Finally, the excavations carried out by Polish archaeologist Krukowski in Caucasus between 1915 and 1918 are worth a mention.

As in the other countries of Europe, these early studies were oriented exclusively to the study of cave stratigraphy and among the structural features only fireplaces were identified.

The period after the Revolution, the 1920s saw the further progress in rockshelter studies coincident with important methodological improvements. Crimea became the focal area for the improvement of field and laboratory studies. Between 1924 and 1926 Gleb Bonch-Osmolovsky directed a remarkable excavation campaign at the Kiik-Koba grotto resulting in the discovery of two Mousterian strata with Neanderthal burials.

Bonch-Osmolovsky played a crucial role in the development of methodology of rockshelter studies in Russia. Having practically no predecessors and being acquainted with the development of prehistoric field

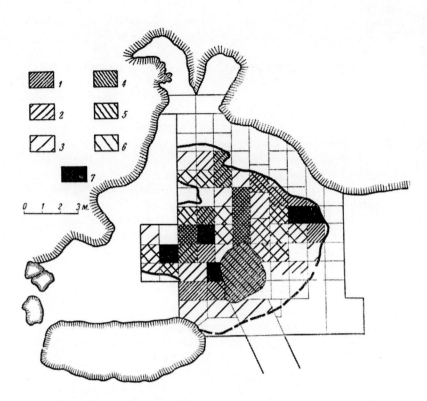

5.2 Spatial distribution of cultural debris in the lower stratum (VI) of the Kiik-Koba grotto, Crimea (after Bonch-Osmolovsky 1940: Figure 25). Lithics per square meter: 1: above 200 pieces; 2: 100 to 200 pieces; 3: up to 100 pieces. Bones per square meter: 4: above 200 g; 5: 100 to 200 g; 6: up to 100 g; 7: above 300 pieces of lithic debris.

archaeology abroad only though literature, he put forward a coherent methodology for interdisciplinary study of these features. His field strategy was based on the excavations by trenches or small areas with profiles each 1 to 2 meters long. The sediments were excavated following the stratigraphy by slices 25 to 30 cm deep, while the structural features were unearthed completely.

The artifacts were plotted, enumerated and their elevation recorded with the information stored on special cards. The description in field notebook was supplemented by photos. All sediments were carefully screened. With aid of tissue paper the imprints of profiles and portions of the cultural layer were made, later these were used for the fabrication of plaster casts for the museum expositions. These excavation techniques allowed for the transformation of the narrow boundaries of strictly stratigraphic approach and the beginning of the study of spatial distributions. Bonch-Osmolovsky (1940:94) emphasized 'the special importance given by the Soviet archaeologists to the study of spatial distribution of findings at the surface of the Paleolithic sites'. In the final monograph on Kiik-Koba one could find schemes of spatial distribution of lithics and bones per square meter for two main cultural layers (**Figure 5.2**). The investigator paid special attention to the study of the relief of the culture-bearing strata and the cave floor, and the identification of small pits. Bonch-Osmolovsky for the first time identified the effect of trampling in vertical distribution of cultural debris associated with the movement of pieces upstream. He also paid attention to the activity of cave hyenas identifying the gnawing marks on bones.

It is worth noting the effect of parallel development which is not uncommon in the history of archaeology. Roughly at the same time as Bonch-Osmolovsky, Denis Peyrony (1932) expressed an interest in the spatial distribution of remains, identifying the Solutrean artificial structure at Fourneau du Diable in 1924. Generally speaking, this intriguing line of inquiry put forward by Bonch-Osmolovsky essentially foreshadowed the methodology of François Bordes (1972) which were to appear in the late 1940s during the exploration campaign at Pech de l'Azé.

Apart from Kiik-Koba, Bonch-Osmolovsky explored several other cave sites (Shaitan-Koba, Kosh-Koba, and Adzhi-Koba), and later the Upper Paleolithic strata at Siuren. Other scholars worked in Crimea, Zabnin and Ernst, studied the Mousterian strata at the Chokurcha grotto in 1928. At the entrance area the huge concentration of mammoth bones covering 18 square meters (not excavated completely) was unearthed.

In the Caucasus the future leader of the Georgian archaeology, Georgiy Nioradze (1933) started the campaign of the excavation of the Upper Paleolithic at Devis-Khvreli (Imeretia).

The 1930s were not so rich in discoveries. During this time the Russian scholars were mainly concerned with excavations of the Upper Paleolithic habitation sites in large horizontal exposures. In Crimea, after the interruption of the activity of Bonch-Osmolovsky and Ernst (both suffered from the political repressions) some campaigns (at Chagorak-Koba, Volchiy Grotto, Bakhchisaraiskaya) were held by Otto Bahder and Dmitri Krainov. At the Volchiy Grotto the huge natural hollow rich in cultural debris located behind the rampart of limestone blocs was unearthed. It was interpreted as remains of a domestic structure built within a naturally protected area.

The activities during the 1930s in the Caucasus were mostly associated with a prominent Soviet prehistorian, Sergey Zamyatnin (1937). From 1936 to 1938 he discovered and excavated the Mousterian assemblages at the Akhshtyrskaya and Navalishenskaya caves located near the Black Sea coast as well as the Mgvimevi rockshelter at Georgia (the last site later was explored by Kiladze-Bedzenishvili). Later Krainov started his own fieldwork at the caves located near Sochi (the Black Sea coastland). Nioradze continued his fieldwork at Sakazhiya.

These years saw an important expansion in the area of cave exploration. Sergey Bibikov discovered scattered Pleistocene remains at the caves located in the Ural Mountains, while Alexey Okladnikov in 1938 and 1939 excavated the first Paleolithic cave site at Central Asia, the Teshik-Tash Grotto with the remains of a destroyed Mousterian burial. All these achievements were connected only with field discoveries. No significant progress occurred in the methodology after Bonch-Osmolovsky.

The period after the World War II was characterized by long-term exploration campaigns, typical for the organization of the Soviet archaeology, and favorable for meticulous study of cave sites. In this paper I confine myself with very brief general remarks omitting long lists of relevant literature which could be found elsewhere.

In Crimea during the late 1940-1950s Formozov and Krainov continued to explore the cave sites. The former excavated the Mousterian assemblage at Staroselie. Vekilova studied the Upper Paleolithic assemblages from the Siuren II rockshelter. Later, from the 1960s on, Kolosov and his students concentrated on the area near the city of Belogorsk where a cluster of Middle Paleolithic sites with foliated bifaces (Zaskalnaya V, VI, IX, Ak-Kaya III, Prolom II, etc.) were explored, some of them yielded burial structures.

The discovery of the famous Kudaro caves in South Ossetia, which yielded deeply stratified Mousterian and Acheulian assemblages, by Vassily Liubin in 1955 was the turning point in the cave site archaeology in Russia. From this time onwards the Kudaro caves became the focal point for the development of excavation techniques summarized in the special paper of Liubin (1990). His methodology was based on the interdisciplinary studies of caves, combination of transverse and longitudinal profiles, wet screening of all sediments, individual plotting of findings. The excavations were oriented toward the maximal micro-stratigraphic resolution, permitting the identification of areas with undisturbed living floors.

The cave sites concentrated in the northwestern Caucasus were the object of long-term studies carried out by Autlev and Formozov. The main area of rockshelter concentration with Mousterian and Upper Paleolithic assemblages is the Borisovskoe gorge investigated by Liubin and Belyaeva. As for the Black Sea coast the 1960s saw a new campaign at the Akhshtyrskaya cave directed by Vekilova, later followed by Chistyakov and Kulakov.

In the Trans-Caucasus area, the Georgian archaeologists were the most active. At the western most part of the republic, at Abkhazia, Soloviev, Tsereteli and Korkiya excavated the cave of Apiancha. In Western Georgia Medeya Nioradze, Tushabramishvili and Berdzenishvili directed the long-term campaigns at the Tskaltsitela River valley (Chakhati, Ortvala, Sakazhiya). The 1970s saw the multidisciplinary study of a concentration of cave sites at Tsutskhvaty directed by Maruashvili. The Paleolithic of the South Georgian Plateau was studied by Grigoliya, Gabuniya and Kikodze. The Acheulian and Mousterian assemblages at the Tsona cave located at high mountains of South Ossetia were the object of exploration carried out by Tushabramishvili.

The fieldwork in other parts of the Caucasus was less intensive. In Azerbaijan the main object of the explorations directed by Guseinov was the cave site of Azykh with rich Acheulian and Mousterian assemblages accompanied by human remains. The Mousterian remains were identified by Dzhafarov and Zeinalov in several other caves. In Armenia the Mousterian assemblages at the cave sites of Erevan I and II, and Lusakert I were investigated by Eritsyan.

Several cave sites are reported from other areas of the Eastern Europe. In Moldavia, the Paleolithic grottoes (Brynzeny I, Vykhvatintsy) were excavated during the 1960 and 1970s by Ketraru, Borziyak, and Anisiutkin. Gladilin explored the Molochnyi Kamen cave at the Trans-Carpatian area (Ukraine). The extreme northeastern part of the European Russia produced only one cave site with the Paleolithic remain, namely Medvezhya (Bear Cave), explored by Guslitser and Kanivets during the 1960s, later followed by Pavlov.

The Ural caves added an important new dimension to the cave studies. From the early 1960s Otto Bahder initiated the investigation of the Pleistocene art at the walls of Kapovaya cave, the first cave art images known from Russia. Shirokov excavated the Upper Paleolithic strata at the Bobylek grotto.

Central Asia remained less investigated in spite of abundance of the Mousterian remains. Lev excavated the Aman-Kutan cave, later Ranov explored Ogzi-Kichik. The main cave site in this area is without doubt Obi-Rakhmat, excavated by Suleimanov followed by Omanzhulov. From 1980 onwards Islamov excavated the Acheulian strata at the Sel-Ungur cave.

The 1950s saw the first attempts to explore the Paleolithic cave sites in remote regions of Siberia. In 1954, Rudenko conducted short-term excavations at the Ust'Kanskaya Cave, later acknowledged as the first Middle Paleolithic site in Siberia. After a long break Okladnikov renewed the exploration of Altai caves (Strashnaya Cave) followed by the new discoveries at the Far East (the Geographical Society Cave), Yakutia (Diuktay) and Yenisei (Dvuglazka).

SOME CONTEMPORARY ACHIEVEMENTS

After a relatively short-term period of stagnation and interruption of several projects during the 'hard times' of the late 1980-early 1990s, the contemporary Russian archaeology embarked on a renewal of active fieldwork in different parts of the country, including cave site exploration. First, the traditional areas of the Caucasus are the object for study by several crews. Among these the large-scale investigation of a complicated network of cave chambers with rich Middle Paleolithic remains at Myshtylagty Lagat (the Weasel Cave) directed by Gidzhrati is worth mentioning. Liubin and Belyaeva renewed the excavations in Armenia (the Pechka grotto). Second, the Ural Mountains continue to be the object of small-scale excavations by Serikov, Kotov, Iurin, etc. as well as continuation of the studies at Kapovaya cave.

The focal area for the cave and rockshelter studies in Russia today is without doubt the Altai mountains (South Siberia). A large interdisciplinary crew of scholars from Novosibirsk directed by Anatoly Derevyanko continues a long-term program of exploration of a rich series of sites containing the remains from Acheulian to the Final Paleolithic (Strashnaya, Kaminnaya, Okladnikov, Denisova, Maloyalomanskaya, Biika I and II, Ust'Kanskaya, Iskra). The data from these sites is yielding rich factual evidence for questions of world prehistory, such as those concerning Middle Paleolithic variability and Middle-to-Upper Paleolithic transition. The unique concentration of multicomponent cave and open-air sites at the northwestern Altai serves now as the key stratigraphic succession for the Pleistocene prehistory of northern and central Asia well comparable with the role of the sites of southwestern France for European prehistory. Apart from the sites located in Siberia, the scholars from Novosibirsk directed several ambitious projects in Central Asia, Mongolia and Uzbekistan. It is worth mentioning that these projects are the area of active international co-operation and a lot of scholars from USA, Japan, South Korea, and Europe take part in these efforts.

CONCLUSIONS

The importance of caves and rockshelters for the development of archaeology is not based solely on the wealth of information they yield on prehistory but also with crucial methodological problems, which Russian scholars share with colleagues from other countries. Among these the cave taphonomy is of prime importance. The role of carnivores versus prehistoric humans in bone assemblage accumulation continues to be a matter of hot debate. Unfortunately there are only few attempts to analyze the faunal assemblages beyond the compiling of simple lists of species and minimal number of individuals. The paleontologist Baryshnikov (Derevyanko et al. 2003) examined the collection of Denisova from this viewpoint. He identified several episodes of the cave chamber occupation by humans and carnivores. Thus the lowermost strata of the cave demonstrate the prevalent use as a cave bear den with only sporadic human intrusions. Later the succession shows the complex interstratification of mostly cave hyena dens (gnaw marks and corrosion of bones, coprolites, etc.) with human occupations. Later the hyenas were supplanted by humans.

Another problem of the identification is man-made structures in cave sites. In the majority of case studies it is far from clear if we are dealing with artificial or natural structures, especially when identifying pavements, concentrations or ramparts made of limestone blocs. In some cases the clear-cut concentrations of cultural debris could be interpreted either as remains of domestic units or simple erosional remnants of culture-bearing strata.

Last but not least is the reconstruction of systems of open-air and cave sites as reflection of prehistoric settlement patterns. In this case it is sometimes very difficult to identify the real function of a cave or rockshelter. One of the obvious examples is Okladnikov cave in the Altai (Derevyanko 1997), where the site consists of a complex network of narrow passages filled by the Pleistocene deposits with abundant Mousterian artifacts and bones. It is not easy to understand the mechanics of site formation and function.

It seems that the progress in dealing with the aforementioned problems largely depends on international co-operation. The active interaction of various national archaeological schools is extremely instrumental in stimulating the advance of Paleolithic studies in Russia, introducing the state-of-the-art Western technology and expertise into the rich Russian prehistoric record.

Acknowledgments. I owe a debt of gratitude to the following archaeologists, who generously supplied me with the information concerning the cave sites: Vassily P. Liubin, Elena V. Belyaeva, Zoia A. Abramova, Anatoly P. Derevyanko, Michael V. Shunkov, Sergey V. Markin, Alexander V. Postnov, Andrey I. Krivoshapkin, Nazim Gidzhrati, and many others. This research has been partly supported by the Program of the Russian Academy of Sciences 'Adaptations of peoples and cultures to environmental, social and technomic transformations'.

REFERENCES

ANUCHIN, D. N. (1879) Otchet ob osmotre doistoricheskikh pamyatnikov Frantsii (in Russian). *Antropologicheskaya vystavka.* Moscow. II, p. 362-381.

BONCH-OSMOLOVSKY, G. A. (1940) *Paleolit Kryma, vyp. 1. Grot Kiik-Koba* (in Russian). Moscow-Leningrad: Izdatelstvo AN SSSR. 266 p.

BORDES, F. (1972) *A Tale of Two Caves*. New York: Harper and Row. 151 p.

DEREVYANKO, A. P. (1997) (ed.). *The Paleolithic of Siberia*. Urbana: University of Illinois Press. 406 p.

DEREVYANKO, A. P. [et al.] (2003) *Prirodnaya sreda i chelovek v paleolite Gornogo Altaya* (in Russian). Novosibirsk: IAEt SO RAN. 448 p.

LIUBIN, V. P. (1990) Stoyanki v skalnykh ubezhischakh: spetsifika i metodika polevykh issledovanii (in Russian). *Kratkie soobscheniya Instituta arkheologii*. Moscow. 202, p. 68-77.

MEREJKOWSKY, C. de. (1884) Station moustérienne en Crimée. *L'Homme*. Paris. 1 (10), p. 300-302.

NIORADZE, G. K. (1933) *Paleoliticheskii chelovek iz Devis-Khvreli* (in Georgian). Trudy Muzeya Gruzii; VI. Tiflis: Muzei Gruzii. 110 p.

OSSOWSKI, G. O. (1895) *O geologicheskom i paleontologicheskom kharaktere pescher iugo-zapadnoi okrainy Evropeiskoi Rossii i smezhnykh s nei oblastei* (in Russian). Tomsk: Tipografiya Makushina. 86 p.

PEYRONY, D. de (1932) *Les gisements préhistoriques de Bourdeilles (Dordogne)*. Archives de l'Institut de Paléontologie humaine; 10. Paris, IPH. 126 p.

SCHUROVSKY, G. E. (1878) Obschaya programma dlya issledovaniya kostenosnykh pescher (in Russian). *Trudy Obschestva liubitelei estestvoznaniya, antropologii i etnografii XVII. Trydy arkheologicheskogo Otdela* 1: 82-88.

ZAMYATNIN, S. N. (1937) *Paleolit Abkhazii* (in Russian). Trudy Instituta abkhazskoi kul'tury. X(1). Sukhumi: IAK AN SSSR. 54 p.

DESERT CAVES AND ROCKSHELTERS IN THE GREAT BASIN OF NORTH AMERICA

C. Melvin AIKENS

Department of Anthropology and Museum of Natural and Cultural History, University of Oregon, Eugene, Oregon 97403-1224, U.S.A.; maikens@uoregon.edu

Abstract. Caves and rockshelters in the Great Basin of North America have played a huge role in the development of archaeological analysis and theory because of the key data they yield on the ecology of hunting-gathering peoples. Many dry caves have provided long stratigraphic sequences and rich assemblages of artifacts, including textile and other normally perishable specimens. The invaluable assemblages of plant and animal remains from caves have especially been keys to understanding the highly mobile lifeway of hunter-gatherers, showing their dependence on spatial variability in the natural environment, and the sociocultural effects of climatically induced changes in local floras and faunas over some 10,000 years of Holocene time. In the history of Great Basin archaeology, four caves sites have been of particular importance in defining both the scientific potentials and the research problems of dry caves. These are Lovelock Cave in Nevada, the Fort Rock and Paisley 5 Mile Point caves in Oregon, and Danger Cave in Utah. Along with the archaeological riches they contained, however, also came stratigraphic and associational problems that are endemic to such sites. This paper offers a brief history of these problems and archaeologists' efforts to overcome them. It urges that future work must be ever more carefully conceived and planned to exact the maximum potential from these precious data sources, which are rapidly being depleted by both scientific and illegal excavations.

Keywords: North America, Great Basin, dry caves, ecology, stratigraphy, contextual problems.

Résumé. Les grottes et abris rocheux dans le Great Basin d'Amérique du Nord on eu un rôle énorme dans le développement de l'analyse et de la théorie archéologiques, car ils fournissent des données essentielles sur l'écologie des peuples chasseurs-récolteurs. Plusieurs caves sèches ont offert de longues sécuences stratigraphiques et de riches collections d'objets artisanaux, dont des pièces de tissu et d'autres spécimens normalement périssables. Les précieuses collections de restes végétaux et animaux présents dans les grottes, particulièrement, ont permis de comprendre le style de vie hautement nomadique des chasseurs-récolteurs; ils ont montré leur dépendance à l'égard de la variabilité spatiale dans leur environnement naturel, et les effets socio-culturels des changements causés par le climat sur la flore et la faune locales pendant environ 10.000 ans de la période de l'Holocène. Dans l'histoire de l'archéologie du Great Basin, quatre grottes ont été particulièrement importantes pour aider à définir, d'une part le potentiel scientifique, d'autre part les problèmes de recherche liés aux caves sèches. Il s'agit de la grotte de Lovelock dans le Nevada, les grottes de Fort Rock et Paisley 5 Mile Point en Oregon, et la grotte de Danger dans l'Utah. En révélant leurs richesses archéologiques, ces grottes ont aussi présent des problèmes stratigraphiques et associationnels communs à ce type de sites. Cet article présente une brève histoire de ces problèmes et des efforts faits par les archéologues pour les surmonter. Il insiste que toute recherche future doit être très attentivement conçue et dirigée afin d'extraire le maximum d'information de ces précieuses sources de donnés, qui sont en train d'être rapidement détruites par les excavations à la fois scientifiques et ill égales.

Mots-clés: Amérique du Nord, Great Basin, caves sèches, écologie, stratigraphie, problèmes contextuels.

The archaeology of dry caves and rockshelters has established a rich and varied record of material culture in the desert west of North America, as well as stratigraphic evidence demonstrating human occupation in the region since the beginning of Holocene time. This paper places in historical and conceptual perspective some of the exciting recent research on archaeology in the arid western United States that is being discussed in the present symposium. The cases selected obviously represent a tiny fraction of the many cave excavations that have been conducted within this region, but they are singled out for the present purpose because of their historical importance in defining and framing key interpretive and methodological issues that remain important today. All age determinations given in this account are based on radiocarbon years.

Basketry, sandals and other artifacts excavated from Lovelock Cave, Nevada in 1909 led to comparisons of the ancient Great Basin culture with that of the Basketmakers of the American Southwest, already quite well known by the turn of the 20th Century (Loud and Harrington 1929). In 1938 sagebrush bark sandals were found in Fort Rock Cave, Oregon, beneath a thick layer of volcanic ash that came from the mid-Holocene eruption of Mount Mazama in the southern Cascade Range. This geological evidence placed early Great Basin occupation much deeper in time than Basketmaker, and we now know the Fort Rock sandal type dates back some 10,000 years (Cressman et al. 1940, 1942; Connolly et al. 1998). Deep, well-stratified deposits at Danger Cave, Utah, briefly tested by archaeologists during 1937-41 and extensively investigated during 1949-53 showed that people first camped more than 10,000 years ago on clean sand left on the cave floor by waters of a final recessional stage of Pleistocene Lake Bonneville, and came back to the cave again and again over succeeding millennia. More than three meters of dry, finely layered culture-bearing deposit slowly accumulated in a stratigraphic record that spans the entire Holocene. The rich inventory and long sequence of artifactual and biotic remains from Danger Cave led to the recognition of an ancient tradition of broad-spectrum hunting and gathering that began to appear some 10,000 years ago and persisted in its essential elements down to the late 19th century (Jennings 1957).

From the beginnings marked by these three signal cave excavations the Great Basin of western North America has become a premier world area for the long-term study of human ecology. It is a topographically and altitudinally varied region where the patchy mosaic distribution of plant and animal resources critically shaped the annual cycle of movements and seasonal aggregations of the native hunting-gathering people. A strong research focus on the long-term trajectory of human ecological change has been well-supported by the rich biotic remains and long cultural sequences preserved in its many dry cave sites, and increasingly by investigations outside the caves into pollen sequences from bog and lake sediments, plant macrofossils from wood rat middens, and erosional and depositional processes along the edges of the region's many Pleistocene lake basins (Grayson 1993).

With all the archaeological blessings bestowed by dry cave sites, however, also come stratigraphic records that are complicated by long histories of disruptive activity by their human and other occupants—and more recently the depredations of irresponsible relic collectors who heedlessly mine and destroy archaeological deposits for the occasional attractive specimens they contain. The various disruptions to which caves are subject often make it hard to understand the precise relationships between artifacts, cultural features, and "ecofacts" found in their deposits, and this has caused archaeologists various kinds of interpretive problems and outright errors. In fact, the history of cave archeology in the arid west of North America has been at the same time a history of archaeologists learning how to dissect and analyze such sites so as to overcome the contextual ambiguities that often arise in an excavation. This is a methodological problem that is of course worldwide, but some of the earliest critical concern with such issues arose from excavations in Great Basin caves. It is interesting to note that the *de facto* learning approach of Great Basin archaeologists usually has been to belatedly recognize mistakes and try subsequently to correct them, or at least do better the next time. It is a learning process that continues to this day.

There are also changes over time in the questions archaeologists seek to investigate. With the wisdom of hindsight it is possible to recognize "mistakes" in earlier work that were not necessarily errors in terms of then-existing standards. They were, rather, failures to imagine in advance all the valuable information that could have been gained by sharper research questions, and procedures of excavation and analysis specifically tailored to them. So much is this syndrome of "making mistakes" an inherent part of archaeological practice and progress that one of the favorite greetings of the celebrated Great Basin archaeologist Jesse D. Jennings when visiting a student or colleague in the field was, "Well, so you're the one making the mistakes on this site." Jennings took it for granted, based on his own very considerable field experience, that mistakes were effectively unavoidable when working through a deposit of unknown structure and content. The present paper gives several Great Basin examples of this peculiarly archaeological learning process and where it is taking us.

Excavations were carried out at Lovelock Cave by L.L. Loud in 1912, by Loud and M.R. Harrington in 1924, and by R.F. Heizer and L. Napton in 1968-69. Professor A.L. Kroeber of the University of California at Berkeley, hearing that important artifacts were coming from a cave being mined for bat guano near Lovelock, Nevada, sent Loud, a guard at the Phoebe Hearst Museum, to make a collection there. Loud spent the late fall and early winter digging alone in Lovelock Cave, gathering a very considerable collection of textile and other artifacts preserved by its dry interior climate. He collected specimens in a workmanlike manner, according to a series of "lots." But he had no training as an archaeologist, and these lots came from large, deep cuts in different parts of the cave that did not provide useful vertical or horizontal control on the provenience of the specimens.

In 1924 Kroeber sent Loud back to Lovelock Cave to work with M.R. Harrington, a well established archaeologist from the Heye Foundation in New York. In the course of this work Harrington excavated one unit stratigraphically and recognized six main occupation levels. These he grouped into Early, Transitional, and Late periods for the cave occupation as a whole, but the great bulk of the Lovelock collection came from other uncontrolled excavations and remains without stratigraphic context. The monograph published in 1929 by Loud and Harrington reported a rich collection of specimens that included among other things basketry, sandals, netting, and even duck decoys that had been made out of rushes and covered with the skins and feathers of actual waterfowl.

The Lovelock Cave textiles showed many broad similarities with those already known from previous excavations in dry Basketmaker and Puebloan rockshelter sites in Utah, Arizona, Colorado, and New Mexico. No traces, however, of Basketmaker/Pueblo cultigens or pottery were present at Lovelock. The collection gave rise to the theory -- subsequently proven erroneous--that the Lovelock Cave culture, and cultures of the Great Basin generally, had come from the Southwest in quite recent times, losing the arts of cultivation, pottery-making, and masonry architecture in the cold, impoverished deserts of the Great Basin. It also became evident that this Lovelock Culture was quite different in its textiles and other distinctive elements from the culture of the Northern Paiutes who occupied the region in ethnographic times. Regrettably, with respect to this latter point, the Lovelock collections overall remained so lacking in stratigraphic control that neither Loud and Harrington, nor subsequent scholars working with the collections, could clearly determine whether Lovelock Culture was abruptly replaced by a Northern Paiute incursion, or whether there was continuity and a gradual transition between the two. The question in fact remains at issue today.

In 1968-69 Robert F. Heizer and his student Lewis K. Napton opened new excavations at Lovelock Cave, hoping to resolve lingering questions and learn more about the site's significance within the region. Unfortunately, by that time the site had been thoroughly ransacked by artifact collectors. Only a few pockets of archaeological deposit could be found, in crevices and among the boulders of the cave bottom, and Heizer and Napton were unable to resolve the old stratigraphic questions. Marvelously, however, they discovered in remnants of the original cave fill many desiccated human coprolites. Through directly radiocarbon-dating these specimens and studying the meal remains they contained, Heizer and Napton (1970) were able to offer a remarkable account of broad-spectrum reliance on lake-marsh resources by the Lovelock Cave people, who had lived there during a period when the lakebed below the site-- mostly dry in modern times--was covered with water. This research sparked great interest and led on to much similar research in the region, establishing a provocative new view of the ecological diversity within Great Basin desert culture that continues to generate productive research today. Heizer and Napton overcame the old stratigraphic problems at Lovelock Cave by directly radiocarbon dating the specimens they studied, and a similar approach to clarifying the development of textiles originally found there is ongoing (Hattori et al. 2000). So much for old methodological errors and continuing efforts to overcome them at Lovelock Cave.

In the late 1930s, University of Oregon Professor Luther S. Cressman, a renounced Episcopal minister with a Ph.D. in Sociology, excavated at a series of caves in south-central Oregon. Like Loud and Harrington, Cressman was another self-taught archaeologist, who records in his 1988 memoir that he learned how to dig by reading the works of Sir Leonard Wooley, and by practice in the field. He also gained a great deal by seeking the collaboration of geologists Ernst Antevs and Howel Williams, and the palynologist Henry P. Hansen, among other specialists from relevant disciplines. Cressman concentrated his attention on cave sites associated with the shorelines of Pleistocene-age lakes-- which are extensive in the Northern Great Basin, -- seeking evidence of people who might have lived there when the lakes stood at those old shorelines. His 1937-38 excavations in Fort Rock Cave revealed more than 70 complete and fragmentary woven sandals made of sagebrush bark, which were buried beneath a thick layer of volcanic tephra. Subsequent research showed the tephra came from the colossal eruption of Mount Mazama that formed the caldera of Crater Lake, 100 km. southwest of Fort Rock Cave. We now know this eruption occurred about 7000 years ago; at the time of Cressman's research, the geologist Howel Williams estimated its age as between 5000 and 10,000 years ago. The sandals themselves are now directly C-14 dated between 9000 and 10,000 years ago (Connolly et al. 1998). The Fort Rock Cave work, a pioneering piece of geo-archaeological research for its time, showed conclusively that the Great Basin desert culture, as originally exemplified by the assemblage from Lovelock Cave, was no late derivative of the Southwestern Basketmaker/Pueblo cultures but in fact far more ancient than they, and likely one of the broadly adapted sources from which they had sprung.

Cressman returned to Fort Rock Cave in 1967-68 for further excavations with his last student, Stephen Bedwell. After 30 years of sporadic looting by artifact collectors, no undisturbed deposits remained in the cave, but tumbled boulders were removed from the apron in front, and in the area that had been thus protected a small pit was dug down to lake gravels from the last high stand of Pluvial Lake Fort Rock, dated to about 13,000 years ago. A mano fragment, a heavily re-sharpened projectile point, and some flakes were found just above the gravels, and charcoal fragments gathered from the bottom of the pit were C-14 dated to 13,200 years ago. Cressman felt that his geo-archaeological approach to demonstrating a Pleistocene presence of "early man" in the Great Basin, which he had initially pioneered more than 30 years before, had been vindicated (Bedwell and Cressman 1970; Cressman 1988). Many professional colleagues, however, did not share his assessment. Because a clear association between the artifacts and the dated charcoal was not documented in detailed notes and photographs, because the radiocarbon date had an unusually large standard deviation, and because the single recovered point resembled a Western Stemmed type elsewhere dated about 8-10,000 years ago, this dating has long been controversial (Haynes 1971).

Cressman also directed work at the Paisley Caves in 1937-38, located in the Summer Lake Basin about 70 km southeast of Fort Rock. There his crew found, at the bottom of a deep excavation amid tumbled roof-fall boulders, the bones of extinct horse and camel, along with a few flaked stone artifacts. These caves had been eroded into the highest shoreline of Pluvial Summer Lake, and Cressman believed that his excavations showed the presence of human hunters there in Pleistocene times. However, because the apparent association between the artifacts and bones was not documented by detailed notes and photographs, this claim has always been treated with skepticism. Accordingly, in the familiar Great Basin way, these caves also were revisited many years later by an archaeologist trying to overcome earlier methodological problems.

In 2002 Dr Dennis L. Jenkins, of the University of Oregon, began a new research program at the Paisley Caves. They, too, had been pillaged by artifact collectors over decades, but by excavating beneath Cressman's earlier back dirt, piled up at the front of the caves, Jenkins succeeded in finding some un-vandalized deposits. At the bottom of deep excavations amid tumbled roof-fall boulders, he, like Cressman, found the bones of extinct animals along with artifacts of stone and woven plant fiber. Again the narrow circumstances of excavation at the very bottom of the cave did not allow direct and clear documentation of the crucial associations in the field, but this time a battery of laboratory tests undreamed of in Cressman's time was brought to bear. This work is still in process, but recent preliminary reports suggest that an analytical effort combining obsidian hydration dating, AMS dating of minute fiber artifacts and animal bone, and DNA testing of human feces from the Paisley Cave 5 deposits is likely to resolve a question that has been hanging for almost 60 years. A single newly reported result is the AMS dating of a human fecal specimen identified by DNA analysis to an age of about 12,300 BP (Jenkins 2006; Willerslev et al. 2006).

My last main example, Danger Cave, lies near the Utah-Nevada border at the western edge of Pluvial Lake Bonneville. As noted above, it occupies a premier place in the history of Great Basin archaeological research. Excavations in 1949-53 by Jesse D. Jennings of the University of Utah recovered a rich record of lithic and perishable artifacts, along with abundant biotic remains, from over three meters of dry and well-stratified deposits that rested on beach sands and gravels from a late low stand of the ancient lake. Some of Willard Libby's first archaeological C-14 assays established the date of initial occupation at over 10,000 BP, and confirmed repeated occupation throughout the Holocene. Jennings' dense monograph integrated in a thoroughgoing way the stratigraphic, radiocarbon, artifactual, floral, faunal, and other data from a single rich and well-controlled site in a *tour de force* never before seen in Great Basin archaeology.

His ecological interpretation, based on the excavated data and guided by published ethnographic and ethno-botanical work on the same areas' 19th century Gosiute people, offered a compelling view of a broadly adapted Great Basin Desert Culture that effectively spanned all of known human time in the region and is still influential today. Jennings' Danger Cave work was a defining moment in Great Basin archaeology, but as in the previous cases, though exceptional for its time it was not perfect. Several significant errors of excavation and interpretation became clear from the critiques of other scholars and the wisdom of hindsight.

Most importantly in the matter of excavation, although the diggers attended rigorously to excavating by the natural stratigraphy of the deposits, the defined stratigraphic units were extremely thick by modern standards. Myriad small or discontinuous depositional events and individual features such as charcoal lenses were visible in the cave profiles, but for purposes of excavation and analysis these features were grouped within larger stratigraphic units marked by continuously traceable layers of spall rock from the cave ceiling. No discrete living surfaces or activity areas, later to become a matter of great interest to archaeologists, were isolated in Jennings' excavations. Further, very few (by modern standards) radiocarbon dates were obtained from the excavation as a whole, and none at all were assayed from unit DIV; it simply was placed in time by the dates from stratigraphic units above and below. Another thing was that the Danger Cave fill had been screened through coarse mesh, and bulk samples of fill were not routinely taken, so that certainly many small specimens from the deposits went uncollected and unanalyzed. In the brief account possible here, the various questions of current interest that are begged by the lack of such detail cannot be outlined, but some at least can be readily enough imagined.

In 1986 and again in 2001-2002 research teams returned to Danger Cave for further excavation and analysis (Rhode et al. 2006; Goebel et al. this volume). One of their objectives was to minutely dissect, record, and collect "all" specimens from a block of undisturbed cave fill, seeking thereby to greatly improve the resolution of the Danger Cave data and rectify some of the errors of omission in Jennings' original excavation. Miraculously, they found a small area left intact by the looters who have struck at the cave in the half-century since Jennings' excavations. The task of analyzing such a prodigious amount of detailed data has proven to be monumental, however, and a decade later the analysis continues.

The results so far in hand have, however, already yielded important new insights into the nature of the Danger Cave occupation. Based on an extremely close analysis of small seeds from the deposit it is now thought, for example that the processing and consumption of such seeds, identified in Jennings' classic analysis as one of the most diagnostic characteristics of the Desert Culture pattern, did not begin at Danger Cave until about 8700 BP. That was some 1600 years after people first began to come there, and a time when other animal and plant resources that had been less laborious to obtain in the vicinity began to diminish as warming climate gradually shrank the wetland that fronted Danger Cave along the old lake margin. A significant number of new radiocarbon dates from Danger Cave also show convincingly that the site was consistently visited throughout the millennia. This lays to rest earlier controversy about long breaks in occupation, postulated to have been caused by mid-postglacial warming and drying, which were possible to entertain given the smaller number of radiocarbon dates presented in Jennings' original report.

It remains to be seen what additional new insights the Danger Cave re-excavation will yield, but one related matter can be mentioned. The renewed research by Madsen, Rhode, and Jones confirmed that the intensive and frequent visitation to which Danger Cave (like many other Great Basin sites) was subject over millennia created palimpsest deposits that made it simply impossible for archaeologists to isolate discrete one-time occupation surfaces or activity areas. This in turn led Madsen and colleagues (Schmitt and Madsen 2005) to seize the opportunity of excavating Camel's Back Cave, in the same general area but at a much less favored locality than Danger Cave, where the sheer rarity of human visitation and long intervening periods of non-occupation made it possible for archaeologists to discriminate and map discrete occupation surfaces in a way that can't be accomplished in most Great Basin sites.

It would be possible to continue in this vein for a long time, adducing other sites, other mistakes and initial failures to recognize important areas of analytical potential, and other promising efforts to overcome the ambiguities of earlier excavations. It would be an interesting excursion, but one that could fill a book. Two major methodological considerations for future research are already clear, however, and I conclude with them.

First, as the above examples show, modern methods of physical and chemical analysis applied directly to artifact specimens and other critical items offer promising ways of overcoming persistent problems of contextual ambiguity in Great Basin cave excavations that have continued to plague our analytic and interpretive efforts. This is an approach that is also being applied, and can be increasingly applied, to museum specimens already in hand and curated for the precise purpose of making further analyses possible. These approaches are expensive, but the results yielded make them worthwhile, and indeed indispensable.

Second, as the above discussion also illustrates, somewhat indirectly, we are increasingly coming to confront the reality that many, and probably most, of the potentially most informative archaeological sites in the Great Basin have already been discovered, and indeed have been "used up" in greater or lesser degree. The attrition is ascribable both to archaeologists groping their way site by site toward more sophisticated questions and forms of analysis, and to heedless artifact collectors and other forces of destruction. As our highest-quality field research base dwindles steadily, we can ill afford further mistakes, failures of imagination, and failures of protective management. We need to get extremely serious about all our future research efforts, focusing them tightly to glean the very most from the potentials that remain to us.

As we already see happening in our fields' best practices generally, methodological procedures and units of excavation and analysis are becoming ever tighter and more precise, requiring us to spend much more time excavating much smaller parts of sites, and to spend much more time and money exhaustively analyzing the materials recovered. It is an excellent trend, and one that is blossoming none too soon.

REFERENCES

BEDWELL, STEPHEN F. (1973) *Fort Rock Basin: Prehistory and Environment.* University of Oregon Books, Eugene.

BEDWELL, STEPHEN F. and LUTHER S. CRESSMAN (1970) *Fort Rock Report.* University of Oregon Anthropological Papers 1, Eugene.

CHAMBERLIN, RALPH V. (1911) Ethno-Botany of the Gosiute Indians of Utah. *Memoirs of the American Anthropological Association* 2(5): 329-405, Lancaster, Pa.

CRESSMAN, LUTHER S. (1988) *A Golden Journey: Memoirs of an Archaeologist.* University of Utah Press, Salt Lake City.

CRESSMAN LUTHER S.; HOWEL WILLIAMS AND ALEX D. KRIEGER (1940) *Early Man in Oregon: Archaeological Studies in the Northern Great Basin.* University of Oregon Monographs, Studies in Anthropology 3, Eugene.

CRESSMAN, LUTHER S.; FRANK C. BAKER, PAUL S. CONGER; HENRY P. HANSEN and ROBERT F. HEIZER (1942) *Archaeological Researches in the Northern Great Basin.* Carnegie Institution of Washington, Publication 538.

CONNOLLY, THOMAS J.; CATHERINE S. FOWLER and WILLIAM J. CANNON (1998) Radiocarbon Evidence Relating to Northern Great Basin Basketry Chronology. *Journal of California and Great Basin Anthropology* 20(1): 88-100.

GRAYSON, DONALD K. (1993) *The Desert's Past: A Natural Prehistory of the Great Basin.* Smithsonian Institution Press, Washington, D.C.

HAYNES, C. VANCE (1971) Time, Environment, and Early Man. *Arctic Anthropology* 8(2): 3-14.

HEIZER, ROBERT F. and LEWIS K. NAPTON (1970) *Archaeological Investigations in Lovelock Cave, Nevada.* University of California Archaeological Research Facility Contributions 10, Berkeley.

JENKINS, DENNIS L. (In Press) Distribution and Dating of Cultural and Paleontological Remains at the Paisley 5 Mile Point Caves (35LK3400) in the Northern Great Basin: an Early Assessment. In *Paleoindian or Paleoarchaic: Great Basin Human Ecology at the Pleistocene-Holocene Transition*, edited by Kelly E. Graf and Dave N. Schmitt, Chapter 4. University of Utah Press, Salt Lake City.

JENNINGS, JESSE D. (1957) Danger Cave. University of Utah Anthropological Papers 27, *Memoirs of the Society for American Archaeology* 14, Salt Lake City.

LOUD, LLEWELLYN L. and MARK R. HARRINGTON (1929) Lovelock Cave. *University of California Publications in American Archaeology and Ethnology* 25(1): 1-183. Berkeley.

RHODE, DAVID; DAVID B. MADSEN and KEVIN T. JONES (2006) Antiquity of Early Holocene Small-Seed Consumption and Processing at Danger Cave. *Antiquity* 80 (2006) 1-12.

SCHMITT, DAVE N. and DAVID B. MADSEN (2005) *Camels Back Cave.* University of Utah Anthropological Papers 125, Salt Lake City.

WILLERSLEV, ESKE; NURIA NAVERAN, TOM GILBERT, JUAN SANCHEZ, DENNIS JENKINS, JONAS BINLADEN, ANDRES GÖTHERSTRÖM, and MICHAEL HOFREITER (2006) Ancient DNA Analyses of pre-Clovis Human Coprolites. Presented at the Great Basin Anthropological Conference, October 19-21, 2006, Las Vegas, Nevada.

ROCKSHELTER ARCHAEOLOGY IN THE MIDDLE TENNESSEE VALLEY OF NORTH AMERICA

Boyce DRISKELL

Department of Anthropology, University of Tennessee, 250 South Stadium Hall, Knoxville, TN, U.S.A. 37996-0720; email: bdriskel@utk.edu.

Abstract. Well known for its numerous archaeological sites testifying to a cultural history rooted in the late Pleistocene, the Middle Tennessee Valley of North Alabama is characterized by karstic topography where sinkholes, caves, and shelters are quite common. Many of these sheltered places were utilized by prehistoric foragers, leaving an important record of life which, unlike most open air sites of the region, has been better protected from processes of destruction. Recent research at Dust Cave combines geomorphology and geochemistry, faunal and floral analyses, and technological studies to interpret data from a series of discrete, organic laden strata within a chronometric framework informed by numerous radiometric determinations. Exploration at Dust Cave is the culmination of rockshelter research in this region which began in the early 1960's and has played a prominent role in development of methods of data retrieval and analysis, and refinement of cultural history and chronology. Current rockshelter research in the region seeks insight into the organization of subsistence and settlement of Late Pleistocene and Early to Middle Holocene foragers and provides a useful forum for discussing forager theory.

Keywords: Late Pleistocene, Paleoindian, Subsistence, North America, Deeply Stratified Deposits, Stratochronological Reconstruction.

Résumé. Bien connu pour ses nombreux sites archaéologiques qui signifie une histoire culturelle enracinée vers la fin du Pléistocène, la Vallée de Tennessee Moyen au nord d'Alabama est charactérisée par une topographie karstique où s'ont trouvés des ponors, des cavernes, et des abris sous roches. Plusieurs de ses endroits abrités ont été utilisés par des fourrageurs préhistoriques, et maintenant contiennent des traces matérielles de la vie qui mieux ont été protégées contre les processus de destruction, contrairement d'autres sites en plein air dans la région. Des recherches récentes à Dust Cave ont intégrées la géomorphologie, la géochimie, les analyses fauniques et florales, et les études technologiques pour interpréter, dans un cadre chronométrique, des données d'une série de strates discrètes et chargées de matières organiques. L'exploration à Dust Cave représente, dans la région, l'apogée des recherches sur les abris sous roches. Ces recherches, qui ont commencé au début des années 1960, ont contribués à l'élaboration des méthodes d'extraction et d'analyses des données, et au raffinement de l'histoire culturelle et de la chronologie. Les recherches courantes sur les abris sous roches dans la région cherchent à fournir une interprétation de la subsistance et des habitations des fourrageurs pendant la fin du Pléistocène, le début de l'Holocène, et l'Holocène moyen, et fournit un cadre utile pour discuter la théorie concernant les fourrageurs.

Mots-clés: Pléistocène supérieur, Paleoindien, Subsistance, Amérique du Nord, Dépôts profondément stratifiés, Reconstitution strato-chronologique.

Well known for its numerous archaeological sites testifying to a cultural history rooted in the late Pleistocene, the middle Tennessee Valley of North Alabama is characterized by karstic topography where sinkholes, caves, and shelters are quite common. Situated in northern Alabama, the middle stretches of the Tennessee River skirts the Cumberland Plateau and dissects the Highland Rim, a plateau of moderate relief nestled along the western edge of the loftier Cumberland Plateau **(Figure 7.1)**. The middle Tennessee Valley is characterized by a rather narrow floodplain, a series of levees, and sluggish tributaries often forming backwater sloughs. The flow of the River is punctuated by Muscle Shoals, the steepest descent in the river system, with a drop of about 40 meters in 55 kilometers.

Over 4,000 caves are listed in the Alabama Cave Survey (2006). These and numerous other sheltered spaces are formed from solution cavities in the local limestones. Additionally, massive sandstone bluffs in some parts of the region provide erosional overhangs utilized as sheltered space.

In antiquity, sheltered spaces in the middle Tennessee Valley were used in three ways. Some caves were utilized for mineral extraction, both by prehistoric and historic peoples (Crothers et al. 2002). A number of caves contain prehistoric art suggesting ritual use of some caves by prehistoric Woodland and, particularly, Mississippian peoples (Simek 2006). Others used the mouths of caves and bluffshelters for domestic purposes including interment of the dead.

While caves and shelters were well known localities for Indian artifacts and burials in the 19th century, perhaps the earliest written description of a sheltered site in the Middle Tennessee Valley by a professional observer was that of Cyrus Thomas in 1890 who briefly described the contents of Hampton Cave near Guntersville, Alabama. Hampton Cave and 10 other caves or shelters were later excavated in the 1930's as part of the large scale salvage excavations associated with reservoir building on the middle Tennessee River during the Great Depression. All were disturbed by looting and exhibited primarily late prehistoric deposits; early Archaic and Paleoindian deposits were not discovered in these excavations.

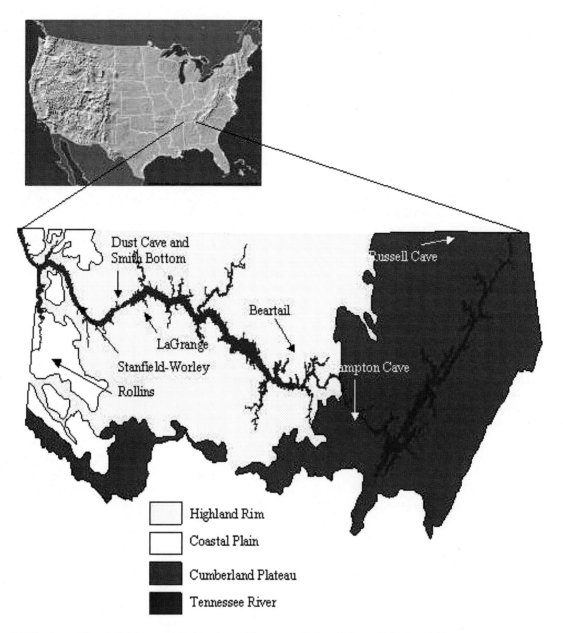

7.1 Physiographic subdivisions and the locations of important sheltered sites within the Middle Tennessee Valley.

After World War II, several avocational archaeology groups were organized in the region. Spurred by western U.S. finds of fluted projectile points associated with extinct Pleistocene animal remains and numerous fluted point finds from eroded, open air sites in the area, these avocationals focused interest on finding local, ancient human counterparts. While surface finds were extremely numerous, no undisturbed Paleoindian deposits had been found in the region that could testify to the chronological or cultural placement of fluted points from the middle Tennessee Valley.

Enthusiastic members of the Chattanooga Chapter of the Tennessee Archaeological Society conducted preliminary excavations at Russell Cave in northeastern Alabama from 1951-1955. Like other sheltered sites excavated earlier in the region, the upper levels of the cave produced late prehistoric remains. Findings at Modoc Shelter in the Illinois Valley to the north suggested that some caves and shelters contained deeply buried, early materials so a consortium of professional and amateur archaeologists, with financial support for the National Geographic Society, probed deeper into the lower deposits of Russell Cave for three additional seasons (1956-1958). Excavations continued in 1962 after federal acquisition of the Cave as a National Monument (Griffin 1974). Even though investigations failed to recover evidence of Paleoindian people, the investigation at Russell Cave highlighted the presence of well-preserved organic remains and the potential for deep, stratified deposits in sheltered sites.

Not to be outdone by the Tennessee Society, the Alabama Archaeological Society, with collaboration from the University of Alabama, sponsored a series of summer student digs referred to as the "North Alabama Project"

(Futato 2004). The Project specifically sought out and investigated sites that might contain Paleoindian remains in stratified context. There was also an undercurrent of intrigue fostered by the presence of crude "pebble tools" in great numbers on many northwest Alabama sites. Called the Lively Complex after the avocational archaeologist who carefully analyzed and typed thousands of these specimens, these simple artifacts resembled tools of the sensational Olduvan industry of east Africa. At the time, prominent members of the Archaeological Society openly wondered if these tools represented a truly ancient, undocumented human past in the middle Tennessee valley; later research confirmed Holocene dating of these artifacts.

The Stanfield Worley Bluff Shelter (DeJarnette, et al. 1962) was the first site targeted. Excavations taking place during the early 1960's revealed a stratified deposit with late prehistoric deposits at the top and a mixed layer of early corner notched and Dalton projectile points at the base. Several other shelters in the middle Tennessee Valley were excavated by the Project over the next several years culminating with the excavation of LaGrange Bluff Shelter in the summer of 1976. Disappointingly, these excavations failed to reveal undisturbed deposits associated with Clovis or other fluted points, but several of the sites contained Dalton bearing basal deposits which clarified the chronological placement of Dalton as a late Paleoindian horizon.

A renewed search for undisturbed, early deposits in sheltered spaces began in 1988 with examination of 20 caves in the Pickwick Reservoir in the western part of the region by staff of the University of Alabama. Twenty caves which showed some archaeological potential (Cobb et al. 1995; Collins 1995) were selected and investigated. The program continued during the summer of 1989 when Dust Cave was one of several caves investigated.

In one of the small test pits at Dust Cave, literally at arm's length, the telltale scraping of towel on flint alerted us to the possible presence of archaeological deposits in the cave; this was quickly confirmed by darker, organic fill and the recovery of debitage. Test units were enlarged and excavations continued. Returning in the summer of 1990, bedrock was eventually reached at a depth of 4.92 meters below datum in the entrance chamber of the cave. Testing continued for several summers following.

At about the same time, the Beartail Rockshelter, located about 100 miles to the east of Dust Cave, came to the attention of archaeologists. Deposits in the shelter showed considerable potential for deeply buried, early remains. Excavated by staff of the University of Alabama over three seasons (1994-1996) with a Legacy Grant from the U.S. Army, the investigations revealed a sequence of prehistoric use dating back to the early Archaic period (Hubbert 1997). Unfortunately, the remarkable preservation of organic remains found at Dust Cave was not duplicated at Beartail Rockshelter so the potential for continued research was limited. At Dust Cave excavations continued for twelve summers, culminating at the end of the summer of 2002.

Dust Cave is remarkable for its deep, complex, and reasonably intact stratigraphy (Driskell 1994, 1996). Informed by 43 14C dates, the sequencing of the Cave's depositional history is quite clear, beginning with initial human use about 10,650 cal. B.C. (Sherwood et al. 2004). In about 3600 cal B.C., headroom became so restrictive that the cave was finally abandoned. Additional detritus from bluffline and cave roof accumulated on the talus and cave floor concealing the archaeological record until recent investigations. Accordingly, Dust Cave contributes substantially to an enhanced precision of regional chronology (**Figure 7.2**) during the Late Pleistocene and early to middle Holocene, providing chronometric collaboration of the several early projectile point horizons.

The complex sedimentary record at Dust Cave has challenged us to experiment with innovative ways to generate data on the Cave's prehistoric occupants. Analysis of structural characteristics of sediments utilizing techniques of micromorphology (Sherwood 2001, Goldberg and Sherwood 1994) has provided most of the data utilized to reconstruct sediment histories. Archaeological chemistry (Homsey 2004; Homsey and Capo 2002, 2006) has aided our understanding of the function of various features found within the cave, including curious fire hardened, prepared red clay surfaces that seem to date from earliest to latest use of the Cave. Sherwood and Chapman (2005) conclude that these prepared and fired clay surfaces are common features of eastern U. S. prehistoric sites but have not been very widely recognized. These features could have functioned as heated cooking surfaces to dry nut meats, roast other plant materials, or possibly cook fish. Examination of the magnetic characteristics of the sediments (Gose 2000; Collins et al. 1994) has assisted in our understanding of the timing of depositional events at Dust Cave.

Generally, in the temperate south, artifacts of perishable materials are poorly preserved. At Dust Cave, however, textile impressions in prepared clay surfaces (Sherwood and Chapman 2005) testify to the presence of a fiber industry. Baskets, bags, lines, nets, and other fiber artifacts were no doubt an important part of subsistence technology from the earliest times, providing various containers, cordage and matting. Bone is well preserved at Dust Cave and bone artifacts, particularly awls, pins, and needles are prominent in the material inventory (Goldman-Finn and Walker 1994). A bone fishhook found at Dust Cave testifies to line fishing early (about 9,000 RYBP) in the Archaic.

Many of the chipped stone artifacts found in the region are made of Ft. Payne chert (Johnson and Meeks 1994), a locally available chert of very high quality. Bifaces

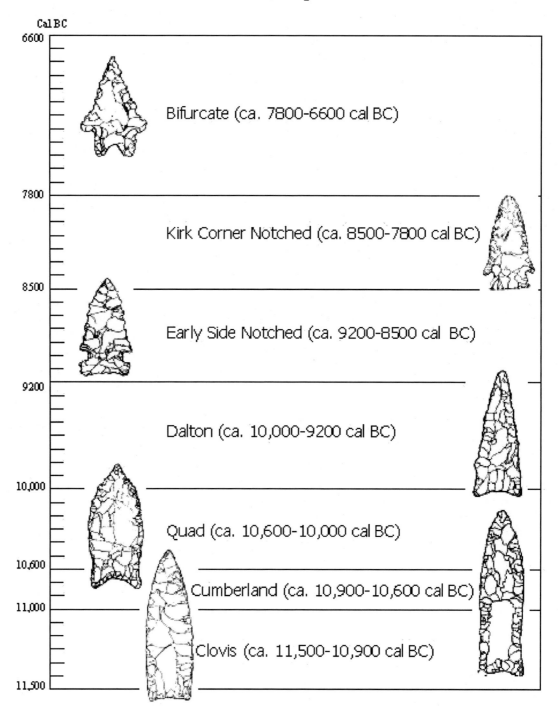

7.2 Suggested chronology for the Paleoindian and Early Archaic periods in the Middle Tennessee Valley (line drawings are from Justice 1987).

dominate the stone tool inventory throughout (Meeks 1994, 1999, 2000; Randall 2002), providing preforms for formal tools and an abundance of flakes for expedient tools. In a coexisting technology, blades struck from prepared blade cores are prevalent in the earlier part of the sequence but precipitously disappear by the end of the Dalton horizon. Other shaped unifaces, such as teardrop and side scrapers, are more prominently represented in the earlier part of the sequence as well, but diminish gradually in proportion to bifacial tools during subsequent Archaic periods.

While some functional studies of the Dust Cave lithic tool assemblage has been attempted (Randall 2002; Meeks 1999, 2000; Walker et al. 2001), only a few artifacts have been examined for microwear. As would be expected, these studies implicate stone tools in hunting pursuits, but more detailed studies are needed to identify tool types for specific tasks and to understand functional, technological and stylistic changes through the seven millennia record of tool use at Dust Cave.

Because of remarkable preservation of faunal and floral remains (Walker et al. 2001), human and animal remains (Hogue 1994; Walker and Parmalee 2004; Walker et al. 2005; Walker 2002; Morey 1994; Parmalee 1994) as well as charred plant materials (Gardner 1994; Hollenbach 2005) are well preserved. Unlike other known early sites in the eastern U.S., large mammals are poorly represented in the collections while aquatic resources are particularly well represented, especially in the earliest deposits (Grover 1994; Walker 1997, 1998, 2000, 2002). While the Dust Cave fauna clearly represent a more reliable sampling of animal exploitation because of exceptional preservation, it is not entirely clear as to whether or not Dust Cave represents a special function site whose refuse was not a very good indicator of the overall subsistence economy of early hunters/gatherers in the middle Tennessee Valley.

Recent research (Hollenbach 2005) examined plant macrofossils from Dust Cave and compared these data and the faunal evidence for subsistence to plant and animal use at several other early, sheltered sites (LaGrange, Rollins, and Stanfield-Worley shelters) in the region. By calculating costs of procurement and season of availability of resources (plant and animal) represented in the archaeological collections at these sites, a model of annual subsistence and settlement patterning for late Paleoindian and early Archaic foragers of the region was developed. In this model, foragers of the area exploited upland nut mast and nearby animal/floral resources in fall, seeds in early winter, floodplain resources including aquatic resources in the winter and spring, and fruits in the summer. Findings strongly support the notion that subsistence and settlement were organized around collectables. Women, children and the elderly probably formed the principal labor force to exploit these resources. Hunting activities were most likely embedded in the collecting cycle, and not the reverse, since prey animals were attracted to the same plant resources (young shoots, mature seeds, nut mast) as their human predators.

While this pattern of subsistence persisted from Late Paleoindian to at least the middle Archaic period in the middle Tennessee Valley, ongoing research by Scott Meeks of the University of Tennessee into the distribution of sites at different times during the late Pleistocene and early Holocene suggests that population densities and perhaps, population concentrations, changed through these first seven millennia of human habitation in the area around Dust Cave.

Unless strictly an artifact of archaeological visibility, Quad peoples reoccupied sites more often and exhibited preferences for upland habitats, particularly sinks, perhaps as a response to the harsher conditions of the Younger-Dryas. As might be expected, population density seemingly increased in the Highland Rim with milder climatic conditions of the Holocene.

Sheltered space in the middle Tennessee Valley has been used for various purposes by humans for about 13,000 years. The earliest professional investigations of caves in the area revealed uses including mortuary sites by late prehistoric peoples, but failed to recognize other uses such as mineral exploitation, ritual practice, or habitation. These functions have all been documented for the area as research in caves and natural shelters has intensified. Because of exceptional preservation within Dust Cave, an unprecedented record of early hunter/gatherer subsistence and technology has been recovered revealing foraging practices in which hunting was probably embedded in, or coordinated with, collecting activities. While population densities and settlement setting may have varied somewhat in conjunction with climatic changes, this lifeway seemingly persisted for many millennia. This pattern is in stark contrast to some models of Paleoindian human organization where hunting is considered to be the primary organizational principle guiding decisions about subsistence and settlement.

REFERENCES

ALABAMA CAVE SURVEY (2006). Web site for the Alabama Cave Survey at http://ourworld.cs.com/alabamacaves/ accessed for number of caves currently listed.

COBB, RICHARD M.; BOYCE N. DRISKELL, and SCOTT MEEKS (1995). Speleoarchaeological Reconnaissance and Test Excavations in the Pickwick Basin. In *Cultural Resources in the Pickwick Reservoir*, edited by CATHERINE C. MEYER. University of Alabama, Division of Archaeology, Report of Investigations 75: 219-261.

COLLINS, MICHAEL B. (1995). Observations on the Geomorphology of Ten Cave/Rockshelter Localities in the Pickwick Basin of Colbert and Lauderdale Counties, Alabama. In *Cultural Resources in the Pickwick Reservoir*, edited by CATHERINE C. MEYER. University of Alabama, Division of Archaeology, Report of Investigations 75:351-376.

COLLINS, M.B.; WULF GOSE and SCOTT SHAW (1994) Preliminary Geomorphological Findings at Dust and Nearby Caves. *Journal of Alabama Archaeology* 40:34-55.

CROTHERS, GEORGE M.; CHARLES F. FAULKNER; JAN F. SIMEK; PATTY JO WATSON and P. WILLEY (2002) Woodland Cave Archaeology in Eastern North America. In: *The Woodland Southeast*, DAVID G. ANDERSON and ROBERT MAINFORT, JR. (eds), University of Alabama Press, Tuscaloosa, pp. 502-524.

DeJARNETTE, DAVID L.; EDWARD B. KURJACK and JAMES W. CAMBRON (1962) Stanfield-Worley

Bluff Shelter Excavations. *Journal of Alabama Archaeology* 8(1 and 2).

DRISKELL, BOYCE N. (1994) Stratigraphy and Chronology at Dust Cave. *Journal of Alabama Archaeology* 40: 17-34.

DRISKELL, BOYCE N. (1996) Stratified Late Pleistocene and Early Holocene Deposits at Dust Cave, Northwest Alabama. In: *The Paleoindian and Early Archaic Southeast,* DAVID ANDERSON and KENNETH SASSAMAN (eds.), University of Alabama Press, Tuscaloosa, pp. 315-330.

FUTATO, EUGENE, (2004) The North Alabama Project: An AAS Excavation Scrapbook. *Journal of Alabama Archaeology* 50(2).

GARDNER, PAUL (1994) Carbonized Plant Remains from Dust Cave. *Journal of Alabama Archaeology* 40:192-211.

GOLDBERG, PAUL AND SARAH SHERWOOD (1994) Micromorphology of Dust Cave Sediments: Some Preliminary Results. *Journal of Alabama Archaeology* 40:56-64.

GOLDMAN-FINN, NURIT and RENEE WALKER (1994) The Dust Cave Bone Tool Assemblage. *Journal of Alabama Archaeology* 40:104-113.

GOSE, WULF A. (2000) Paleomagnetic Studies of Burned Rocks. *Journal of Archaeological Science* 27:409-421.

GRIFFIN, JOHN W. (1974) *Investigations in Russell Cave.* Publications in Archaeology 13. National Park Service, Washington, D.C.

GROVER, JENNIFER (1994) Faunal Remains from Dust Cave. Journal of Alabama Archaeology 40:114-131.

HOGUE, S. HOMES (1994) Human Skeletal Remains from Dust Cave. *Journal of Alabama Archaeology* 40:170-188.

HOLLENBACH, KANDACE D. (2005) *Gathering in the Late Paleoindian and Early Archaic Periods in the Middle Tennessee River Valley, Northwest Alabama.* Ph.D. dissertation, Department of Anthropology, University of North Carolina, Chapel Hill.

HOMSEY, LARA K. (2004) *The Form, Function and Organization of Anthropogenic Deposits at Dust Cave, Alabama.* Ph. D. Dissertation, Department of Anthropology, University of Pittsburgh.

HOMSEY, LARA K. and ROSEMARY C. CAPO (2002) Geochemical Identification of Feature Function and Activity Areas at a Late Paleoindian through Middle Archaic Archaeological Site: Dust Cave, Al. 2002 *Geological Society of America Abstracts with Programs*, p. 51-5.

HOMSEY, LARA K. and ROSEMARY C. CAPO (2006) Integrating Geochemistry and Micromorphology to Interpret Feature Use at Dust Cave, a Paleoindian through Middle Archaic Site in Northwest Alabama, *Geoarchaeology* 21(3): 237-269.

HUBBERT, CHARLES M. (1997) *The Beartail Rockshelter Legacy Project.* Technical report of the Office of Archaeological Services, University of Alabama, Tuscaloosa, Alabama.

JOHNSON, HUNTER and SCOTT MEEKS (1994) Source Areas and Prehistoric Use of Fort Payne Chert. *Journal of Alabama Archaeology* 40:65-76.

JUSTICE, NOEL D. (1987) *Stone Age Spear and Arrow Points of the Midcontinental and Eastern United States.* Indiana University Press, Bloomington.

MEEKS, SCOTT C. (1994) Lithic Artifacts from Dust Cave. *Journal of Alabama Archaeology* 40:77-103.

MEEKS, SCOTT C. (1999) The Function of Stone Tools in Prehistoric Exchange Systems: A Look at Benton Interaction in the Mid-South. In *Raw Materials and Exchange in the Mid-South,* EVAN PEACOCK and SAMUEL O. BROOKES, eds. Archaeological Report 29, Mississippi Department of Archives and History, pp 29-43.

MEEKS, SCOTT C. (2000) *The Use and Function of Late Middle Archaic Projectile Points in the Midsouth.* University of Alabama Office of Archaeological Services, Report of Investigations 77. Moundville, Alabama.

MOREY, DARCY F. (1994) Canis Remains from Dust Cave. *Journal of Alabama Archaeology* 40:160-169.

PARMALEE, PAUL (1994) Freshwater Mussels from Dust and Smith Bottom Caves, Alabama. *Journal of Alabama Archaeology* 40:132-159.

RANDALL, ASA (2002) *Technofunctional Variation in Early Side-Notched Hafted Bifaces: A View from the Middle Tennessee River Valley in Northwest Alabama.* Thesis submitted for partial fulfillment of the M.A in Anthropology, the University of Florida.

SHERWOOD, SARAH C. (2001) *The Geoarchaeology of Dust Cave: A Late Paleoindian through Middle Archaic site in the Western Middle Tennessee River Valley.* Ph. D. Dissertation, Department of Anthropology, University of Tennessee, Knoxville.

SHERWOOD, SARAH C. and JEFFERSON CHAPMAN (2005) Identification and Potential Significance of Early Holocene Prepared Clay Surfaces: Examples from Dust Cave and Icehouse Bottom. *Southeastern. Archaeology* 24(1): 70-82.

SHERWOOD, SARAH C.; BOYCE N. DRISKELL; ASA R. RANDALL and SCOTT C. MEETS (2004) Chronology and Stratigraphy at Dust Cave, Alabama. *American Antiquity* 69(3):533-554.

SIMEK, JAN (2006) Personal communication concerning research on cave art in the middle southeast, USA, July 2006.

WALKER, RENEE B. (1997) Late Paleoindian Faunal Remains from Dust Cave, Alabama. *Current Research in the Pleistocene* 14:85-87.

WALKER, RENEE B. (1998) *The Late Paleoindian Through Middle Archaic Faunal Evidence from Dust Cave, Alabama.* Ph. D. dissertation, Department of Anthropology, University of Tennessee, Knoxville.

WALKER, RENEE B. (2000) *Subsistence Strategies at Dust Cave: Changes from the Late Paleoindian through Middle Archaic Occupations.* University of Alabama Office of Archaeological Services, Report of Investigations 78.

WALKER, RENEE B. (2002) Early Holocene Ecological Adaptations in North Alabama. In: *Culture,*

Environment, and Conservation in the Appalachian South, BENITA J. HOWELL (ed). University of Illinois Press, Urbana, pp. 21-41.

WALKER, RENEE B. AND PAUL W. PARMALEE (2004) A Noteworthy Cache of Branta canadensis at Dust Cave, Northwestern Alabama. *Journal of Alabama Archaeology* 50(1):18-35.

WALKER, RENEE; KANDACE DETWILER; SCOTT MEEKS and BOYCE DRISKELL (2001) Berries, Bones, and Blades: Reconstructing Late Paleoindian Subsistence Economy at Dust Cave, Alabama. *Midcontinental Journal of Archaeology* 26(2):169-198.

WALKER, RENEE B.; DARCY F. MOREY and JOHN H. RELETHFORD (2005) Early and Mid-Holocene Dogs in Southeastern North America: Examples from Dust Cave. *Southeastern Archaeology* 24(1): 83-92.

ROCKSHELTER OF THE MIDDLE ROCKY MOUNTAINS: 70 YEARS OF RESEARCH

Marcel KORNFELD

George C. Frison Institute, Department 3431, 1000 East University Avenue, University of Wyoming, Laramie, 82071 USA; anpro1@uwyo.edu

Abstract. The Middle Rocky Mountains abound in rockshelters many of which were utilized by prehistoric inhabitants, arguably since the earliest human occupations of the Americas. The shelters in this region vary in size, morphology, bedrock geology, sedimentation, time depth of human use, tempo and periodicity of use, and in many other characteristics. The shelters played a significant role in developing regional cultural chronology, which although considerably developed at this time, continues to unfold with new chronologically diagnostic artifacts from shelters holding a key to understanding. Presently we are reexamining previously investigated shelters with cutting edge research strategies, namely interdisciplinary in nature, as well as initiating studies of newly discovered shelters. Both are yielding significant information about rockshelter use, variability, and links to other components of regional settlement.

Keywords: Rockshelters, North America, Rocky Mountains, Research History, New Approaches

Résumé. Middle Rocky Mountains abondent dans abris que beaucoup d'entre lequel a été utilisé par les habitants préhistoriques, sans doute depuis les occupations humaines les plus premières des Amériques. Les abris dans cette région varient dans la taille, la morphologie, la géologie de fondement, la sédimentation, la profondeur de temps d'usage humain, le tempo et la périodicité d'usage, et dans beaucoup de caractéristiques autres. Les abris ont joué un rôle significatif dans la chronologie culturelle régionale et en voie de développement, qui bien que considérablement développé, continue en ce moment à déplier avec les nouveaux objets chronologiquement diagnostiques des abris tenant une clef à la compréhension. En ce moment nous réexaminons abris précédemment des examinés avec les stratégies de recherche d'avant-garde, à savoir interdisciplinaire dans la nature, de même qu'inaugurer les études d'abris récemment découverts. Les deux produisent l'information significative de l'usage des abris, la variabilité, et les liens aux autres composants de règlement régional.

Mots-clés: Abri, Amérique du nord, Rocky Mountains, L'Histoire Recherche, les Nouvelles Approches

Numerous shelters in the Bighorn region (**Figure 8.1**) have been investigated over the past 60 years and provide a rich source of information about prehistory, yet not all have been discovered and many of the excavated ones are potentially sources for additional information about prehistoric peoples and environments. The purpose of this paper is a brief overview of closed sites in the Middle Rocky Mountains, focusing on the western foothills of the Bighorns, followed by a brief discussion of our ongoing investigations.

Except for a few outliers the Bighorn Mountains are the eastern boundary of the Central Rocky Mountain chain. As with all of the Rocky Mountain ranges, the Bighorns formed during the Cenozoic Laramide Orogeny. The eastern and western flanks of the Bighorns slope down steeply from a relatively flat plateau-like upland to the basins. This area plus several kilometers in either direction is generally referred to as the foothills. The foothills are carved by numerous deep canyons, exposing the entire set of geological formations of the Bighorns. Many of these formations are limestones or otherwise conducive to erosion that forms rockshelters and caves throughout the region (Sando 1974).

BIGHORN REGION SHELTERS

It is impossible to provide a complete review of the Bighorn Region Shelters. I wish to mention several significant sites in the region, than spend much of this review on the shelters of the western flanks of the Bighorn Mountains proper. I organize this discussion chronologically, the first discoveries followed by Culture-Historical Archeology concerned with building a chronology, followed by the Wyoming Archaeological Society and other amateur investigations of the 1950s and the 1960s. I then turn to studies by the Office of the Wyoming State Archaeologist in the 1970s and 80s and the contemporary studies from the massive CRM works of the same time period. This roughly parallels the history of American archaeology (Willey and Sabloff 1993).

First Exploration and Description (or Exploratory-Descriptive Period)

Dinwoody Cave on the Wind River Indian Reservation was excavated and recorded by Works Projects Administration crews under the direction of Tom Sowers, the State Supervisor of Survey, from December 1938 to March 1939 (Sowers 1941). Sowers indicates that only one third of the cave was excavated and this produced 649 specimens. As a student of E.B. Renaud, Sowers presents basic statistics on the entire assemblage. Unfortunately no real interpretation follows. These collections were returned to the Wind River Indian Reservations sometimes after analysis and their storage location forgotten until recently. Their re-discovery was

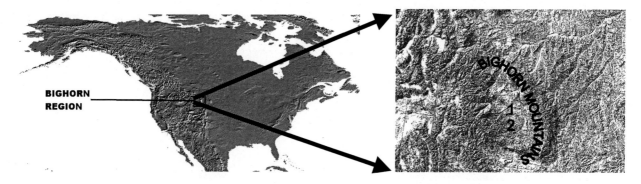

8.1 Location of the Middle Rocky Mountains (left) and the Bighorn Region (right). (left adapted from www.ngdc.noaa.gov/seg/topo/img/nasm2.gif; right modified www.nationalatlas.gov/natlas/print.cfm?bgo...).

however short-lived and ended with reburial by the Wind River Indian Reservations (Reher personal communication 2003), resulting in their permanent loss to science and scientific inquiry.

Birdshead Cave was excavated as a part of the Boysen Reservoir project, although the cave is far from the reservoir flood waters (Bliss 1950). In those days (1940s), research under salvage project that was thought to bear significant information on regional prehistory, could still be conducted outside of the immediate impact zone (e.g., Kornfeld et al. 1995). Birdshead Cave produced a stratified sequence of chronologically diagnostic artifacts, perishable materials, pottery, steatite, and hearth features. Feathers, arrow shafts, cordage, and rabbit skin robes were all part of the perishables. Seven cultural occupations were identified in the shelter and Bliss mentions that possibly nine may be present. Bliss compared the diagnostics to some other western sequences, but insufficient regional data was yet available to develop a regional chronology, although this was Bliss's expressed goal (Bliss 1950:187). Nevertheless, Birdshead provided significant background for the next stratified shelter to be excavated or interpreted, in particular Pictograph Cave.

Exploratory-Descriptive to Culture History-Chronology

Pictograph Cave story is somewhat happier than Dinwoody and well known to most Plains prehistorians, although there are controversies regarding the excavation and write-up of the methods (Frison-Forbis letter exchange late 1980s-early 1990s; Frison, personal communication 1990s). The cave contained rock images, ornaments, harpoons, and a variety of manufacturing debris and of course a cultural sequence exemplified by projectile points. The cave played a major role in the construction of the first Northwest Plains cultural chronology (Mulloy 1958) that still forms the basis of Northwest Plains cultural sequence (e.g., Frison 1991; Reeves 1983). The majority of the collections and notes, however, perished in a fire and are no longer available for study. Of significance to the cultural chronology was Mulloy's interpretation of a cultural hiatus between the Early Prehistoric period and the Early Middle Period.

WAS and other Avocationals

The excitement of the development of regional chronology spawned much work by the newly founded Wyoming Archaeological Society (WAS) and other avocationals, confirming and refining the sequence proposed by Mulloy, but still largely operating in the exploratory-descriptive mode. Although not in chronological order of investigation, this includes at least such sites as Leigh Cave, Spring Creek Cave, Sweem-Taylor, NA301, JO301, JO303, and many others.

Leigh Cave in Ten Sleep Canyon (Frison and Huseas 1968) produced Middle Plains Archaic and earlier components. Most significant, however, are the perishable artifacts, especially the food residues. The cave produced a 2 cm thick layer of allium bulb peels (wild onion) and a significant amount of roasted Mormon crickets (Anabrus simplex). Along with other data coming out at the time (Mummy, Bighorn Canyon Caves, etc.), Leigh Cave provided for a discussion of a broader and archaic pattern of prehistoric subsistence (Frison 1991).

A wealth of perishable material was also recovered form Spring Creek Cave (Frison 1965). Here, along with Middle Plains Archaic, Late Plains Archaic, and Late Prehistoric diagnostics, in particularly the latter, was a wealth of wood and fiber objects (see Larson, this volumer Fig 21.3). Spring Creek provides a unique opportunity to study hunting technology such as arrow shafts, foreshafts, and atl-atls, in addition to wood pegs and several other types of wood objects. Finally fiber material shows several types of manufacturing techniques used for cordage as well as the raw materials, plants, used for these objects/tools (Frison 1968), in other words, evidence for prehistoric plant use. The manufacturing techniques are also claimed to have ethnic implications (Frison 1968; Adovasio et al. 1982).

Other significant sites investigated by WAS and associated individuals include Wedding of the Waters, Daugherty, Trapper, and Pow-wow shelter, among others. Wedding of the Waters produced Middle Plains Archaic, Early Plains Archaic, and Late Paleoindian components and may have deeper cultural components below a massive roof fall (Frison, personal communication 2004). Daugherty Cave, one of the largest in the Bighorns, has mainly Late period occupations (Late Plains Archaic, Late Prehistoric and Protohistoric). Of interest are incised limestone tablets, a hafted biface, and an immense quantity of perishable material such as moccasins, basketry, cordage, and so on.

Other early investigations carried on were at Trapper Creek (Anonymous 1960). Although, much time was spent on the rock images and their recording, several 5 foot squares were excavated, enough to demonstrate a Late Prehistoric occupation and recover some perishable material.

Lastly, several rockshelters were investigated by the Girls Scouts in the Teen Sleep area. Sometimes done under professional guidance and resulting in publications (Mack 1971). These add significantly to the Bighorn region shelter database.

WAS Summary. Much of the WAS work provided badly needed descriptions of the archeological materials that had been known by the local collectors, who have long extracted perishables from these or similar shelters and that adorn many ranch houses throughout the west. However, this material was unavailable in professional literature of the Bighorn region. More significantly it began to build a database for closing the hiatus in occupation proposed by earlier studies (Mulloy 1958). The latter was to play a significant role in the concept of the Altithermal refugium for Plains peoples confronted by the deteriorating climatic conditions of the early/mid-Holocene (Antevs 1955).

More Culture-History: Refining the Chronology of the Hiatus and Early Prehistoric Period

The 1960s saw the investigation of several major shelters that synthesized significant aspects of the regional cultural chronology, namely Mummy Cave and Bighorn Canyon shelters. Mummy Cave provided the most stratified sequence of deposits up to that time. The sequence stretched from 9200BP to the Protohistoric (about 200 bp; measured in radiocarbon years BP) in 38 cultural occupations (Husted and Edgar 2002). Most significant was the Early Plains Archaic and Late Paleoindian (later to be called Foothill-Mountain) evidence, for the first time showing Late Paleoindian complexes unlike those known from the Plains to the east, or even the nearby Horner site (Jepsen 1953; Frison and Todd 1987) and in secure chronological position. Its faunal evidence was also at odds with Plains complexes, showing use of bighorn sheep and a slew of other small and medium mammal species (Hughes 2003).

On the heels of the Mummy Cave investigation and improving on many of its results is the Bighorn Canyon study. Specifically Husted's (1969) work at Bottleneck,

Mangus, and Sorenson rockshelters defined the basic late Paleoindian, Foothill-Mountain complexes and refined their chronological positions. Home to Pryor Stemmed, Lanceolate (now Alder; see Davis et al. 1989), and Lovell Constricted projectile points, the shelters also confirmed the medium mammal (i.e., bighorn sheep, etc.) utilization thus providing confirmation for alternative lifeways in the Bighorn region. Perhaps most significantly for Husted, the Mummy Cave, Bighorn Shelters, and other nearby sites provided a comparative base for developing the concept of the Western Macrotradition, a Rocky Mountain based cultural entity, but including the Great Basin, and Plateau areas, with a 10,000 year (measured in radiocarbon years BP) time depth (Husted 1969, 1995; Husted and Mallory n.d.; cf. Lahrens 1976).

Culture History and Adaptation: OWSA and CRM in the 1970s and 1980s

Although refinement of culture sequences remained significant, the import of the New Archeology (Binford 1962) began to be felt with the next series of shelter studies. At the expense of seeing science conservatively as an orderly progression that increases the knowledge base, rather than a radical departure from the past that rejects past knowledge claims in light of new paradigms (Kuhn 1970), the developed chronological sequences allowed for asking new questions based on cultural adaptation and evolution as organizing principles.

One of the first studies of this period was at Medicine Lodge Creek. The site at the confluence of Dry and "Wet" Medicine Lodge Creeks (**Figure 8.2**) is perhaps the most significant site not only in the Bighorn region, but in the Northwest Plains, and indeed has implications for the entire North American continent. The site is the location of magnificent rock images, but its significance lies in the long cultural chronology and numerous unique artifacts and features (**Figure 8.3**) (Frison, 1991)

Medicine Lodge Creek is the location of most cultural components of any Bighorn region shelter (n=13). Of course this depends on how you cut the time pie. Current chronologies are designed to maximize the distinctions within the temporal sequence, whether this is meaningful to cultural processes, time and tempo of evolution, is another question. Nevertheless, it is easier to lump than to divide, thus if divisions are present they can always be lumped if questions and methods warrant.

Although smaller and with shorter cultural sequences, Bush, Carter, Little Canyon Creek, Paint Rock V, Rice, Schiffer, Southsider, Eagle, and Wortham shelters provide a wealth of data for Bighorn Region prehistory, Foothill-Mountain cultural sequences, adaptive strategies, and cultural evolution.

At Bush Shelter, a 9300 year sequence was established with a rich assemblage of chipped stone, suggesting a major regional biface production locality (Miller 1988).

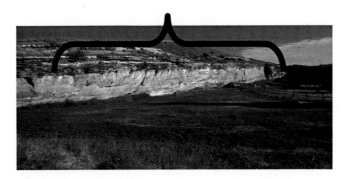

8.2 Medicine Lodge Creek Site. (site is located below the cliff between the brackets).

8.3 Rock images at Medicine Lodge Creek Site.

At the nearby Little Canyon Creek (Shaw 1982), the lithic production is limited, but both Bush and Little Canyon Creek contained an assemblage of similar bone awls and needles (at Little Canyon Creek). The bone tools are highly suggestive of leather working and the manufacturing of clothing. Analysis of faunal remains from several rock shelters and other upland sites allowed Shaw to suggest seasonality of foothill occupations and develop a preliminary model for the Bighorn region settlement system.

Further evidence of Late Paleoindian chronologies was recovered at Paint Rock V, Schiffer, and Southsider shelters. Paint Rock V and Schiffer contained notable Pryor Stemmed components, while Schiffer also yielded Late Paleoindian storage facilities. The latter, along with similar features at Medicine Lodge Creek, the faunal remains at Mummy Cave, the Bighorn Canyon shelters, and other sites, provide significant support for the Foothill-Mountain adaptive strategies of the Bighorn region during the early Holocene. Although the general characteristics of such strategies have been known for some time (e.g., Walker 1975; Galvan 1976; Frison 1976; Frison and Grey 1980; Frison, Wilson and Wilson 1976), we are only now beginning to reap the full benefits of these earlier syntheses (e.g., Hughes 2003; Hutter 2001; Bryan 2006). The latter three studies are beginning to show the links between the paleoclimatic episodes and foraging behavior as well as show periods of resource intensification, when either the resources already in use began to be processed more intensively or smaller animals were incorporated into diets.

Among other notable shelters is Eagle on Little Mountain in the northern part of the Bighorns (see Finley, this volume). Test excavations in the shelter showed a long stratigraphic sequence from middle Paleoindian to Late Prehistoric periods (Chomko 1982, 1990). Several undated features precede the earliest radiocarbon date of 9775. On the other end of the chronological spectrum is Wortham Shelter another significant Little Mountain site (Greer 1978). The shelter yielded shallow deposits with two cultural components, several features, and a large quantity of perishable materials. Of particular note are the 322 arrow shafts or fragments, demonstrating several types of notching styles. The shelter also yielded a significant faunal assemblage. Although over a dozen species are represented, bison is extremely common, unlike at most shelters.

Summary of Shelters

We have mentioned only a few of the shelters directly, however, about 49 shelters in the Bighorn region have had significant investigation (**Table 8.1**). At the 49 shelters investigated, between one and 13 cultural complexes are represented for a minimum of 151 components and for an even greater number of occupations. There are 15 rockshelters in this sample (approximately 1/3) with a single identified complex and one shelter (approximately 2%) with 13 complexes (Medicine Lodge Creek).

Perhaps of significance, however, are the shelters with multiple complexes. The 12 shelters with greater than five complexes represented are: Medicine Lodge Creek, Granite Creek, Wedding of the Waters, Eagle, Two Moon, BA, Little Canyon Creek, Ditch Creek, Sorenson, Mummy, Birdshead, and Pictograph. In addition to the long chronological sequences these shelters yielded large artifact inventories. Out of these 12 shelters all but three (Eagle, Two Moon, and BA) have had extensive excavations, have been looted, or have been otherwise destroyed. Of the three that have not, Two Moon and BA, are currently being tested (e.g., Beers et al. 2005). On the

NAME	NUMBER	PROTO	LP	AVONLEA	LPA	PELICAN	MPA	HANNA	DUNCAN	MCKEAN	OXBOW	EPA	LPI	PRYOR	LOVELL	ALDER	CODY	FOLSOM	GOSHEN	CLOVIS	TOTAL
Granite Creek			1		1	1	1		1	1		1		1		1					9
Spring Creek			1		1				1												3
Wedding of the Waters					1	1			1		1	1									5
Leigh Cave												1	1								4
Eagle Shelter	48BH657		1		1		1	1	1			1									6
Two Moon Sthelter					1		1					1		1				1			5
BA Cave				1	1		1			1				1							5
Grayhound Shelter					1											1					2
	48JO1197	1																			1
	48JO1254		1																		1
	48SH287																				1
Daugherty Cave		1	1		1																3
	48JO5		1																		2
	48JO1271-1?																				1
	48BH1078						1					1									2
Hibernation Shelter	48SH468-2		1		1		1														3
Beehive	48BH346						1					1									2
	48BH1830											1									1
	48JO301																				1
Rock Creek	48JO315		1		1																2
	48SH288				1																2
	48WA663A											1		1							2
	48WA663B											1		1							2
Carter Cave	48WA365						1					1									2
Rice Cave	48WA363												1								1
	48SH290-1																				1
	48WA393						1	1										1			3
Paint Rock V													1								1
Schiffer Cave	48JO314	1	1		1	1	1		1	1		1	1	1		1	1		1	1	13
Medicine Lodge Creek	48BH499		1		1	1															4
Bush Shelter			1		1		1	1	1												5
Little Canyon Creek Cave							1														1
Wortham Shelter			1	1	1			1		1		1	1								8
Ditch Creek Shelter	48BH364		1		1	1															4
Southsider Shelter	48BH305											1									1
Trap Hole			1		1			1	1	1		1	1	1							8
Bottleneck Cave			1		1									1	1	1					4
Mangus Shelter			1		1		1	1	1			1		1							7
Sorenson Shelter			1		1																2
Noname Shelter							1		1	1		1	1	1							6
Pictograph Cave			1		1	1	1	1		1		1			1		1				8
Mummy Cave			1		1		1						1			1					5
Bridshead Cave		1																			1
Kaufman Cave			1																		1
Trapper Cave																					0
Juniper Cave																					0
Sweem-Taylor	48JO303																				
Wet Medicine Lodge																					
TOTAL		4	21	2	20	5	18	6	8	6	1	23	7	9	3	6	4	2	3	3	151

8.1. A preliminary tabulation of Bighorn region shelters and components.

PROTO= Protohistoric, LP=Late Prehistoric, LPA=Late Plains Archaic, MPA=Middle Plains Archaic, EPA=Early Plains Archaic, LPI=Late Paleoindian

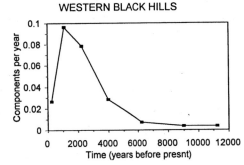

8.4 Temporal distribution of site densities in the Bighorn region (right), compared to the western Black Hills (left). Site density is calculated per time period (Early Paleoindian, Late Paleoindian, Early Archaic, Middle Archaic, Late Archaic, Late Prehistoric, and Protohistoric).

Prehistoric (n=3), then Late Paleoindian and Middle Plains Archaic (n=2), while Early Paleoindian, Paleoindian, Late Plains Archaic, and Protohistoric are represented by one case each. The small sample, however, currently renders any implications of this data unsupportable.

As the Bighorn shelters document significant human use since the first inhabitants of the Americas entered the region. That occupation suggests an increased use of the region through prehistory, until the Protohistoric crash (**Figure 8.4**). This type of shelter use intensity compares favorably to that of other Plains areas. In the western Black Hills, inclusive of large areas of the Powder River Basin and High Plains to the south, for example, the same use intensity is documented (Kornfeld 2003). The shelter data pose several questions, one of which is where is the Altithermal refugium? Earlier works have suggested that there had been a significant abandonment of Plains areas east of the Rockies during the Altithermal and a concomitant consolidation in the western uplands of the Rockies during this hot and dry climatic episode. If such is the case, there should be a significant increase, a spike, in occupation density during the Early Plains Archaic. In fact, in the Bighorn region, including rockshelters, there is no such increase (see **Figure 8.4**). There is, however, a slight increase between the Early Paleoindian and Late Paleoindian and an abrupt increase between Early Plains Archaic and Middle Plains Archaic. The point is that there is significant human occupation of the Bighorn region shelters and we are only now beginning to get the necessary quantity of data to be able to interpret its significance to prehistoric cultural systematics, and evolutionary and adaptive strategies of prehistoric peoples.

WHY CONTINUE RESEARCH OR WHAT WE DON'T YET KNOW ABOUT SHELTERS?

The reviewed rockshelter research leaves many unanswered question about the shelters and their human occupants. Most of the rockshelter investigations in the Bighorn region were conducted 30-40 years ago (e.g., Frison 1962; Husted 1969; Mulloy 1958; Wedel et al. 1968), with some studies conducted over 50 years ago (e.g., Bliss 1950). Archaeological theory and methods have substantially increased the quantity and quality of the data and interpretation potential over this period of time (e.g., Binford 1983; Binford and Binford 1968; Hodder 1997; Renfrew and Bahn 2000). Revisiting previously investigated shelters as well as initiating studies at newly discovered shelters has the potential to enhance our understanding of Bighorn region rockshelter use and thus the region's prehistory.

There are many simple questions about closed site archeology left by previous investigations. Such questions address: shelter formation and sedimentation, cultural occupation relative to geologic deposition, deposit architecture, paleoenvironmental conditions, paleoenvironmental association of cultural components, as well as why some shelters were occupied while others were not. Do the components represent single or multiple occupations, that is, palimpsests? What is the nature of these occupations, periodicity, re-occupation tempo, and so on? It is unlikely that these questions, important for understanding shelter use and their relationship to other sites as well as the cultural systems, can be answered with any of the extant data (previously excavated shelters).

Another relevant question is whether bedrock was reached in the excavated shelters and whether deeper archaeological deposits are present. If older human occupations are present and not known, then our interpretation about human use of the shelters and the region is skewed by the false negative (i.e., the shelters were not occupied). For instance, we may think that the shelters either were not used, or were used only sporadically prior to a certain time, while in fact the shelters are simply not completely excavated. Clearly this may introduce bias into the interpretation of regional adaptive strategies. A corollary of this issue, fluted point use of shelters in North America, has been a matter of debate for some time (e.g., Collins 1991; Kelly and Todd 1988; Walthall 1998). The debate revolves around whether the lack of prehistoric cultural materials in rockshelters at this early time of American prehistory (greater than 10,000 radiocarbon BP) is a function of the fact that: 1) shelters were not used; 2) shelter deposits have eroded away; or 3) shelter deposits no longer lie in the shelters themselves, but in aprons further outside because the shelters have collapsed. To evaluate these and other possibilities, the entire shelter chronostratigraphy must be dated and excavated down to bedrock, and shelter aprons must be tested.

To deal with some of the unanswered questions, and in the lacuna of rockshelter research in the Bighorns from the late 1970s through the early 1990s, we have initiated a Bighorn shelter study about 15 years ago. As a part of this research we are revisiting many of the previously excavated shelters and conducting additional studies in them, including re-recording the profiles in greater detail, collecting granulometric, paleoenvironmental, chronologic, and other samples, and surveying, recording, and testing shelters throughout the region.

This study, however, will not do much for the lack of understanding of shelter sedimentation, deposit architecture, and cultural occupation relative to sedimentation. To investigate such questions we have initiated state of the art excavation and recording procedures at two shelters, which was yet further upgraded this year with the McPherron and Dibble (2002) EDMwin software.

In our work at two Black Mountain sites all the bone and chipped stone larger than 2 cm and all the eboulis larger than 10 cm is mapped three dimensionally and positional data is recorded. This is the first time in the Middle Rocky Mountains, and perhaps in any closed site in North America, that this sort of detailed fieldwork has been carried on.

Thus the research goals of the Bighorn region shelter studies are to synthesize our understanding of the occupation of the region's rockshelters by including the additional studies of chronostratigraphy, sediments, formational history of individual shelters, and the formational history of each shelter's fill. In addition, archaeological collections are being re-examined with new methods that may shed additional light on rockshelter use (e.g., Hughes 2003; Hutter 2001). Such studies will allow for a more robust synthesis of Bighorn region rockshelter occupation and the region's prehistoric human economy, ecology, and adaptation.

PRELIMINARY RESULTS

Within these goals, our ongoing research has focused on two areas with shelter reconnaissance throughout the Bighorns. At Paint Rock Canyon, the primary effort has been a survey for closed sites, with over 180 shelters systematically recorded, of which at least 17% have surface cultural material (Larson, this volume).

In examining our data several observations can be made: 1) We can classify closed Bighorn sites into four categories, perhaps with some implications for cultural or

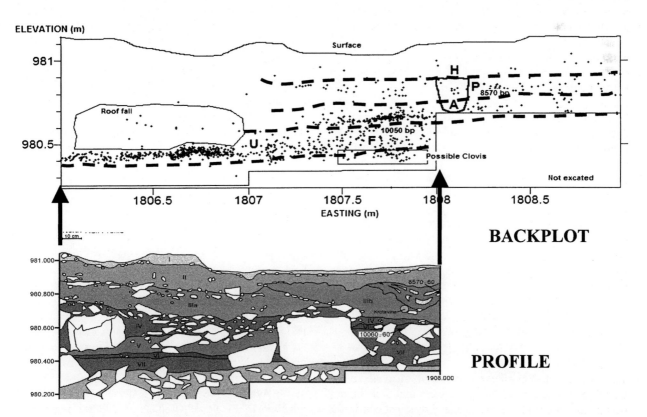

8.5 Backplot (top) and stratigraphic section (bottom) of Two Moon shelter.

natural site formation processes: caves, rockshelters, overhangs, and boulder shelters; 2) Four common types of shelter formation processes in the Bighorns, account for the size and shape of the cavity: paleokarst solutions, active karsts, erosion/exfoliation, and colluviation; and 3) Many more closed sites are in the region, perhaps numbering in the 1000s and perhaps as many with human occupation.

Turning to our detailed site scale studies (**Figure 8.5**), Two Moon Shelter and BA Cave have demonstrated significant variation in shelter sedimentation. While Two Moon is filled primarily with an allogenic fluvial deposit and autogenic eboulis, BA Cave is a combination of allogenic eolian sediment also containing abundant eboulis (Finley 2001 and this volume). On the other hand, while BA sediments from surface to about 1 m span the last 4000 years, and overlay a relatively thin layer possibly reaching to 12,000 radiocarbon BP years ago, at Two Moon the past 8000 radiocarbon BP years is encased in a thin surface layer, some places less than a few centimeters thick, while 8000-12,000 radiocarbon BPyear old deposits are in the underlying 40± centimeters.

Two Moon Shelter provides a good example of complexity of site formation and vertical site structure that might be expected at other closed sites of the Bighorns. A sagittal diagrams of point provenienced chipped stone shows three components from 8000 radiocarbon BP to greater than 10,500 radiocarbon BP years ago, including Pryor Stemmed, Agate Basin, and Folsom. Preliminary analysis of the lower components shows statistically significant differences in artifact positional data between cultural layers and features. In one example, the inclinations are shown to be greater in the area (quad- 50x50 cm) of a hearth than in its surroundings. There may be several possible behavioral implications of these differences.

Of more importance are the recent results of faunal analysis of BA Cave. Unlike Two Moon, BA contains an abundance of faunal remains, heavily fragmented by processing (Bryan 2006). The excavation technique, a combination of stratigraphic and spit control, together with paleoenvironmental reconstruction (Finley 2001 and this volume), has allowed for the first time at a Middle Rocky Mountain Shelter and one of the rare instances in North America, a link between the state of the paleoenvironment and cultural processes. Bryan has recently determined that there is a strong correlation between faunal use and the condition of the resource with a broader spectrum diet indicated during poor conditions. Only with further fine scale excavation at BA Cave and other sites will we be able to fully appreciate the implications of these results.

CONCLUSION

In conclusion, the closed sites in the Middle Rocky Mountain, although mined for perishable and chronologically diagnostic artifacts, stratigraphic, and chronostratigraphic sequences are an untapped source of much richer and more robust information for understanding prehistory of this region. Many more of these kinds of sites are present than we have imagined just a few years ago and renewed and reinvigorated investigations are beginning to yield results. We can expect much more in the future both in a way of understanding closed sites and (Northwest Plains and Rocky Mountain) prehistory.

Acknowledgements. This paper is the product of over a dozen years of research in the Bighorn region. Many people have enhanced this endeavor, one of the primary being George Frison and Mike Bies. Through countless hours of discussion and visiting of many of the sites mentioned in the text I have grown to appreciate the Bighorn region for the clues it holds to the past. I particularly want to thank George Frison for encouraging me to move into the Bighorns to investigate prehistory and rockshelters in particular. I also thank him for several of the photographs used in this paper. I thank Mike Bies for the funding that he has provided through the Bureau of Land Management and years of camaraderie. Without his assistance our continued research in the Bighorns could not continue.

REFERENCES

ADOVASIO, J.M.; R.L. ANDREWS; and C.S. FOWLER (1982). Some observations on the putative Fremont "Presence" in southern Idaho. *Plains Anthropologist* 27(95):19-27.

ANONYMOUS (1960). Trapper Creek Site. *The Wyoming Archaeologist* III(8 and 9):3.

ANTEVS, E. (1955). Geologic-climatic dating in the west. *American Antiquity* 20(4):317-335.

BEERS, J., M. KORNFELD, and G.C. FRISON (2005). *The Black Mountain Archeological District: Preliminary Results of the 2004 Field Studies*. Technical Report No. 33, George C. Frison Institute of Archaeology and Anthropology, University of Wyoming, Laramie.

BINFORD, L.R. (1962). Archaeology as anthropology. *American Antiquity* 28(2):217-225.

BINFORD, L.R. (1983). *In Pursuit of the Past*. Thames and Hudson, New York.

BINFORD, S.R. and L.R. BINFORD, editors (1968). *New Perspectives in Archeology*. Aldine Publishing Company, Chicago.

BLACK, KEVIN (1991). Archaic continuity in the Colorado Rockies: the Mountain Tradition. *Plains Anthropologist* 36(133):1-29.

BLISS, W. L. (1950). Birdshead Cave, a stratified site in Wind River Basin, Wyoming. *American Antiquity* 15(3):187-196.

BRYAN, KARINA M. (2006). *Periodic Late Holocene Human Resource Intensification in the Bighorn Mountains: Evidence from BA Cave*. Unpublished MA Thesis, Department of Anthropology, University of Wyoming, Laramie.

CHOMKO, S. A. (1982). Eagle Shelter, Bighorn County, Wyoming. *Archaeology in Montana* 23(1):27-42.

CHOMKO, S. A. (1990). Chronometric dates from eagle shelter, Big Horn County, Wyoming. *Archaeology in Montana* 31(2):51-55.

COLLINS, M. (1991). Rockshelters and the early archaeological record in the Americas. In *The First Americans: Search and Research*, edited by T.D. DILLEHAY and D.J. MELTZER, pp. 157-182. CRC Press, Boca Raton, Florida.

DAVIS, LESLIE B.; STEPHEN A. AABERG; WILLIAM P. ECKERLE; JOHN W. FISHER; and SALLY T. GREISER (1989). Montane Paleoindian Occupation of the Barton Gulch Site, Ruby Valley, Southwestern Montana. *Current Research in the Pleistocene* 6:7-9.

FINLEY, JUDSON (2001). *Late Holocene Environments and Rockshelter Formation Processes in the Bighorn Mountains, Wyoming*. Unpublished MA thesis, Department of Anthropology, Washington State University, Pullman.

FRISON, G. C. (1962). Wedding of the Waters cave: a stratified site in the Bighorn Basin of northern Wyoming. *Plains Anthropologist* 7(18):246-265.

FRISON, G. C. (1968). Daugherty Cave, Wyoming. *Plains Anthropologist* 13(42):253-295.

FRISON, G. C. (1965). Spring Creek Cave, Wyoming. *American Antiquity* 31(1):81-94.

FRISON, G. C. (1976). The chronology of Paleo-Indian and Altithermal Period groups in the Bighorn Basin, Wyoming. In *Culture change and continuity: essays in honor of James Bennett Griffin*, edited by CHARLES E. CLELAND, pp. 147-173. Academic Press, New York.

FRISON, G. C. (1991). *Prehistoric Hunters of the High Plains*. 2nd ed. Academic Press, San Diego.

FRISON, G. C. and D. C. GREY (1980). Pryor Stemmed, a specialized Paleo-Indian ecological adaptation. *Plains Anthropologist* 25(87):27-46.

FRISON, G. C. and M. HUSEAS (1968). Leigh Cave, Wyoming, Site 48WA304. *Wyoming Archaeologist* 11(3):20-33.

FRISON, G.C. and L.C. TODD (1987). *The Horner Site: the type site of the Cody cultural complex*. Academic Press, Orlando.

FRISON, G. C.; M. WILSON and D. J. WILSON (1976). Fossil bison and artifacts from an Early Altithermal Period arroyo trap in Wyoming. *American Antiquity* 41(1):28-57.

GALVAN, M. E. (1976). *The Vegetative Ecology of the Medicine Lodge Creek Site: An Approach to Archaic Subsistence Problems*. Unpublished MA thesis, Department of Anthropology, University of Wyoming, Laramie.

GREER, J. W. (1978). Wortham Shelter: an Avonlea site in the Bighorn River Canyon, Wyoming. *Archaeology in Montana* 19(3):1-104.

HODDER, IAN (1997). 'Always momentary, fluid and flexible': towards a reflexive excavation methodology. *Antiquity* 71(273):691-700

HUGHES, S. (2003). *Beyond The Altithermal: The Role of Climate Change in the Prehistoric Adaptations of Northwestern Wyoming*. Unpublished Ph.D. dissertation, Department of Anthropology, University of Wyoming, Laramie.

HUSTED, W. M. (1969) *Bighorn Canyon Archeology*. Publications in Salvage Archaeology, No. 12. River Basin Surveys, Museum of Natural History, Smithsonian Institution. Lincoln, Nebraska.

HUSTED, W. M. (1995) The Western Macrotradition twenty-seven years later. *Archaeology in Montana* 36(1):37-92.

HUSTED, W. M. and R. EDGAR (2002). *The Archaeology of Mummy Cave, Wyoming: An Introduction to Shoshonean Prehistory*. National Park Service, Midwest Archaeological Center Special Report No. 4 and Southeast Archaeological Center Technical Reports Series No. 9. Lincoln, Nebraska.

HUSTED, W.M. and OSCAR MALLORY (n.d.). The Western Macrotradition. Manuscript in preparation at River Basin Surveys, Smithsonian Institute, Lincoln.

HUTTER, P. (2001). Assessment of Changing Diet-breadth at Southsider Shelter. Unpublished Master's Thesis, Department of Anthropology, University of Wyoming.

JEPSEN, G.L. (1953). Ancient buffalo hunters. *Princeton Alumni Weekly* 53(25):10-12.

KELLY, R.L., and L.C. TODD (1988). Coming into the country: Early Paleoindian hunting and mobility. *American Antiquity* 53:231-244.

KORNFELD, M. (2003). *Affluent Foragers of the North American Plains*. BAR International Series 1106, Oxford.

KORNFELD, M.; G.C. FRISON; and M.L. LARSON (1995). *Keyhole Reservoir Archaeology: Glimpses of the Past fro Northeast Wyoming*. Occasional Papers on Wyoming Archaeology, No. 5. Office of the Wyoming State Archaeologist, Department of Anthropology, University of Wyoming, Laramie.

KUHN, T.S. (1970). *The Structure of Scientific Revolutions*. 2nd edition. The University of Chicago Press, London.

LAGESON, D. and D. SPEARING (1988). *Roadside Geology of Wyoming*. Mountain Press, Missoula, Montana.

LAHRENS, L. (1976). *The Myers-Hindman Site: An Exploratory Study of Human Occupation Patterns in the Upper Yellowstone Valley from 7000 B.C. to A.D. 1200*. Anthropologos Researches International Incorporated, Livingston, Montana.

McPHERRON, SHANNON P. and HAROLD L. DIBBLE (2002). *Using Computers in Archaeology*. McGraw-Hill, Boston.

MACK, JOANNE M. (1971). *Archaeological Investigations in the Bighorn Basin, Wyoming*. Unpublished MA thesis, Department of Anthropology, University of Wyoming, Laramie.

MILLER, K. (1988). *A comparative analysis of cultural materials from two Bighorn Mountains archaeological sites: a record of 10,000 years of occupation.* Unpublished Master's Thesis, Department of Anthropology, University of Wyoming, Laramie.

MULLOY, W. (1958). *A Preliminary Historical Outline for the Northwestern Plains.* University of Wyoming Publications XXII (1). Laramie.

RAND MCNALLY (1974). *The International Atlas.* Rand McNally and Company, Chicago.

REEVES, B.O.K. (1983). *Cultural Change in the Northern Plains: 1000 B.C.-A.D. 1000.* Archaeological Survey of Alberta, Occasional Paper No. 20, Alberta Culture, Edmonton, Alberta.

RENFREW, C, and P. BAHN (2000). *Archaeology Theories Methods and Practice.* 3rd edition. Thames and Hudson, New York.

SANDO, W.J. (1974). Ancient solution phenomena in the Madison Limestone (Mississippian) of north-central Wyoming. *Journal of Research of the U.S. Geological Survey* 2:133-141.

SHAW, L. C. (1982). *Early Plains Archaic Procurement Systems during the Altithermal: the Wyoming Evidence.* Unpublished MA thesis, Department of Anthropology, University of Wyoming, Laramie.

SOWERS, T. C. (1941). *The Wyoming Archaeological Survey, A Report.* Federal Works Agency Work Projects Administration, State of Wyoming. University of Wyoming, Laramie, Wyoming.

WALKER, D.N. (1975). *A Cultural and Ecological Analysis of the Vertebrate Fauna from the Medicine Lodge Creek Site (48BH499).* Unpublished MA thesis, Department of Anthropology, University of Wyoming, Laramie.

WALTHALL, John A. (1998). Rockshelters and hunter-gatherer adaptation to the Pleistocene/Holocene transition. *American Antiquity* 63(2):223-238.

WEDEL, W.R.; W.M. HUSTED, and J. MOSS (1968). Mummy Cave: prehistoric record from the Rocky Mountains of Wyoming. *Science* 160:184-186.

WILLEY, G. R. and J. A. SABLOFF (1993). *A History of American Archaeology.* 3rd edition. W. H. Freeman, San Francisco, CA.

www.ngdc.noaa.gov/seg/topo/img/nasm2.gif (accessed December 21, 2006).

A CENTURY OF BASKETMAKER II ROCKSHELTER ARCHAEOLOGY IN THE AMERICAN SOUTHWEST: THE TRANSITION TO FARMING ACROSS THE COLORADO PLATEAU

Francis E. SMILEY and Susan GREGG SMILEY

Department of Anthropology, Northern Arizona University, Flagstaff, AZ, 86011-5200 U.S.A.; francis.smiley@nau.edu

Abstract. The Basketmakers II peoples are the earliest farmers on the Colorado Plateau, and the Basketmaker archaeological remains provide a remarkable window on the foundation of Pueblo societies in Utah, Arizona, Colorado, and New Mexico. Over the last 120 years, the excellent preservation at Basketmaker rockshelters has provided a superb laboratory for studying the transition from hunting and gathering to agriculture. We sketch the geography of Basketmaker II archaeology and summarize the results of over a century of chronometric research. Then we trace the evolution of storage technology and the construction of small, surface and semi-subterranean dwellings.

Résumé. Les Basketmakers sont les fermiers les plus premiers sur le Plateau du Colorado, et leurs restes archéologiques fournissent une fenêtre remarquable sur la fondation de sociétés de Pueblo dans Utah, Arizona, Colorado, et Nouveau Mexique. Durant ces 120 dernières années, la préservation remarquable aux abris de rocher de Basketmaker a fourni un laboratoire excellent pour étudier la transition de chasser et le rassemblement à l'agriculture. Nous esquissons la géographie de Basketmaker II archéologie et résumons les résultats de par-dessus un siècle de recherche de chronometric. Alors nous traçons l'évolution de technologie de stockage et la construction de petit, la surface et les demeures à demi souterraines.

This article provides a brief, wide-ranging, and selective overview of the pan-regional rockshelter archaeology of early farming peoples on the Colorado Plateau of the northern American Southwest. The period during which early farming groups used rockshelters as domiciles, storage, and burial facilities is called the White Dog Phase. We do not attempt a detailed treatment of the White Dog Phase in the limited space available. Detailed treatments can be readily found in Smiley (2002), Matson (1991), and Wills (1988).

Just prior to and just after the turn of the twentieth century, the Wetherill brothers, a host of adventurers and treasure hunters, and professional academic archaeologists like Alfred Vincent Kidder and Samuel Guernsey discovered remarkable sites and developed an interpretive framework for early southwestern agriculture in the northern Southwest. The sites universally lay in the grand and capacious rockshelters in the deeply incised canyons of the Colorado Plateau. Because these earliest Plateau farmers lived in and used sheltered sites in arid environments, we have one of the most complete material records of any early farming society in the world.

Early investigators viewed the early farmers as vaguely defined "elder brothers" of the Cliff-Dwellers (Prudden 1897). The initial explorations occurred in southeastern Utah in the Grand Gulch area. These "elder brothers" used the rockshelters of the Colorado Plateau as domiciles, as storage facilities, and as burial sites for their dead. Like the cliff dwellers in the same rockshelters and canyon complexes, the elder brothers grew corn and squash. In contrast to the cliff dwellers, they had neither ceramics nor above-ground architecture. They did, however, make elegant baskets, use the atlatl and dart, and manufacture a range of textiles.

The "Basket Makers," as the earliest excavators called the cave-dwelling farmers, appeared to predate the cliff-dwelling Puebloan peoples, but to what extent no one knew. The Basket Makers have come to be known as the "Basketmaker II" peoples and the early portion of Basketmaker cultural development has come to be called the White Dog Phase. Basketmaker II peoples clearly comprise a key transitional cultural phenomenon in the development of agriculture and sedentism in the American Southwest.

On the one hand, the Basketmaker peoples of the northern Southwest farmed, stored surpluses, and ceremoniously buried their dead in domestic context—all traits we usually associate with small-scale, tribal, agricultural societies. On the other hand, the early Basketmakers appear to occupy a position defined in social-organizational, settlement, and demographic terms that lies closer to structures usually ascribed to band-level hunting and gathering societies. The Basketmakers clearly fall somewhere between the band-level, nonfood-producing adaptation of the Archaic peoples and that of the agricultural, tribal, Puebloan societies (Smiley 2000).

A BASKETMAKER II SATELLITE PERSPECTIVE

Now, let us develop grand regional perspective on Basketmaker II archaeology using "...photorealistic, mathematical simulations created from satellite data ... (Bowen 2006)."

9.1 Photorealistic simulation of the northern American Southwest showing the Colorado Plateau and the approximate region of Basketmaker II occupations from southern Nevada on the west to southwestern Colorado and northwestern New Mexico on the east. The timing of archaeological research in the four major research foci: Grand Gulch/Comb Ridge - 1890-1900 and 1970-2006; Marsh Pass /Black Mesa - 1914-1920s and 1970s-1980s; Durango area - 1950s; and Canyon del Muerto in the Canyon de Chelly complex - 1920s. Aerial composite image courtesy of Dr. William A. Bowen - California Geographical Survey - http:\\geogdata.csun.edu.

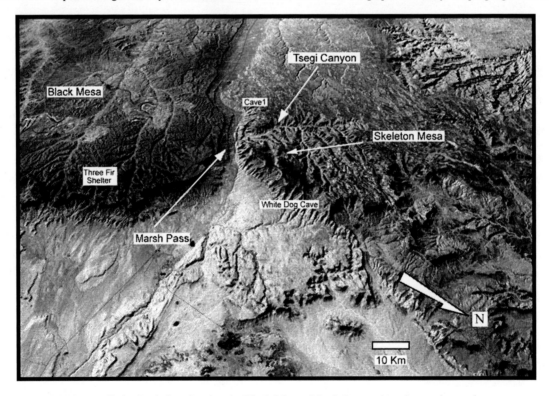

9.2 Photorealistic simulation showing the Black Mesa - Marsh Pass region view to the southwest. Aerial composite image courtesy of Dr. William A. Bowen - California Geographical Survey - http://geogdata.csun.edu.

The Basketmaker II peoples inhabited, for the most part, the southern and middle Colorado Plateau (**Figure 9.1**). Black Mesa, Marsh Pass, and Skeleton Mesa provide examples of the extensive complexes of deeply incised canyons that are ubiquitous across the Colorado Plateau (**Figure 9.2**). A second glance (**Figure 9.1**) shows four primary Basketmaker II site concentrations – not the only, by any stretch of the imagination – but the primary research foci across more than a century of archaeological research. The primary foci of research consist of the Grand Gulch/Comb Ridge area of southeastern Utah, the shelters in Hidden Valley near Durango, Colorado, a large number of rockshelters in the Black Mesa/Marsh Pass region of northeastern Arizona, and rockshelters in Canyon del Muerto of the Canyon de Chelly complex in eastern Arizona.

For lack of space, we focus discussion on the Grand Gulch/Comb Ridge, Marsh Pass/Black Mesa, and Canyon de Chelly areas in this paper. In approximate chronological order of research emphasis, we examine the sites occupied by the first farmers in the northern Southwest.

Grand Gulch

The earliest Basketmaker II archaeology – and some of the earliest archaeological investigations in the Southwest – occurred in southeastern Utah in Grand Gulch, part of the vast and deeply incised canyon complex of tributaries to the Colorado River. The Basketmaker II interpretive framework began to develop in the 1890s during the period of discovery and frenzied initial excavations in Grand Gulch and other areas of southeastern Utah (Prudden 1897; Pepper 1902). The excavations were swift and, at least for archaeologists of our time, depressingly thorough. The early excavators coveted the rich assemblages of basketry and textiles associated with well-preserved human burials. The early excavations and sites yielded the full range of materials representing virtually all activities: from mummified human remains to clothing to the materials of agriculture, technology, trade, and art. Southeastern Utah including Grand Gulch, Cedar Mesa, and Comb Ridge are all major research foci for Basketmaker II archaeology at various times over the past century (**Figure 9.3**).

During the initial period of interest in the Basketmakers a century ago excavators came to understand the stratigraphic position of the Basketmaker remains below the remains of cliff dwellings in the great rockshelters of the Plateau (Kidder and Guernsey 1919). The subposition of Basketmaker remains demonstrated the antiquity of both human occupation in the New World and something of the sequence of the development of agricultural ways of life.

Marsh Pass

Jumping ahead now about two decades to the period 1914-1923, A. V. Kidder and Samuel Guernsey began exploring the archaeology of northern Arizona, particularly in the area of Marsh. Marsh Pass forms a narrow defile between the Black Mesa massif on the southeast and Skeleton Mesa on the northwest (**Figure 9.2**). Note the deeply incised canyon systems in the east-facing scarp of Skeleton Mesa as well as the adjacent Tsegi Canyon labyrinth of canyons. The canyons contained many shelters inhabited by the Basketmakers and produced the well-known assemblages recovered by Kidder and Guernsey (1919; Guernsey and Kidder 1921). White Dog Cave and Kinboko Cave 1, the best known of the numerous sites excavated by Kidder and Guernsey, yielded phenomenally well-preserved archaeological assemblages. **Figure 9.4** is a view of White Dog Cave named for the mummified carcass of a canine recovered in a human burial.

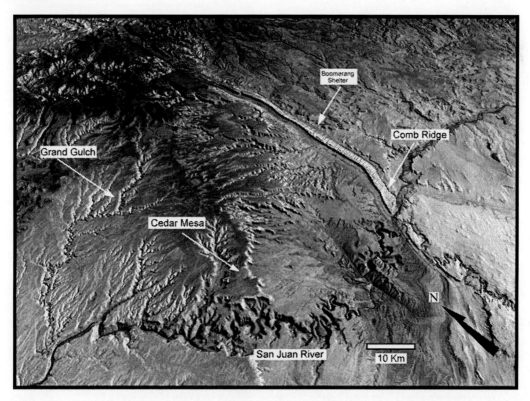

9.3 Photorealistic simulation showing the Grand Gulch - Cedar Mesa - Comb Ridge region. Aerial composite image courtesy of Dr. William A. Bowen - California Geographical Survey - http:\\geogdata.csun.edu.

9.4 White Dog Cave near Kayenta, AZ. Photo by F. Smiley.

Another famous Marsh Pass shelter, Cave 1 in House Canyon, lies several kilometers to the southwest of White Dog Cave. Cave 1 produced a range of materials from human burials to storage facilities. The illustrative work in Kidder's and Guernsey's monographs gives a remarkable glimpse of the range and state of preservation of Basketmaker materials from the March Pass Area and established a standard of archaeological reporting that remains impressive today. Among remains of early farming societies the sites of the March Pass region yielded the famous sunflower cache, many examples of twined string bag technology, quantities of excellent baskets, fur and feather twined robes, and the tools and weaponry of the Basketmakers of the Marsh Pass Shelters. One other noteworthy material turned out to be ubiquitous in the caves of the Basketmakers: *Zea mays*. The excavators noted the presence of corn as complete cobs, kernels, and the remnants of corn plants in virtually all excavated deposits. Because the Basketmakers appeared to have had a developed culture and well-entrenched lifeways, Kidder gave them the designation "Basketmaker II" assuming that archaeologists would one day discover incipient cultivators who would turn out to be the precursors of the Basketmakers (Kidder 1927).

Black Mesa

Cheek-by-jewel with Marsh Pass (**Figure 9.2**) the Black Mesa massif served as a locus of Basketmaker research in the late 1970s and 1980s (Smiley 1985, 2002). Black Mesa is a huge landform measuring about 80 km east/west and about 120 km north/south. Three Fir Shelter, located on northern Black Mesa (**Figure 9.2**) provided some landmark radiometric and settlement information that clarified questions about Basketmaker chronometry, rockshelter use-life, and site function for shelters across the Marsh Pass/Black Mesa region (Smiley and Parry 1990; Smiley 2002).

Three Fir Shelter, high in the Douglas fir stands of northern Black Mesa, gave evidence of very early corn farming on the Colorado Plateau as early as 3600 years BP. The sandy, capacious floor of the shelter had been intensively honeycombed with storage and thermal pits (**Figure 9.5**). In addition, the site yielded a small surface structure nestled among a complex of storage features. Few other surface structures have been found in Basketmaker II shelters and the Three Fir surface

9.5 Interior of Three Fir Shelter on northern Black Mesa. Photo by W. J. Parry.

structure was difficult to identify during excavation. Accordingly, the paucity of reported structures in other Basketmaker II sites may be the result of the rapid and often uncontrolled excavation methods used across the Colorado Plateau up to the 1970s. Not only was Three Fir Shelter occupied by corn farming peoples by about 3600 BP, but the occupation lasted until about 1800 BP (Smiley and Parry 1990). The work on Black Mesa and in Three Fir Shelter in particular has demonstrated that the Basketmakers arrived far earlier and used sheltered sites far longer than previously thought (Smiley 1985; Berry 1982). The site-use trajectory appears to parallel that of Bat Cave to the south (Wills 1988), also occupied possibly as early as 4000 BP by farming populations.

Canyon Del Muerto

To examine Basketmaker II archaeology in eastern Arizona in the Canyon de Chelly complex (**Figure 9.1**) we step back from the 1970s and 1980s excavations on Black Mesa to the 1920s when Earl Morris conducted his landmark archaeological explorations. Morris examined a number of Canyon del Muerto sites, perhaps the most spectacular of which is Tseahatso or Big Cave (**Figure 9.6**). The sheltered portion of the site extends for nearly 100 meters. Like many Colorado Plateau rockshelters, Tseahatso is so large that one gets no inkling of perspective from distant shots and when one stands within such shelters, the size and architecture are overwhelming. The sheltered sites in Canyon del Muerto produced remarkable assemblages of perishables: basketry, textiles, extraordinarily beautifully crafted sandals, and human burials. Morris' work revealed the range of Basketmaker funerary style and custom – from the lavish funerary treatment apparent in the Chief's Grave in Tseahatso (Morris 1925) to burials with almost no grave goods, the variation in practice remains an intriguing research area today.

9.7 Fish Mouth Cave in the Comb Ridge. View to the southwest, photo by F. Smiley.

Comb Ridge

Finally, we arrive back at the northern reach of Basketmaker archaeological settlement and research in the Comb Ridge-Cedar Mesa- Grand Gulch region. The early work in Grand Gulch in the 1890s put the Basketmakers on the map and generated untold excitement to dig the Southwest. In the 1970s and 1980s a great deal of research took place on Cedar Mesa and in Grand Gulch (Matson 1991; Matson and Lipe 1978).

By the 1990s, Smiley and Robins (1997) wondered if anything might remain after a century of predations by treasure seekers and casual excavators. Smiley and Robins began to look at rockshelters in the east-trending canyons of Comb Ridge, the sinuous and magnificent geologic feature that stretches north from the San Juan River (**Figure 9.3**). Perhaps the best known of the Comb shelters is Fish Mouth Cave (**Figure 9.7**). Fish Mouth Cave, like Tseahatso in Canyon del Muerto, is so large and has such a huge sheltered area that only a spherical stitched image in the form of a video can possibly do the site justice.

Smiley and Robins examined several Comb shelters, among them, Boomerang Shelter, a great, curving sheltered space under the Comb nearly 100 meters long and 50 meters deep (**Figure 9.8** and **Figure 9.9**). The crew look like ants and give some idea of the scale of the shelter.

Although Boomerang Shelter and every other Comb shelter Smiley and Robins examined appeared to have been hit by B-52 bombing sorties, the result of a century of looting, test excavations demonstrated that the materials left on the backdirt by pothunters could be used to give a remarkably accurate chronometric reading for the sites and that the sites had long use-life profiles similar to those of Three Fir Shelter and Bat Cave (Smiley 1997). In addition, Smiley and Robins found

9.6 Tseahatso (Big Cave) in Canyon del Muerto in eastern Arizona, photo by F. Smiley.

9.8 Boomerang Shelter in Comb Ridge. View to the south, photo by F. Smiley.

9.9 Interior of Boomerang Shelter showing the main prehistoric habitation area. View to the southeast. Note crew members in distance, photo by F. Smiley.

intact deep cultural strata in the shelters. Finding intact cultural strata of Basketmaker II age in Boomerang Shelter provided a major pleasant surprise since many of the Basketmaker II rockshelter occupations across the northern Southwest exhibit little stratigraphy. Three Fir Shelter with approximately 20 cm of cultural deposits that contain nearly 2000 years of human use supplies an apt example. Accordingly, Smiley and Robins found that the cultural and scientific values of these ancestral Puebloan occupations remain exceedingly high. The looters had disturbed only the upper 70 cm of approximately 2 meters of Basketmaker II stratigraphy.

The stratigraphy at Boomerang gave a clear sequence of occupation and of the development of storage technology during the period of Basketmaker use. The Boomerang Shelter profile showed the sequential development of Basketmaker II storage technology from small earthen pits to larger pits and eventually to small slab-lined cists and large slab-lined and mortared storage cists late in the sequence (Smiley and Robins 2005).

CONCLUSION

The Basketmaker II groups across the northern Southwest during the White Dog phase, the early Neolithic of the northern Southwest, clearly appear transitional. They act in many respects like tribal farmers and in many other important respects like mobile hunter-gatherer band societies. Clearly, the Basketmakers are what Smiley (2000) has termed, *infra-tribal* cultures. As such, they incorporate new behaviors, technologies, ideologies, and social forms; as well as farming, storage, violence on a level not previously evidenced, regional adaptations, and a very clearly altered human/land relationship as evidenced in particular by their funerary customs that set them unequivocally apart from the earlier Archaic peoples.

In this sketch of Basketmaker archaeology and Basketmaker shelters we have had to leave out a great deal. Perhaps the most interesting aspect of Basketmaker II adaptation to the stochastic climate and difficult edaphic conditions of the Colorado Plateau is the relatively high mobility that the groups appear to have maintained across nearly two millennia of subsistence farming. Accordingly, the Basketmakers add a great deal of variability to our store of knowledge on the behaviors and adaptations of early farmers around the world.

Acknowledgements. We wish to thank Marcel Kornfeld for the opportunity for and the help with our participation in the symposium and in this volume. We are indebted to William Bowen of the California Geographical Survey - (http://geogdata.csun.edu) for his above-and-beyond-the-call help in providing the photorealistic simulation images of the Southwest. Thanks to F. Smiley's research partners William J. Parry and Michael R. Robins for the efforts they have contributed over the years in the courses of the various projects mentioned here. We the authors bear responsibility for any errors or omissions.

REFERENCES

BERRY, Michael S.(1982) *Time, Space, and Transition in Anasazi Prehistory*. University of Utah Press, Salt Lake City.

BOWEN, William (2006) *Technical Addendum to Arizona Atlas of Panoramic Aerial Images* http://130.166.124.2/az_panorama_atlas/page16/page16.html.

GUERNSEY, S. J.; KIDDER, A. V. (1921) Basket Maker Caves of Northeastern Arizona: Report on the Explorations, 1916-1917. *Papers of the Peabody Museum of American Archaeology and Ethnology* Vol. 8, No. 2. Harvard University, Cambridge.

KIDDER, A. V. (1927) Southwestern Archaeological Conference. *Science* 66:489-491.

KIDDER, A. V.; GUERNSEY, S. J. (1919) *Archaeological Explorations in Northeastern Arizona*. Bureau of American Ethnology Bulletin No. 65. Smithsonian Institution, Washington, D.C.

MATSON, R. G. (1991) *The Origins of Southwestern Agriculture*. University of Arizona Press, Tucson.

MATSON, R. G.; LIPE, W. D. (1978) Settlement Patterns on Cedar Mesa: Boom and Bust on the Northern Periphery. In, *Investigations of the Southwestern Anthropological Research Group*, edited by ROBERT C. EULER and GEORGE J. GUMERMAN, pp. 1-12. Museum of Northern Arizona, Flagstaff.

MORRIS, E. H. (1925) Exploring the Canyon of Death. *National Geographic* 48(3):262-300.

PEPPER, G. H. (1902) The Ancient Basketmakers of Southeastern Utah. *American Museum Journal* Vol. II, No. 4; Guide Leaflet No. 6, New York.

PRUDDEN, M. T. (1897) An Elder Brother to the Cliff-Dwellers. *Harper's New Monthly Magazine* 95(565):56-62.

SMILEY, F. E. (2002) The First Farmers: The White Dog and Lolomai Phases In *Prehistoric Culture Change on the Colorado Plateaus*, F. E. SMILEY and SHIRLEY L. POWELL (eds.) University of Arizona Press, Tucson.

SMILEY, F. E. (2000) Infra-Tribal Systems and Cultural Evolutionary Studies of Early Agriculture on the Colorado Plateau. Paper presented at the 65th Annual Meeting of the Society for American Archaeology, Philadelphia.

SMILEY, F. E. (1997) Toward Chronometric Resolution for Early Agriculture. In *Early Farmers in the Northern Southwest: Papers on Chronometry, Social Dynamics, and Ecology*, pp. 13-42. Animas-La Plata Archaeological Research Paper No. 7., edited by FRANCIS E. SMILEY and MICHAEL R. ROBINS.

SMILEY, F. E. (1985) *The Chronometrics of Early Agricultural Sites in Northeastern Arizona: Approaches to the Interpretation of Radiocarbon Dates*. University Microfilms, Ann Arbor.

SMILEY, F. E.; PARRY, W. J. (1990) Early, Intensive, and Rapid: Rethinking the Agricultural Transition in the Northern Southwest. Paper presented at the 55th Annual Meeting of the Society for American Archaeology, Las Vegas, Nevada.

SMILEY, F. E.; ROBINS, M. R. (2005) Help for the Looted Rockshelters of the Colorado Plateau in a New Century of Archaeology: New Basketmaker II Research on the Great Comb Ridge. In *The Colorado Plateau II,* edited by CHARLES VAN RIPER III and DAVID J. MATSON. University of Arizona Press, Tucson.

SMILEY, F. E.; ROBINS, M. R. (1997) *Early Farmers in the Northern Southwest: Papers on Chronometry, Social Dynamics, and Ecology*. Animas-La Plata Archaeological Research Paper No. 7., edited by FRANCIS E. SMILEY and MICHAEL R. ROBINS, Northern Arizona University, Flagstaff.

SMILEY, F. E.; ROBINS, M. R. (2005) Help for the Looted Rockshelters of the Colorado Plateau in a New Century of Archaeology: New Basketmaker II Research on the Great Comb Ridge. In *The Colorado Plateau II,* edited by CHARLES VAN RIPER III and DAVID J. MATSON. University of Arizona Press, Tucson, with Michael Robins.

WILLS, W. H. (1988) *Early Prehistoric Agriculture in the American Southwest*. School of American Research Press, Santa Fe.

SUBTERRANEAN CAVES, THEIR MORPHOLOGY AND ARCHAEOLOGICAL CONTENT: THE MORTUARY CAVES IN COAHUILA, MEXICO

Leticia GONZÁLEZ ARRATIA

Museo Regional de La Laguna, Instituto Nacional de Antropología e Historia, México; legoar@yahoo.com

Abstract. Geomorphological elements of subterranean caves in the Coahuila desert, selected by prehispanic hunter-gatherers to deposit their dead, repeat consistently several natural characteristics. Also cultural traits such as the flexed position of the corpses and the items associated with it. A hypothesis about the reasons that the desert people selected subterranean caves as cemeteries is that the morphological aspects of these caves functioned as a metaphor of the mother's womb. The first half of the paper describes the morphological elements as well as the problems of doing archaeology in this type of context and the second part, the archaeological material and interpretation.

Keywords: Subterranean caves, entrance, shaft, floor, chamber, mortuary ritual, metaphor, flexed skeletons.

Résumé. Certain elements geomorphologiques des caves subterranees du desert de Coahuila, au Mexique, ont etée selecciónee para les chasseurs cuillers prehispaniques pour depositer leur morts. Ces caves ont des characteristiques naturelles qui se repeat comme l'entrance, le tunnel, et les chambres. Les items culturelles et la position flexee des corps se repeat aussi. Une hypothèse sur la raison que le peuple du desert ont choisit subterranean caves comme cemmeteries peut etre ce que la morphologie des caves ont functionee comme une metaphor du ventre de la mère. La première partie de cette essay descrit les elements morphologique de ces caves, et aussi les problèmes de faire achaeologie dans cet type de context. La second part, descrit le materielle archaeologique et son interpretation.

Mots-Clés: caves subterranees, entrée, tunnel, chambre, ritual mortuaire, metaphor, corps flexee.

Few caves or shelters were professionally explored, excavated and studied in the history of archaeology of the Coahuila desert in Northern Mexico. Only two archaeologists, both pioneers, worked in caves and shelters. The first was Walter W. Taylor who excavated several habitation caves and burial shelters in 1941 in central Coahuila in the Cuatro Cienegas region. The other one was Luis Aveleyra Arroyo de Anda who in 1953 and 1954, explored Candelaria Cave, a subterranean cave used as a cemetery by ancient prehispanic hunter-gatherer people, in the Laguna Region in Southwest Coahuila, México.

THE SUBTERRANEAN CAVE: A HISTORY OF DISCOVERIES AND LOOTING

At least since about 1845, prehispanic inhumation in subterranean caves were reported to have taken place in limestone mountains and hills in the Laguna region, and at the time they were reported, all had been looted, including Candelaria cave (**Table 10.1**). In all cases the caves contained more than one corpse, all were arranged in a flexed position and wrapped either with petates or with weaved mantles. When referring to the number of corpses found in Lugar de los Sepulcros Cave (Avila about 1845:6) writes many. Edward Palmer in 1880 informed that before Coyote Cave was looted, its discoverer assured him there had been very many bundles (Palmer 1880). Candelaria cave reported 113 skulls which accounts for at least 113 individuals (Romano 2005:27). In Tres Manantiales Cave, there were at least ten individuals inhumed (Mansilla et. al 2006).

CHRONOLOGY

Few radiocarbon dates have been obtained in subterranean caves. The most reliable are three: two from Candelaria which yielded dates between 1095 and 1315 B.C. (Aveleyra, 1964:129) and one for Coyote Cave (1010 to 1020 B.C.) (González Arratia 2006:60). According to studies of archaeological material from Candelaria, Coyote and Tres Manantiales cave which are

1845 (ca.)	Sierra de San Lorenzo (Lugar de los Sepulcros Cave)	José Ma. Avila
1880	Sierra de San Lorenzo (Coyote Cave)	Edward Palmer
1953-1954	Sierra de la Candelaria (Candelaria-cave)	Luis Aveleyra Arroyo de Anda
2002	Sierra de Tlahualilo (Tres Manantiales Cave)	Looted and reported by ejidatarios (peasants)

Table 10.1. Reported inhumations in subterranean caves.

Site	Measurements		
	Entrance hole (mts.)	Shaft (mts.)	Chamber (mts.)
Lugar de los Sepulcros	7.50 m. x 4.17 m. (irregular hole)	16.7 mts.	
Coyote	No data	13.65 to 16.38 mts.	8.36 to 12.54 square mts.
Candelaria	1.22 m.	9.0 m. long	9 m. long 10 m. height
Tres Manantiales	1.0 m. x 0.70 cms. (irregular hole)		7 mts. width, 3 mts. length 2 mts. high.

Table 10.2. Cave measurements and characteristics.

very similar in all three except for the final wrapping, it was concluded that this type of inhumation in subterranean caves might have been practiced in the Laguna region at least during more than two hundred years, mainly the XIIIth century B.C.

DESCRIBING THE CAVES

In order to briefly describe natural characteristics of subterranean cave, I have divided them into several components as follows: 1) entrance (hole); 2) shaft or tunnel; 3) floor; 4) subterranean chamber(s) (**Table 10.2**). The entrance is usually described as a hole (irregular) or a very narrow crevice, which leads to a long shaft such as a chimney shaft. In Candelaria Cave it measures 9 m in length; in Coyote cave 13 to 16 m in length. Sometimes the hole faces the sky so it is not easily visible from the distance and sometimes it is not even visible when close up.

Inside, chambers are not large and usually more than one is connected to the first chamber (the one the shaft leads to) either laterally through a narrow horizontal passage or beneath it through a hole. In the case of Candelaria Cave the following description was made by a geologist:

> The entrance is roughly circular and very small, some 4 ft. across: it opens horizontally on to a vertical shaft some 30 feet deep, so that ropes or, better still, a rope ladder, are absolutely essential… At the bottom of this shaft the floor of the cave slants in accordance with the dip of the strata. Some important rock falls have taken place, crushing a certain amount of material; and work inside the cave is fraught with considerable danger. (Martínez del Río, 1953:215)

At least four of the reported archaeological subterranean caves show these same features as may be appreciated by the commentaries of their explorers:

In the XIX century Avila (about1845) refers to the entrance of Lugar de los Sepulcros Cave, as a deep hole displaying at its end a series of chambers. In order to descend, it was necessary to fasten several ropes (Avila about 1845:465-466). Once inside, they found several chambers.

In 1880, Edward Palmer descends through a hole to Coyote cave, using a long rope, at the end of which he finds a chamber and underneath it, another one (Palmer, 1880:1). The floor of Candelaria cave was reached by employing a rope ladder. They explored the main chamber and another one, noting that there were "many" other chambers which were not explored.

Tres Manantiales Cave that was explored by my team, (after its initial looting), is also composed of several chambers. The floor is a mixture of fallen rocks, dust and animal's excrements. We also found human coprolites.

ARCHAEOLOGICAL CONTENT OF THE CAVES

All of the mentioned caves contained bundles which included a flexed skeleton wrapped either in a woven tile (Candelaria, Coyote, Lugar de los Sepulcros) or a petate (Tres Manantiales) and tied in the first case either with bands and cords, or only with cords in the second case. Inside the bundles there were different instruments and adornments, on the outside were larger artifacts such as bows, large baskets, cradles, leaves of prickly pear and agaves. Personal testimonies and physical anthropological reports inform that they were systematically placed and included women, men and children (Avila about 1845; Palmer 1880; Romano 2006; Mansilla et. al. 2006). **Table 10.3** is a list of most of the archaeological artifacts found inside and outside the bundles. Only the ones that appear in more than one cave are on the list.

HOW DO THESE CAVERNS FORM?

The caverns form through the forces of subterranean water flow such as creeks or springs, in the interior of mountains composed of sedimentary soluble rocks for example limestone, dolomite, gypsum, etc. The water dissolves the rock and leaves empty spaces (holes, hollows) since its constituents turn out in solution and are carried by the liquid out of the cavern. These hollows are called subterranean caves (Zumberge 1971:150).

Archaeological object	Caves			
	Candelaria	Tres Manantiales	Coyote	L. de los Sepulcros
Deer antlers	x	x		
Atlatl	x		x	x
Bow	x	x		
Baskets, coiling technique	x	x	x	
Calabazo		x	x	
Shell, seeds and bones ornaments (necklace, bracelet)	x	x	x	x
Cordage in various degrees of thickness	x	x	x	x
Wooded handled stone knive.	x	x	x	
Cradle frame	x	x	x	x
Large mat of thin sticks closely bound together	x	x		x
Fiber skirts	x		x	
Shafts and arrow shafts	x	x		
Rodent's incisive on a wooden handle	x	x		
Woven textile	x		x	x
Rabbit stick	x	x		
Digging stick	x	x		
Diagonal reed matting	x	x		
Decorated matting	x	x		
Piece of fur	x		x	
Stone projectile points	x		x	x
Awl	x	x		
Scraper on a wooden handle	x	x		
Sandals	x	x	x	x
Yahual	x	x	x	x

Table 10.3. Artifacts found in subterranean burial caves.

CHARACTERISTICS OF THE ENVIRONMENT OF SUBTERRANEAN CAVES AS REPORTED FOR COAHUILA

The subterranean caves have a number of common characteristics, that include at least the following: 1) Big and small rocks from the roof lying on the floor; 2) Fine powder make up the floor; 3) Almost total darkness; 4) Lack of ventilation. All explorers mention the fact that big rocks were covering all or part of the floor and thus destroyed many mortuary bundles. In the case of Candelaria, the archaeologist mention that merely picking up artifacts and bundles produced so much powder that at times they could hardly see anything.

CONCLUSION

There is substantial evidence that prehispanic people in southern México (Mesoamerica), venerated mountains and caves (Limón, 1990). But my hypothesis is that hunting and gatherer peoples in the desert of northern Mexico established a link between inhumation and caves reinterpreting the subterranean cave as a metaphor of the mother's womb.

The archaeological findings I have mentioned above, point to an intentional selection of caves with shafts and underground chamber or chambers. In all cases skeletons were flexed in a fetal position, and turned into a bundle by wrapping them up with several types of textiles. In other places I have mentioned the complexities involved in the preparation, transportation and deposition of the corpses (González Arratia 2002:60-63). Here, I would like to point out only about the relationship between the characteristics of the subterranean cave, the mortuary bundle and the flexed position. The flexed position is one of the most important characteristics since all corpses so far found in these types of caves, whether women, men or children, exhibit it.

One interpretation of this position could be that it refers to the fetal stage, previous to birth, and a reminder of the ties between the infant and the mother which partially end at the moment of birth. Once dead, this position might indicate the disposition or obligation of the spirit of the individual to return to a point similar to that of departure. According to prehispanic legends and myths, almost any cave would do, but I propose that the subterranean caves with a shafts come closer to the metaphor of the mother's womb since this type of cave is also associated to water.

In order to complete the cycle of returning to the earth's womb, the individual should traverse a road full of dangers, and the subterranean cave would also represent the shortest way to the current of water which leads to the expected goal, a perfect conclusion of a cycle which apparently starts in the mother's womb. But there are several practical and methodological problems and even risks that archaeologists face to explore, collect and excavate this type of archaeological context.

From a methodological point of view, there are two main problems. One is that all caves known so far, have been looted, the bundles opened and the archaeological context altered. Thus, in most cases, there is a lack of direct association between a specific skeleton and the artifacts or/and between skeletons. Another problem is that according to reports or testimonies, so far all bundles were found on the surface and no internment has been reported. Therefore, in the future, subterranean caves should be excavated regardless its physical characteristics which include difficulty in getting inside, lack of natural light, small and low size of chambers, big fallen rocks which probably hide part of the archaeological record, and a floor composed of very fine soil which could be indicating poor conditions for a good stratigraphy without forgetting that removing it even minimally produces several practical problems.

REFERENCES

AVELEYRA ARROYO de ANDA, LUIS, (1964). Sobre dos fechas de radiocarbono 14 para la Cueva de la Candelaria, Coahuila, in *Anales de Antropología*. México. Vol. 1. p. 125-130.

AVELEYRA, ARROYO de ANDA, LUIS; MANUEL MALDONADO KOERDEL and PABLO MARTINEZ del RIO, (1956). *Cueva de la Candelaria, Vol. I,* Memorias del Instituto Nacional de Antropologia e Historia V, Instituto Nacional de Antropología e Historia, Secretaría de Educación Pública, México, México.

AVILA, JOSÉ MARIA, (abaout 1845). *Tres Días de Paseo in Album Mex,* Tomo 1, México, p.465-468.

GONZÁLEZ ARRATIA, LETICIA (1999). *La arqueología de Coahuila y sus fuentes bibliográficas,* Instituto Nacional de Antropología e Historia, México.

GONZÁLEZ ARRATIA, LETICIA, (1999). *Museo Regional de la Laguna y la Cueva de la Candelaria,* Instituto Nacional de Antropología e Historia, México.

GONZÁLEZ ARRATIA, LETICIA (2004). La cultura del desierto y una de sus tradiciones simbólicas: el ritual mortuorio, in SALAS QUINTANAL, H.; PÉREZ-TAYLOR, RAFAEL, eds. - *Desierto y fronteras. El norte de México y otros contextos culturales. V Coloquio Paul Kirchhoff.*, p.367-386, Universidad Nacional Autónoma de México, México.

GONZÁLEZ ARRATIA, LETICIA, (2006). *La exploración de Edward Palmer en varias cuevas mortuorias en Coahuila en el siglo XIX*, Instituto Nacional de Antropología e Historia, México.

LIMÓN OLVERA, SILVIA (1990). *Las Cuevas y el Mito de Origen.* Los casos inca y mexica, Consejo General para la Cultura y las Artes, México.

MANSILLA, JOSEFINA e Ilán Leboreira, *Informe sobre los restos óseos de la Cueva de Tres Manantiales*, [manuscript], 2006, Dirección de Antropología Física, INAH, México.

MARTÍMEZ del RIO, PABLO (1953), A preliminary Report on the Mortuary Cave of Coahuila, Mexico, in *Bulletin of the Texas Archaeology Society,* The Texas Archaeological Society, Austin, Vol. 24, (pp. 208-216)

PALMER, EDWARD, *Notes on the Coahuila Caves,* [manuscript], 1880, Alfred Tozzer Library, Peabody Museum, Massachusetts. (13 pages)

ROMANO PACHECO, ARTURO (2005). *Los restos óseos humanos de la cueva de La Candelaria, Coahuila.* Craneología, Instituto Nacional de Antropología e Historia, México.

ZUMBERGE, JAMES H. (1971). *Geología Elemental*, C.E.C.S.A., México.

PART II – CURRENT RESEARCH IN EURASIA AND AFRICA

ROCKSHELTER STUDIES IN SOUTHWESTERN IBERIA: THE CASE OF VALE BOI (ALGARVE, SOUTHERN PORTUGAL)

Nuno BICHO[1], Mary C. STINER[2], Delminda MOURA[3], and Armando LUCENA[1]

[1]FCHS, Universidade do Algarve, campus de Gambelas, 8000 Faro, Portugal; nbicho@ualg.pt
[2]Department of Anthropology, University of Arizona, Tucson, AZ, 85721-0030, U.S.A.
[3]FCMA, Universidade do Algarve, campus de Gambelas, 8000 Faro, Portugal

Abstract. Rockshelters have been most important in the study of prehistoric archaeology, particularly for the Paleolithic period. In Portugal, however, very few rockshelters have been found, tested and excavated that shed light on the Paleolithic occupation of the Western edge of Europe. This paper will focus on the case of the rockshelter of Vale Boi (Algarve, Southern Portugal), a site with Upper Paleolithic and early Neolithic occupations. Interpretation of the site and their human occupations will be discussed and interpreted based on various aspects, specifically site formation processes on slope deposits, intra-site spatial organization, and excavation techniques.

Résumé. Les abris ont été les plus importants dans l'étude de l'archéologie préhistorique, en particulier dans ce qui concerne des temps de Paléolithique. Au Portugal, cependant, peu d'abris ont été trouvés, examinés et excavés fournissant l'information sur le Paléolithique du bord occidental de l'Europe. Cet article se concentrera sur cas de l'Abri de Vale Boi (Algarve, Portugal méridional), un gisement avec Paléolithique supérieur et Néolithique ancien. L'interprétation de le gisement sera discutée et interprétée basé sur de divers aspects, spécifiquement processus de formation archéologique, sur des dépôts de pente, intra-spatial organisation et techniques d'excavation.

The rockshelter of Vale Boi, Algarve, Portugal, was discovered in 1998 during survey within the research project of *Human Paleolithic occupation of Algarve* (funded by Fundação para a Ciência e Tecnologia, projects PRAXIS/PCHS/C/HAR/70/96). This project took place between 1997 and 2000, and had as its main objective the construction of a chrono-cultural framework for the Paleolithic and Epipaleolithic of southern Portugal. This was an area where very little data were known for those periods (**Figure 11.1**). Thus, the project consisted mainly of survey, testing and radiocarbon dating. More than 65 sites were located, some of which were tested and dated. The most important of these sites is Vale Boi. The initial investigative team was composed by geomorphologists, zooarchaeologists and archaeologists, among them C. Reid Ferring, Mary Stiner, John Lindly and Nuno Bicho. Since then, the research team has changed due to various earlier commitments (e.g., Reid Ferring is presently working in the Dmanisi Project), but still various projects were carried out, mostly to deal with the site of Vale Boi (project POCTI/HAR/37543/2001 funded by FCT).

Vale Boi was found while the team was surveying the fluvial contexts in the county of Vila do Bispo. The site is located on the left side of the river named Vale Boi, in front of the small town with the same name (**Figure 11.2**). The area is marked by the presence of limestone ridges (named locally *Barrocal*) that border the schist region to the north. The Atlantic shore is about 2 km to the south and is characterized by a series of sandy inlets, usually with very good access. The Paleolithic human occupation is present in one of the limestone slopes, topped by a rockshelter with a 20 m high face. The alluvial plain is some 40 m below and in between there is a sequence of terraces of which some were occupied by the Upper Paleolithic hunter-gatherers.

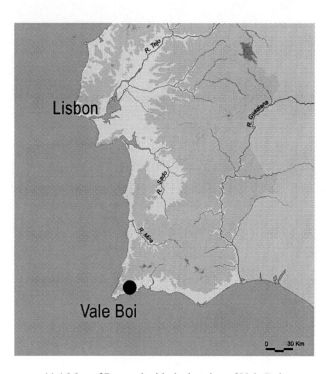

11.1 Map of Portugal with the location of Vale Boi.

At the time of the site discovery, there were artifacts spread over the surface, including quartz and flint artifacts, bones and marine shells, suggesting a Magdalenian occupation on more than 1000 m². After testing and excavation the archaeological area is now estimated to be over 10,000 m², covering the whole slope.

The first two 1m² tests (G25 and Z27) were carried out in 2000 in two areas with high artifact concentration on the site surface. These tests were located in mid slope on flattish areas of two different terraces. The higher one, G25, showed a long sequence starting in early Gravettian and lasting, without interruptions, to early Magdalenian.

11.2 General view of the site of Vale Boi.

The second test, Z27 had two main levels, one Solutrean and one Gravettian.

In 2001 8 m² were opened adjacent to G25, while Z27 was included in an area with a total of 6 m². Two new areas were tested as well and are now under excavation. Located respectively at the base of the slope (designated by *Terrace*) and at the top, in the rockshelter area (known as *Rockshelter*).

METHODOLOGY OF EXCAVATION

The whole grid system is based on a single datum with the arbitrary coordinates of 100.00 m north and 100.00 m east. The absolute elevation is 34.50 meters a.s.l., obtained from a topographic bench mark located about 1 km southwest of the site. The excavation was carried out using an EDM system with two total stations, a GTS 226 from TopCon and a SET330RK3 from Sokkia. There are three grid areas, corresponding to each of the three areas of excavation, the *Rockshelter*, the *Slope*, and the *Terrace* (**Figure 11.3**).

The 1 square meter tests were excavated by 5 cm artificial spits, while the excavation followed a more complex system. The basic vertical units were major geological layers labeled alphabetically starting at the top. Each of these was excavated by 5 cm spits, designated by numbers, starting on the top with level 1. Within each spit, all artifacts and faunal remains larger than 2 cm were three dimensionally plotted. The exceptions in terms of size are bladelets, complete retouched tools, whole shells or adornments that are always plotted. All plotted pieces have a sequential number that corresponds to the shot number in the total station. Tags are numbered previously and bagged with the artifact, bone or shell as they are recovered and also are three dimensionally plotted. If the number on a tag does not correspond to the number in the total station, a new reading is obtained corresponding to a new tag, and the older tag is kept for the excavation records. This method is used because there is no electricity at the site for computer equipment (which of course could be resolved with electric generators or batteries, but since it is in a public open area, no equipment can be left at the site while the team is not present). The tags are made on 90 g tracing paper so they are resistant to handling, humidity and washing. Pre-printed tags include the site ID code (VB), the year of excavation, and the three dimensional shot number. To these are added by hand the date of collection as well as the excavation unit, layer and spit. These are registered because they simplify the following analysis of the artifacts and faunal remains.

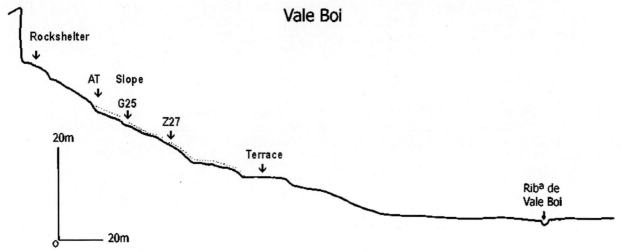

11.3 Schematic section of the site.

The horizontal control also uses the so called bucket approach (McPherron and Dibble 2002). This is a system where a 10 l bucket is used as excavation unit. For each full bucket of sediment there is a three dimensional EDM shot for the mid point where the sediment was recovered. Each bucket is given a new tag. This is placed in the bag with the materials recovered in the screen (all sediment is screened using a 3 mm mesh) from that sediment bucket. In practical terms, this means that each 5 cm spit in each one square meter is divided in, on average, 8 to 10 smaller areas comprised by the volume of a single bucket with specific artifacts and faunal remains relating to a three dimensional reference point. The data base collected with the total station is labeled according to the type of reference points that is obtained: bucket (b – balde in Portuguese), bone (o - osso), shell (c – concha), lithic (l – lítico), Radiocarbon sample (ca – carbono), and, finally, topographic reading (t – topografia).

THE SLOPE AREA

The slope area has been the subject of four different 1 square meter tests, of which two were expanded into larger areas. The location of each test was chosen due to the local topography and the concentration (or lack of) of artifacts. Thus, areas of higher concentration of surface artifacts were tested (G25 and Z27), corresponding to flatter areas of the slope to determine the presence of in situ materials. The other two tests (AZ20 and AT21) were carried out higher in the slope in areas where there were no artifacts on the surface to establish the limits of the human occupation.

The results in every single case were better than expected. In the G25 area, the stratigraphy is still not completely known, although it seems that the lower section is essentially sterile with extremely rare small bone fragments and no artifacts. This section reaches at least 2.5 meters below present surface and it is characterized by very fine reddish sediment (clays and silts) with very angular limestone clasts of small dimensions (less than 5 cm). At about 1.5 m below surface there is a break and a new unit starts where the Gravettian is present with very high quantities of artifacts and faunal remains. The fine sediments are mostly composed of silts and the color is darker with shades of brown. The frequency of *éboulis* is lower than below, but on the other hand there is some soil development attested by the presence of small iron and manganese nodules and concretions, that frequently have adhered to the artifacts and bones.

This layer can reach a meter in thickness and is topped by a heterogeneous layer of medium to large size limestone blocks, corresponding to the beginning of the Last Glacial Maximum, some 21,000 years ago (**Table 11.1**). Above, the sediment characteristics are the same as below the blocks, but with Solutrean artifacts and fauna. Above this is a thin darker layer, marked by frequent roots and humus, and lithic artifacts that seem to date to early Magdalenian. It should be noted that the sediment package with archaeology is marked by the complete absence of sterile zones and that the archaeological materials are dense through out the deposit.

Preliminary sedimentological analysis indicates that there is no significant alterations to the vertical position of the initial sediment. This is also confirmed by the artifacts and faunal remains. Although everything is highly fragmented, this is not due to site formation processes, but because it is a midden, where the daily garbage was dumped. Bone fragments show green fractures in every case, and there are very few complete bones (either long bones or other types). This is the use of grinding stone building technique to obtain grease and other proteins and minerals from the marrow and spongy areas of the bones (Stiner 2003). This same technique also demands a frequent use of boiling stones, which resulted in the case of Vale Boi in many fragmented quartz lithic remains, also present in this area. Shells are also frequently broken, and in the case of large specimens such as scallops (*Pecten maximus*) refitting was carried out very

Period	Area	Lab Code	Material	Result
Early Neolithic	Terrace	Wk-17843	bone	6020 ± 35*
Early Neolithic	Terrace	Wk-17842	bone	6095 ± 40*
Early Neolithic	Terrace	Wk-17030	bone	6035 ± 40*
Early Neolithic	Terrace	OxA-13445	bone	6040 ± 35*
Early Neolithic	Terrace	TO-12197	Human bone	7500 ± 90*
Late Solutrean	Slope	Wk-12131	bone	17,634±110
Early Solutrean	Rockshelter	Wk-17840	charcoal	20,340±160
Proto-solutrean	Slope	Wk-12130	bone	18,410±165**
Late Gravettian	Slope	Wk-16415	shell	21,830±195
Late Gravettian	Slope	Wk-13686	bones	22,470±235
Early Gravettian	Slope	Wk-12132	charcoal	24,300±205
Early Gravettian	Slope	Wk-16414	shell	23,995±230
Early Gravettian	Slope	Wk-17841	shell	24,560±570

Table 11.1. AMS dates from Vale Boi. (* in Carvalho et al, in press. ** Since the % of N (.18) in this sample is very low, the result should be considered as a minimum date.)

11.4 Group of bone tools from the Gravettian and Solutrean middens.

11.5 Solutrean points found in the Midden.

successfully with small fragments coming from one, or in rarer cases two adjacent units, usually from a single spit, but in certain cases from two continuous vertical spits. The same fact was also found in the case of various bone points (**Figure 11.4**), of which two were entirely refitted (one with two fragments, and another with more than 10 fragments). No refitting has been attempted so far for lithic artifacts (**Figure 11.5**). Finally, another indicator of the good preservation of this area is the fact that there is a wide range of sizes in the lithic materials and in the small fauna, including numerous very small elements less than 5 mm in size, proving that there was no size sorting in the deposit. Artifacts and fauna show a wide diversity of directions of deposition and are basically flat following the inclination of the base of the deposit.

The second area, around Z27, is very different. The total thickness of the deposit above bedrock is between 1.2 m and 0.5 m. There are only two main layers one with Solutrean and another with Gravettian. The top layer is not more than 40 cm thick, with the faunal remains extremely rare and fairly physically and chemically worn. This is probably the result of the action of a fluvial regime with running water altering the chemical characteristics of the bone as well as dragging them down slope.

The lower level, with abundant faunal remains and lithic materials, seems to be slightly better preserved since the erosion did not completely destroy the fauna. However, there is still some evidence of bone surface deterioration. Also, no refitting was obtained, since there are no shells or bone tools. In addition, it is clear that size sorting took place in this area with small-sized lithic artifacts being very rare. Also noticed during excavation was the vertical and oblique position of the artifacts following the dip and strike of the slope.

The function of the Z27 area, very similar to that of G25, was likely a garbage dump with the formation of a midden, perhaps contemporaneous in both areas. But because of its location, in a more inclined stretch and further down slope than G25, the site formation processes resulted in significant alteration of the archaeological record. Even though the morphological characteristics of both locales are very similar, a semi-circular depression in the limestone bedrock filled with Paleolithic archaeological remains of very different kinds.

The two tests (AZ20 and AT21) located on the slope further up from G25, showed a very distinct pattern. Unlike the two areas just described, the tests did not have a compact midden. In fact, there was a single archaeological horizon in each test, of about 10 cm thick, not very rich, but there were both faunal remains and lithic artifacts. Also unlike the other two areas, the bones (there were no shells) were not as fragmented and there were a few whole bones, though faunal remains, in general, are not as common as down slope. While the lower test has a single Solutrean level, the higher one is characterized by a Gravettian occupation. The geological context also seems somewhat different, since the deposit is marked by the presence of very large limestone blocks, instead of the small clasts seen down slope.

THE TERRACE

One of the goals of testing after the excavation around Z27 and G25 was to determine the extent of the site. Thus, a series of tests were carried out in the lower section of the site, referred to as the *Terrace*. This is the second step up from the alluvial plain, where very few artifacts were present on the surface. Thus, the team thought that those artifacts had just rolled down (up slope the archaeological horizons are right at the surface) and no *in situ* materials would be found. The first test, J20, had some lithic artifacts on the surface spits, but their frequency increased as the depth increased and about 40 cm below surface there was a well preserved early Neolithic level, dated to about 6000 BP. This had not only habitation features (Carvalho et al. in press), but also ceramics, lithic materials and a wide diversity of fauna (birds, herbivores, both domestic and wild, marine shells, and fish) indicating that the economy was still very much a broad spectrum diet.

At the base of the Holocene sediments, there was a break with a clear disconformity. Initially, it was thought that the only Holocene human occupation was the Neolithic. However, the AMS date of a single human tooth found at the edge of the excavation area showed that there was human presence at the site during Mesolithic times. Unfortunately, the Mesolithic archaeological context was eroded away, probably together with the sediment package that contained it, leaving only that isolated human tooth (Carvalho et al. in press).

The Pleistocene section seems to be very similar from top to bottom, with only an increase in the clay content towards the bottom of the sequence. It is also marked by different levels of limestone blocks that result from both the slope and the erosion of the local bedrock that is present at the surface in the northwest corner of the excavation. There are various archaeological levels present in the sequence, with at least two Solutrean and two Gravettian horizons. Faunal remains are rare in the Solutrean horizons, more so in the east side of the excavated area, while as one approaches the exposed bedrock in the northeast corner, the amount of bones increases. Here, very much like in the AZ20 and AT21 tests in the top part of the slope, bones are frequently whole. This pattern of good preservation in the northeast corner is likely the result of the proximity to the limestone bedrock that helped to preserve the faunal remains. These are very common in the lower archaeological horizons dated to the Gravettian, and also very close to bed rock.

Unfortunately, the upper deposit is marked by clear intrusions with artifacts coming from upslope, and it will be very difficult in the case of the top Solutrean horizon to separate the intrusive artifacts from those that are *in situ*, since they do not present any morphological or damage differences, not even at the level of inclination of deposition.

THE ROCKSHELTER

The stratigraphy in the tests AZ20 and AT21 in the upper section of the slope suggested that, perhaps, the rockshelter would reveal only sterile deposits. The reason for this was the clear decrease of artifact and faunal remains in those two tests, as one goes up slope. Also, the fact that while the archaeological horizon in AT was found at about 50 cm below surface and at AZ, further up slope, the human occupation was already at about 1 m below surface, indicated that, if there was archaeology in the rockshelter, it would probably be so deep that excavation would be nearly impossible. In any case, within the main plan to identify and understand the extent and limits of the human occupation, two tests were carried out, one against the rock face (T1) and the other (T2), slightly lower about 10 meters down slope, on an area where the inclination was not to steep. Since the surface had very large blocks from rockfall, the tests, unlike those done in the two lower areas of the site, were 2 m^2 along the slope, so it would be possible to excavate around the blocks and remove them to the lower section in the direction of the slope.

Test T1 was dug to 2.5 m below surface. The top 50 cm were very late Holocene, loose sandy sediment with roots and no artifacts but for a couple of metal caps from recent bottles. Below was a well developed and very hard breccia 1.5 m thick. Finally, below, was fine sediment, mixed with small-sized limestone clasts and a couple of artifacts and shells, appearing to be mixed and part of a archaeological horizon with unknown chronology. The probability that there is not far from T1 a true archaeological level is very high, but the difficulty to excavate through 1.5 m of breccia to reach it at 2.5 m of depth where no heavy machinery can go would turn the excavation into a nearly impossible task.

Fortunately, T2 revealed itself much easier to deal with. The stratigraphy was similar to that of T1, but the thickness of each layer was considerably less. Thus, the artifact layer was present at a maximum depth of 1 m below surface. The top of archaeological horizon presented a slight inclination, essentially parallel to that of the local surface, but as it was excavated, the base was practically leveled.

Unlike the other areas, the archaeology of T2 is marked by the presence of many whole bones, very few fractured artifacts, and also by fairly thin, distinctive levels, sometimes separated by thin sterile horizons. In addition, the frequency of broken quartz fragments is virtually absent when compared with the extensive amounts found in the middens and in the Terrace area.

Recovered from T2 were three Solutrean levels, and as the area was expanded in the direction of the slope another level (Magdalenian) was found resting on top of the Solutrean sequence. It was decided to open in this direction for two main reasons. First, the archaeological level was closer to the surface in the western section. And second it would be easier to remove the very large blocks, some weighing over 100 kg by pushing and rolling them in the direction of the slope. Since the bedrock has not been reached, the complete stratigraphy is not yet known and it may be as long as in other areas. In fact, it may be longer, since together with the Gravettian materials from the midden there are a few artifacts that are clearly of Mousterian age and are likely coming from up slope.

The characteristics of the deposit (that is the accumulation of thin, distinctive archaeological horizons), rare broken artifacts, including bifacial Solutrean points, whole bones, a wide range of marine shell species used as ornaments and the lack of compact artifacts and sediments forming a midden indicated a very different situation from the slope section, that of a living space. And in fact, such hypothesis come to be confirmed since in the last days of the 2006 field season a hearth was found against the south cut. The idea of a living zone had been suggested before by the presence of engraved small, thin schist plaque with at least three animals, of which two are easily identifiable: a horse and an aurochs.

Finally, this area also revealed another very interesting aspect. Unit T2 was started in 2004. In 2005, the vegetation cover down slope from the test was cleared to make access easier and so the excavation area could be expanded. With the removal of the bushes, the team uncovered a ring of large blocks that were clearly the collapse of the overhang of the shelter. The general shape of the ring seemed to indicate that the shelter could be just the entrance to a cave. In the last day of the 2005 field season, in the eastern side of the excavation, a large crack more than 20 cm wide and a meter deep appeared, separating clearly the outside or entrance sedimentary deposits (corresponding to a rockshelter) and the inside cave deposits with smaller angular clasts, no large boulders, and with a much higher degree of carbonate deposition. This area has not yet been excavated since it was found in 2005.

CONCLUSION

Except for references to chronology and to a few artifacts or techniques, very little cultural interpretation was attempted. The reason is that the main topic of the paper and of the present volume is the setting and context of rockshelters and not of the Upper Paleolithic or the Neolithic. Still, a few ideas should be presented.

The diet is one of the more interesting aspects found in Vale Boi. A broad spectrum diet is present from the beginning of the human occupation, that is in the Gravettian (Bicho et al. 2003; Stiner 2003; Manne et al. 2005), with a wide range of species, large and medium herbivores (such as the horse, aurochs, ibex, red deer, and wild ass), lagomorphs, birds, shellfish (in very large quantities during the Gravettian) and marine mammals (**Table 11.2**). It should be noted that the marine resources are very important from very early on, aspect that is also seen in other sites of Portugal (Bicho 2004; Bicho and Stiner 2006; Bicho and Haws in press). Still the composition of the fauna is very much a terrestrial assemblage, much like at other sites in Portugal (Bicho et al. 2006; Zilhão 1997), and it does not change through time. Although there are some frequency variations, likely due to changes in the local and regional vegetation cover and temperature. The better example seems to be the rabbits in Portuguese Estremadura that tend to decrease in the colder phases in the Late Pleistocene (Bicho et al. 2004), or the case of horse and aurochs that become more frequent in Vale Boi with the development of the LGM that caused a more open vegetation cover around the site (Bicho et al. 2003; Bicho and Stiner 2006). Another interesting aspect is the sharp decrease in the use of marine shells starting in the early LGM (Bicho et al. 2003; Bicho and Stiner 2006). This is likely not the result of a change in diet or gathering habits, but it is due to a change in the location of the shore. With the LGM and the regression of the Atlantic, the paleoshore moves further away from Vale Boi, probably no more than 20 km (Bicho 2004). The result was that only important

Name	species	notes
limpet	*Patella vulgata* and *Patella ulyssiponensis*	
mussel	*Mytilus edulus/galloprovincialis*	
clam	*Ruditapes decussates*	
cockle	*Cerastoderma edule*	
salops	*Pecten maximus*	adornment? tool?
	Littorina obtusata	adornment
	Littorina mariae	adornment
	Dentalium sp.	adornment
birds	indet. Sp.	mid size
rabbit	*Oryctolagus cuniculus*	
fox	*Vulpes* sp.	probably *V. vulpes*
linx	*Felis pardina*?	Large-bodied, probably Iberian linx
red deer	*Cervus elaphus*	
aurochs	*Bos primigenius*	
ibex	*Capra* sp.	Probably *C. pyrenaica*, large-bodied
wild boar	*Sus scrofa*	
horse	*Equus caballus*	small-bodied
wild ass	*Equus hydruntinus*	

Table 11.2. Faunal List from G25 and Z27 (slope area).

items, such as shells for adornments were frequently brought in to the site (Bicho et al. 2004), while those species used for diet very rarely reached Vale Boi.

The adornments present in the Gravettian and in the Solutrean show a diversity and a frequency that are not known in other parts of Portugal, at least for those phases of the Upper Paleolithic (Bicho et al. 2004; Vanhaeren and d'Errico 2002; Calapez 2003). On the other hand, they tend to show a similar pattern seen in Mediterranean Spain for the same period (Soler Mayor 2002), a pattern that is also present in the bone tool assemblage and bifacial Solutrean points. Bone tools are abundant at Vale Boi in the Gravettian and Solutrean. In fact, there are more bone tools in Vale Boi during those periods, than in the rest of Portugal (Bicho et al. 2004). With regard to bifacial Solutrean armatures, the same pattern is visible (Bicho 2006), the stylistic characteristics and the type diversity in Vale Boi are in everything identical to that of the Parpalló area (Fullola 1985; Villaverde 1994), and the presence of a Solutrean engraved plaquette also separates Vale Boi from the traditional Solutrean of the Portuguese Estremadura and clusters it with the Solutrean from Mediterranean Spain.

On a final note on the history of early prehistoric rockshelter studies in Portugal is that very little has been done on this topic. In fact, besides Vale Boi, still subject of excavations, there are only two rockshelters excavated and published in Portugal: Lagar Velho also known as Lapedo (Zilhão and Trinkaus 2002) and of Paleolithic age; and Pena d'Água with Neolithic deposits (Carvalho 1998). The tendency in Portugal, since the 19th century has been always the excavation of caves, and in fact there are about a dozen excavated caves in central Portugal with early prehistoric deposits (Zilhão 1997). Only recently have rockshelters started to be explored and thought to be important during surveys. Likely, in the next couple of decades more rockshelters will be discovered and excavated in Portugal.

REFERENCES

BICHO, N. (2003) A importância dos recursos aquáticos na economia dos caçadores-recolectores do Paleolítico e Epipaleolítico do Algarve. *Xelb*, 4:11-26.

BICHO, N. (2006) *Fashion and glamour: weaponry and beads as territorial markers in Southern Iberia*. Paper presented at the XVth UISPP Congress, Lisbon.

BICHO, N.; M. STINER; J. LINDLY; C.R. FERRING; J. CORREIA (2003) Preliminary results from the Upper Paleolithic site of Vale Boi, southwestern Portugal. *Journal of Iberian Archaeology*, 5:51-66.

BICHO, N, J. HAWS (in press) At the land's end: marine resources and the importance of fluctuations in the coast line in the prehistoric hunter-gatherer economy of Portugal. *Quaternary Science Review*.

BICHO, N.; J. HAWS; B. HOCKETT (2006) Two sides of the same coin – rocks, bones and site function of Picareiro Cave, Central Portugal. *Journal of Anthropological Archaeology*, 25:485-499.

BICHO, N.; M. STINER (2006) Gravettian coastal adaptations from Vale Boi, Algarve (Portugal). In *La cuenca mediterránea durante el Paleolítico Superior. Teunión de la VIII Comisión del Paleolítico Superior*. J. SANCHADRIAN, A. BELÉN MARQUEA AND J. FULLOLA y PERICOT (eds.), pp. 92-107. Fundación Cueva de Nerja: Nerja.

BICHO, N.; M. STINER; J. LINDLY (2004) Shell Ornaments, bone tools and long distance connections in the Upper Paleolithic of Southern Portugal. In *La Spiritualité*, M. OTTE (ed.), pp.71-80. Liege: ERAUL.

CALLAPEZ, P. (2003) Moluscos marinhos e fluviais do Paleolítico Superior da Gruta do Caldeirão (Tomar, Portugal): evidências de ordem sistemática, paleobiológica e paleobiogeográfica. *Revista Portuguesa de Arqueologia*, 6(1):5-15.

CARVALHO, A. (1998) O Abrigo da Pena d'Água (Rexaldia, Torres Novas): resultado dos trabalhos de 1992-1997. *Revista Portuguesa de Arqueologia*, 1(2):39-72.

CARVALHO, A.; DEAN, R; BICHO, N.; FIGUEIRAL I.; PETCHEY, F.; DAVIS, S.; JACKES; M., LUBELL D.; BEUKENS, R.; MORALES, A.; ROSELLÓ, E. (in press) O Neolítico antigo de Vale Boi (Algarve, Portugal). Primeiros resultados. In *IV Congreso del Neolítico en la Península Ibérica*. Alicante: Museo de Arqueología de Alicante.

FULLOLA, J. (1985) Les pieces à ailerons et pédoncule comme élément différential du Solutréen ibérique. In *La signification culturelle des industries lithiques*, M. OTTE (ed.), pp. 339-352. Oxford: BAR.

MANNE, T.; STINER, M.; BICHO, N. (2005) Evidence for Resource Intensification in Algarve (Portugal) During the Upper Paleolithic. In *Animais na Préhistória e Arqueologia da Península Ibérica. Actas do IV Congresso de Arqueologia Peninsular*, pp. 145-158. Universidade do Algarve: Faro.

McPHERRON, S. and H. DIBBLE (2002) *Using computers in archaeology. A practical guide*. Boston: McGraw-Hill Mayfield.

SOLER MAYOR, B. (2002) Adorna, imagen y comunicación. In *De Neandertales a Cromañones. El inicio del poblamiento humano en las tierras valencianas*, V. Villaverde (ed.), pp.367-376. Valencia: Universitat de València.

STINER, M. (2003) Zooarchaeological evidence for resource intensification in Algarve, southern Portugal. *Promontoria*, 1:27-61.

VANHAEREN, M.; F. D'ERRICO (2002) The body ornaments associated with the burial. In *Portrait of the Artist as a Child. The Gravettian Human Skeleton from the Abrigo do Lagar Velho and its Archaeological Context*, J. ZILHÃO and E. TRINKAUS (eds.), pp. 154-186. Lisboa: IPA.

VILLAVERDE, V. (1994) Le Solutréen de faciès ibérique: caractéristiques industrielles et artistiques. In *Le Solutréen en Péninsule ibérique*, pp. 11-29. Mâcon: Musée Départamental de Solutré.

ZILHÃO, J. (1997) *O Paleolítico Superior da Estremadura Portuguesa*. Lisbon: Edições Colibri.

ZILHÃO, J.; E. TRINKAUS (eds.) (2002) *Portrait of the Artist as a Child. The Gravettian Human Skeleton from the Abrigo do Lagar Velho and its Archaeological Context*. Lisbon: IPA.

EARLY TARDIGLACIAL HUMAN USES OF EL MIRÓN CAVE (CANTABRIA, SPAIN)

Lawrence Guy STRAUS and Manuel González MORALES*

*Department of Anthropology, University of New Mexico, Albuquerque, NM 87131, USA; lstraus@unm.edu and Instituto de Investigaciones Prehistóricas, Universidad de Cantabria, 39071 Santander, SPAIN; morales@unican.es

Abstract. Like many large caves with strategic locations and favorable orientations, El Mirón, in the Cantabrian Cordillera of northern Spain, was used for a wide variety of purposes in the course of repeated human occupations during 41,000 years. Among the most substantial of the Paleolithic occupations were those of the early Magdalenian (17-13 kya). These are characterized by dense midden deposits replete with artifactual and faunal assemblages, attesting to intensive exploitation of both local and extra-local catchment areas and to a multitude of activities from hunting and fishing to butchery and food processing, to knapping, sewing, artistic creation, etc. There are also many hearths, often filled with fire-cracked rocks, as is common in Magdalenian sites.

Keywords: El Mirón Cave, Cantabrian Spain, Early Magdalenian, Hearths.

Résumé. Comme beaucoup de grandes grottes dans des endroits stratégiques et avec des orientations favorables, El Mirón dans la Cordillère Cantabrique du Nord de l'Espagne, fut employée pour une ample variété de fins au travers de maintes occupations humaines pendant 41.000 ans. Parmi les occupations plus substantielles furent celles du Magdalénien ancien (17-13 ka). Elles sont caractérisées par des dépôts de résidus chargées de restes d'industries et de faunes qui attestent l'exploitation intensive des zones locaux et non-locaux et une multitude d'activités depuis la chasse et la pêche à la boucherie et la cuisine, à la taille de pierre, la couture, la création artistique, etc. Il y a aussi de nombreux foyers, beaucoup d'eux remplis de galets fracturés par le feu, ce qui sont communs parmi les gisements magdaléniens.

Mots-Clés: La Grotte de El Mirón, L'Espagne Cantabrique, Magdalénien ancien; Foyers.

Certain caves have been repeatedly chosen for human use throughout prehistory. When present in a landscape, caves and rockshelters are extraordinarily attractive to humans, whether they are Pleistocene foragers or Holocene farmers, and whether the caves are used for short-term logistical bivouacs such as hunting, fishing or gathering camps; long-term, multi-purpose, residential base loci; field stations; or farmhouses/barns. Why specific caves were frequently used can in part be imagined by commonsense, although clearly an understanding of changes in environments is important to any detailed reconstruction of the changing roles of caves over time as places in the human landscape. A combination of location and physical characteristics (e.g., size, orientation, shape of mouth, internal topography, solar illumination, humidity, drafts) is likely to be critical in the human decisions about occupying a cave short- and/or long-term (see Montagnari Kokelj, this volume). Cave mouths, the areas most heavily used by people, are dynamic structures. Their walls, ceilings and floors, as well as roof-fall blocks, entrance overhangs and exterior talus slopes change greatly over time, creating new possibilities and/or problems for potential human occupants in terms of living space, accessibility, light and commodity. Climate changes can affect the livability of caves by modifying air currents, ceiling drips, surface water flow and ponding.

Caves experiencing "optimal" conditions from the standpoint of human groups with a particular set of demographic, social and economic characteristics and shelter needs stand to witness large-scale, repetitive occupation. Given relative structural stability, caves can be used in redundant fashion by successive human groups taking advantage of (or working with) the same physical characteristics of the cavities in question (e.g., overhangs, "site furniture" such as blocks, exterior terraces and other level areas, wall alcoves, ceiling drips, streams, ponds, muddy areas) to organize their various kinds of activity areas. Activities that were likely to have been more-or-less incompatible with one another (e.g., hide-scraping and stone-knapping; cooking and sleeping; defecation and eating) would have been spatially segregated along "redundant" lines, as long as the cave configuration remained similar over time. Some activities required more light, more broad, clear, level spaces, less draft, better access to areas where noxious or cumbersome waste could be discarded, etc., than others, and humans were likely to organize their use of space within caves accordingly. Thus, the human use of caves is the result of a combination of environmental, structural and socio-economic factors. Of the many exterior-opening caves in any given region, only a relatively few were so favorable over the long term as to house major human occupations for extended periods of time (see Straus 1979, 1990, 1997). Such a cave was El Mirón, at 43°15' North, in the Cantabrian Cordillera, between Santander and Bilbao in northern Atlantic Spain. Our excavations since 1996 have revealed evidence of human occupations from at least the late Middle Paleolithic to the late Middle Ages– dated by 60 ^{14}C assays between 41,000 BP- A.D.1400 and indeed late into the 20th century, when the cave vestibule continued to shelter both livestock and people (e.g., Straus and González Morales 2003a, b, 2005; Straus et al. 2001).

EL MIRÓN CAVE

Archeologically known for over a century and surrounded by famous cave art loci, Mirón is located at 260 m above present sea level and about 100 m above the valley floor on the face of a dramatic, cirque-like cliff mid-way up Monte Pando, at the eastern end of the Valle de Ruesga the deeply entrenched upper-middle course of the Río Asón. The cave dominates the gorges of two Asón tributaries and, facing due west, has a panoramic view of all three valleys and the cordilleran slopes above them. Directly in the cave's axis is the impressive, 1000 m-high, pyramidal-shaped Pico San Vicente, across the Calera and Gándara gorges at the eastern end of the steep, jagged Sierra del Hornijo. The Asón is the main avenue of communication in eastern Cantabria between the Atlantic coast and a relatively low pass that in turn leads up to the high Castilian meseta. Via its eastern tributary, the Carranza, the Asón also traditionally provided the easiest access between Cantabria and the Basque Country, regions which are otherwise separated by a mountain chain that extends north to the shore. The shore is 25 km from Mirón via the Asón. During the Last Glacial, the distance would have been about 7-12 km further north than at present.

The Mirón mouth is large, 16 m wide by about 20 m high, and visible from great distances. As landmarks, San Vicente and the Mirón mouth are remarkable. The 30 m-deep Mirón vestibule is sunlit and spacious, with a flat, 13 m-high ceiling, a level ground surface, and a 7-12 m width. The accessible, but dark, inner cave extends another 100 m from the NE corner of the vestibule.

The ceiling of the cave seems to have been remarkably stable, now doubt formed by exceptionally hard limestone overlying softer limestone in which Mirón formed. There are few indications of major episodes of "recent" rockfall. Similarly, there are few areas of major drips: a couple in the center of the vestibule and one in the middle of the inner cave, with active calcium carbonate precipitation. Thus the overall configuration of the cave would seem to have changed little since the Tardiglacial, except for the fact that the ground surface of the cave was about 2-3 m lower than at present–based on our excavations in the vestibule front, center and rear. In the rear, where the strata slope down toward the cave mouth, following the underlying geometry of the eroded face of ancient alluvia that fill the inner cave, erosion (both natural and human) has likely stripped away the Holocene deposits. The early Magdalenian deposits of the vestibule rear and ramp areas seem to have been banked up against the scoured, planed-down surface of the underlying alluvia–either atop an intervening wedge of Solutrean, early Upper Paleolithic (EUP) and late Middle Paleolithic (LMP) levels or directly.

THE EARLY MAGDALENIAN OCCUPATIONS OF EL MIRÓN

Although the Upper Magdalenian and Azilian culture-stratigraphic units are represented throughout the vestibule, with ^{14}C dates between 13,000-10,300 BP (final Tardiglacial), the levels are poor and attest to short, spatially discrete occupations of the cave. Similarly, the Solutrean occupations (19,000-17,000 BP the Last Glacial Maximum) may have been low-intensity and discontinuous in nature. Even more ephemeral were the widely separated visits to the cave during the LMP/EUP (41,000-27,000 BP the Interpleniglacial) and Mesolithic (9500-8500 BP the Preboreal). After the Mesolithic, the rhythm of occupational intensity increased greatly in the Neolithic, Chalcolithic and Bronze Age (5800-3200 BP, mid-Holocene).

It is the Early (i.e., "Initial," "Lower" and "Middle") Magdalenian that interests us here (González Morales and Straus 2005). This period, dated by 25 ^{14}C assays between 17,000-13,000, witnessed the most intensive, repeated UP occupations of El Mirón. At the level of archeological visibility, these deposits, constituting a distinctive stratigraphic horizon (levels 17-14 in the vestibule front; levels 312-310 in the mid-vestibule; levels 119-108 in the rear), seem to represent continuous use of the cave. This means that in excavating them, we can discern no interruptions in occupation, no sterile lenses, no hiati. Indeed the distinctions we make between "levels" are often problematic. The top and bottom of the horizon are, however, stratigraphically sharply marked. The lack of stratigraphic gaps within the horizon does not mean that early Magdalenian people were sedentary. Indeed this would be highly improbable for any forager society dependent on exhaustible, mobile game and fish species. The dense nature of these "midden" deposits does suggest very frequent, intensive human residences in the cave during the early Tardiglacial. In addition to the massive nature of the early Magdalenian "palimpsest" in the Mirón vestibule (80 cm-thick in the front, 90 cm in the mid-vestibule trench, 100 cm in the rear), there are clear indicators that human use of the cave had been much more extensive in this period. In a niche ("A") in the south wall of the cave in the narrow "ramp" area, a few cm above the steeply sloping eroded surface of the ancient alluvia, we excavated an intact, bone- and artifact- rich fill of dark, "chocolate" brown silt-loam very similar to the richest sediments of the early Magdalenian levels in the vestibule. Although no diagnostic artifacts were found, a ^{14}C assay on a bone from Niche A yielded a date of 16,600±90 BP–consistent with the initial Magdalenian. Much further back, at 25 m east of the top of the ramp, in the totally dark inner cave, we excavated a small sondage in the base of an exploratory trench dug in the 1950s. At the base of this pit we found blades, a blade core and charcoal chunks, one of which was dated to 14,620±80 BP.

It appears that in the early Magdalenian, humans not only occupied Mirón very frequently and "massively," but also fully. One can imagine that very different activities would have been conducted at the sunlit front of the ample vestibule, in the darker, cooler vestibule rear, on the steeply sloping ramp, and in the completely dark, humid inner cave. The similarity in color, texture, organic composition, and extreme abundance of bones, lithic artifacts, fire-cracked rocks and charcoal among the deposits of the vestibule front, middle and rear and the "ramp" area niche is striking, and suggests that the intensive nature of the occupations in the period between 17,000-14,500 BP was similar within different areas of the cave–a sheltered space so large as to provide room for humans to "spread out" and to spatially structure their many activities each time they came back to Mirón. We know that those activities ranged from the artistic (stratigraphically dated rock engravings and engraved deer shoulder blades) to the "mundane" (e.g., butchering areas, cooking areas around sometimes-formal, constructed hearths, stone-knapping areas, possible weapon re-arming areas, sewing areas, areas requiring ad hoc surface "paving"). Other areas with ochre stains and lumps (including "crayons") could have been the residues of either artistic or practical (e.g., hide-preparation, microlith "gluing") activities. That activity areas may have been systematically structured is suggested by the presence in early Magdalenian times of a possible stone wall separating the (less fully sunlit) rear of the vestibule from the middle and front. The sheer amounts and diversity of heavily fragmented ibex and red deer bone (plus remains of other ungulates and of salmon), knapping debris, lithic and osseous tools (e.g., scrapers, burins, perforators, knives, needles, awls), weapon elements (backed bladelets, antler sagaies), personal ornaments (perforated teeth, shells and stones), fire-cracked rocks, lumps and patches of ochre throughout this vast cave are suggestive of frequent, large-scale human occupations, by groups consisting of possibly several families, including women, children and elderly, not just male hunting parties. Residues of limited-function ibex hunting camps could in fact be "mixed-in" with the massive residential palimpsests, but they would be virtually impossible to discern, given the overall "noise" of the compounded, multi-purpose, larger-group occupation residues. What can be said about these repeated early Magdalenian occupations of Mirón, despite the palimpsest nature of their residues, is that the composition of their "trash" did vary in space and through time.

The oldest post-Solutrean assemblages from the vestibule rear are characterized by the significant presence of "macroliths" made on local non-flint raw materials (mudstone, quartzite, limestone), which contrast sharply with assemblages dominated by microlithic bladelets (many backed) made on excellent-quality, non-local flint that characterize the later early Magdalenian levels. By "macroliths" we mean large flakes, flake cores, sidescrapers, denticulates, notches and even choppers. The significance of the macrolithic-rich "initial Magdalenian" assemblages (about 17-16,000 BP) might be as an indication of relatively restricted human mobility, with a high degree of strictly local lithic supplying. (Even this statement has to be nuanced, because some of the earliest post-Solutrean levels contain both local lithic macroliths mainly large flakes and cores, and abundant artifacts, including microdébitage, of excellent quality non-local flint. The lowest of these levels recently yielded a perforated pendant engraved with a horse-head image on a piece of slate. Slate is a material that does not occur commonly in Cantabria or the Spanish Basque Region, but can be found to the west in Asturias and to the east in the Pyrenees, a fact which might suggests long-distance contacts). In contrast, the classic Lower Magdalenian occupations have very few macroliths on local non-flint materials, but are massively dominated by non-local flints that attest to either trade or travel to sources of superb Upper Cretaceous flint along the present-day sea cliffs of west Vizcaya and east Cantabria (which, in the Tardiglacial, would have risen above a narrow, then-dry continental shelf) at distances of about 30-60 km from Mirón. Indeed, the world of the classic "Cantabrian" Lower Magdalenian is territorially defined by very distinctive and similar striated engravings of red deer hinds on red deer scapulae and on cave walls found mainly within the east-central area of the Province of Cantabria (50x25 km). Overall, a trend of expansion in use-territories is suggested for the Magdalenian, and, in fact, in the Middle Magdalenian there are indications of broader (extra-regional)-scale social contacts, notably the presence of clearly "French Pyrenean" art, ornamental objects (contours découpés, rondelles) in several Cantabrian and Asturian sites. Mirón, despite its inland, montane location, participated in these trends because it is near the best avenue of access to the Basque Region (Straus et al. 2002a, b).

THE HEARTHS OF THE EARLY MAGDALENIAN

The early Magdalenian occupation surfaces in the vestibule rear (9-10 m^2 excavated) are replete with fire-cracked rocks (fractured, blackened/reddened sandstone and quartzite cobbles, plus thermally altered limestone éboulis), very abundant blackened bones, crazed and pot-lidded flint flakes, specks of charcoal and charcoal-stained sediments, patches of ash, ochre stains, etc. Such evidence is widely distributed, but dissection of the levels (108-119) over the years has revealed definite concentrations, some with undoubted, albeit simple structures. Detailed analyses of the hearths and their contents are being conducted by Y. Nakazawa as part of his Ph.D. dissertation. The quantities of artifacts and bones are far greater than those shown in the figures (which are only the larger items). Nakazawa's work will include spatial studies of distributions of fire-cracked rocks, lithic artifacts and faunal remains in and around the hearths.

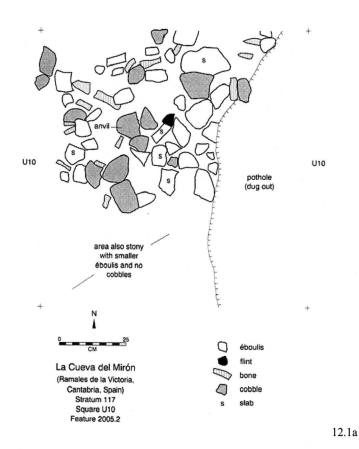

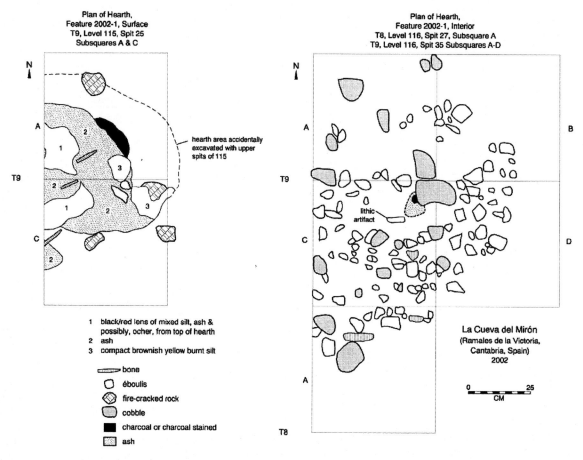

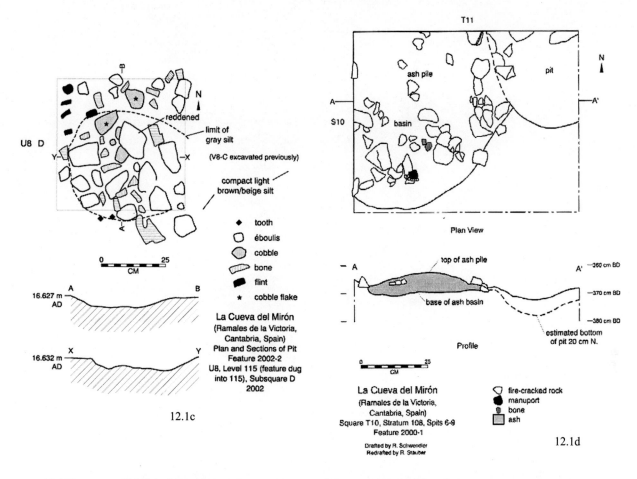

12.1 Features at El Mirón 2000-2005. a) Feature 2005.2, b) Feature 2002.1, c) Feature 2002.2, d) Feature 2000.1.

Aside from an enigmatic "hole" (Feature 2006.2, "Level" 119.1) cut into Level 119.2 from 119, the oldest evident feature uncovered so far in the vestibule rear (Feature 2005.2) is in Level 117, square U10 (**Figure 12.1a**). This consists of over a dozen cobbles (most relatively large, and some fire-cracked), 5 slabs ("plaquettes") and several medium-large-size éboulis together with a "halo" of long bone fragments. There are flakes throughout, a large flint chunk and an anvil stone in the middle of the concentration. The full original extent of this cobble or slab concentration is unknown, since the eastern side was removed earlier by looters and the northern end is under an unexcavated profile, but the uncovered area amounts to about two-thirds of a square meter. There is no charcoal and it is unlikely that this is actually a hearth, even if the fire-cracked cobbles could have come from the cleaning-out of (a) nearby hearth(s). Given this feature's location in the downslope area of Level 117, with fine, clayey silt, it is possible that it represents an expedient "pavement" to make dry and useable an otherwise possibly muddy area.

Immediately overlying Level 116 revealed three features. The clearest is Feature 2002.1, located in square T9 and a small area of T8 (**Figure 12.1b**). The surface appeared as a number of patches of ash and burnt silt in a roughly circular area demarcated by cobbles and éboulis. Beneath the surface, there were masses of small éboulis, about a dozen large cobbles and several small ones, small numbers of lithics and bones, around a small central "core" of ash and charcoal at the base of a shallow basin. This oval feature measured 1.75-0.75 m. Feature 2004.3 is also in Level 116, and horizontally adjacent to, but stratigraphically below Feature 2002.1 in square T10 and along the northern edge of T9. The center is a small depression, which is surrounded by about a dozen large cobbles (plus a few small ones), a half-dozen long-bone fragments lying flat, plus one oriented vertically, and several flint artifacts. There is no evidence that this was a hearth, rather possibly a posthole surrounded by supporting rocks. The third Level 116 feature (2002.3) was found in the narrow area of half-square V7. The southern half of the square is under the profile, while square V8 to the north was excavated as an exploratory 1 m² test pit in the initial stages of work in the vestibule rear. The feature consists of a small ash patch ringed by éboulis and large-medium sized cobbles, an adjacent charcoal-stained area, and several bones and flints (including a core). This appears to represent about a quarter of a hearth that had been built in shallow natural

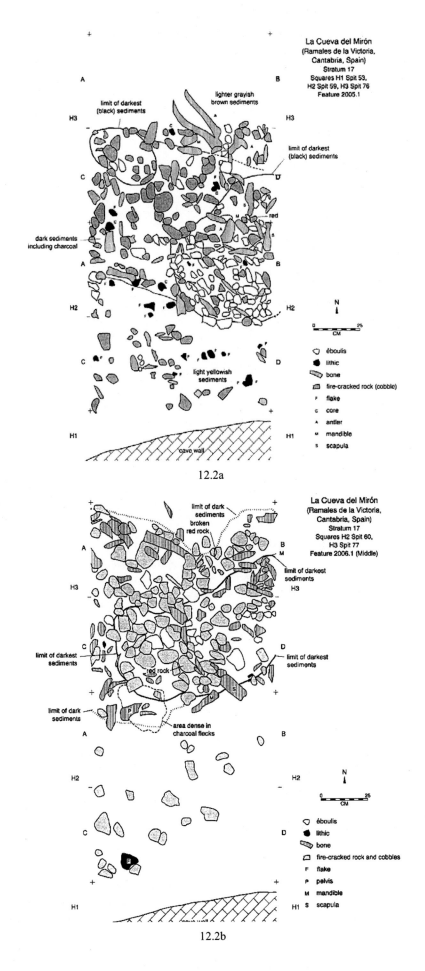

12.2a

12.2b

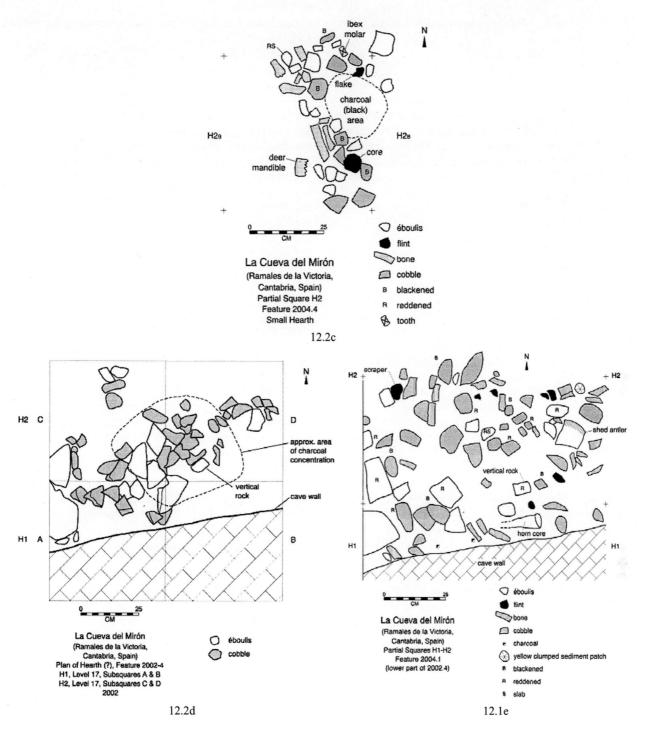

12.2 Features at El Mirón 2004-2005. a) Feature 2005.1, b) Feature 2006.1, c) Feature 2004.4, d) Feature 2002.4, e: Feature 2004.1.

or artificial depression. This "pit" seems to have cut down from Level 116 into 117.

Feature 2002.2 is a small, but clear hearth pit dug down from Level 115 into the top of 116. It is located in square U8D, plus a small part of the northern edge of U7B (**Figure 12.1c**). The feature thus covered slightly more than 0.25 m². It is defined by an oval-shape area of gray (ashy?) silt that coincides with a mass of large, medium and small éboulis, one large and two small cobbles, two huge cobble flakes, numerous large bones and several flint artifacts (mainly at the northwest edge). The largest limestone block (in the center of the feature) is reddened. The ashy silt, rocks, bones and flints are in a shallow basin (about 2-4 cm deep).

The first (i.e., stratigraphically highest) early Magdalenian feature to be found was a clear hearth (2000.1) (**Figure 12.1d**). It was located in Level 108, square T10, in the northwest corner of the Corral excavation area. Part of the feature continues into unexcavated squares S10 to the west and T11 to the north. This elaborate feature continued through four excavation spits and consists of an ash pile within a shallow basin adjacent to a small pit. The full diameter of neither can be accurately given since they both continue under the stratigraphic profiles. The known area of the ash-filled basin measures about 65 x 70 cm and that of the small pit, about 35-40 cm. The basin contains a large number of apparently fire-cracked rocks, especially concentrated along the southern and eastern peripheries, plus large bones and a manuport. The pit was at least 8 cm deep, but possibly more, since it is hard to determine where its original top had been. The ash pile was maximally about 8 cm thick. The pit was virtually devoid of rocks. It is unclear whether one had cut into the other, but it is most likely that the rock-laden hearth and the adjacent pit were built and used together.

The mid-vestibule trench is only 1 m wide, and the early Magdalenian levels were reached in only three sondages, the largest of which, square P6, is a 1 m^2 pit. This pit cut through the full Magdalenian stratigraphic sequence, by far the richest cultural stratum of which, Level 312 (a massive palimpsest corresponding to the whole of the Lower Magdalenian), is densely packed with unstructured hearth debris (charcoal, fire-cracked rocks, blackened bones, fire-crazed flints, etc.). However, no actual feature was observed in level 312. In P6, subsquare D, we did find a small concentration of éboulis and large-medium cobbles (n=4). This feature (2001.2) extends under the south stratigraphic profile, so it was larger than the exposed area of about 30 x 50 cm. The rocks seem to have been in a very shallow basin that cut (on a slight slope) from level 308 into 309. The feature thus dates to the Upper Magdalenian. There were few artifacts or bones, and charcoal may have washed away or percolated downward. Whether this had been a small hearth or an expedient, localized pavement cannot be readily determined.

Rich, early Magdalenian level 17 in the 9-10 m^2 Cabin area (vestibule front) has yielded evidence of several features. The lowest to be found so far is 2005.1 (**Figure 12.2a**). This major hearth continued all the way to the base of Level 17/top of Level 18 as Feature 2006.1 (**Figure 12.2b**). It is in squares H2 and H3 (the center being in H3C+D) and consists of a denser than "usual" mass of éboulis and fire-cracked cobbles at the center of which is a 75 x 60 cm concentration of black, charcoal-rich sediments, that continued eastward into I2 and I3. There is a "halo" of large lithic artifacts (mainly flakes and cores) around the hearth per se and impressive quantities of (red deer and ibex) bones (sometimes "solid" masses) both inside and outside it, including a large rack of red deer antlers immediately to the northwest of the hearth area. There are numerous mandibles. In the upper part of the feature, there is a second, smaller (35 x 25 cm), oval patch of black sediments filled with fire-cracked cobbles in H3C, to the northwest of the main hearth. It is not evident that either charcoal-rich patch actually fills a dug-out pit; the fires may simply have been built on the living surface. Upon reaching the feature base (atop light grey, éboulis-rich Level 18) it became apparent to us that the hearth had first been laid in a slight depression in the "virgin" surface as it existed when people of Lower Magdalenian age first occupied this area of the cave. The area between the two upper charcoal-rich patches is also rich in charcoal, but not as dark. The lower part of the hearth is somewhat smaller, but more uniform in terms of its black clayey silt matrix that is rich in not only charcoal dust, but also actual lumps. Some of the lumps were selected for ^{14}C dating. Analysis of the rock fill (by Nakazawa and Straus) revealed the presence of many blackened/burned and/or fire-cracked sandstone cobbles of standardized size and numerous anvils (some of which are also fire-cracked). This hearth seems to have been reused repeatedly over a long period of time during the formation of the massive Level 17 "midden" deposit. Adjacent to it, about 1-1.5 m to the north, was a cluster of ibex crania with horn cores, associated with several geometrically engraved, quadrangular-section antler point fragments.

Another small, sub-circular patch of black, charcoal-rich sediments was found higher up in level 17 of square H2, subsquare B: feature 2004.4 (**Figure 12.2c**). Measuring about 20 x 25 cm, this patch (itself devoid of large rocks), is surrounded by fire-cracked cobbles, slabs and éboulis, as well as large bones, teeth and a pair of large flint artifacts (one of which is a core). This seems to have been a small hearth, apparently build on the living surface.

Even higher in level 17 in squares H1 and H2 was feature 2002.4/2004.1 (**Figures 12.2d** and **Figure 12.2e**). This is a mass of fire-cracked cobbles, slabs and éboulis, at the center of which was a subcircular concentration of charcoal measuring about 40 x 70 cm. The rocks are both within and around the hearth. The concentration has considerable depth. There are many large bones around the hearth, including an ibex horn core and a shed red deer antler. There is also a "halo" of large flint artifacts (including a scraper) to the north and east of the hearth. (The southern boundary of the feature is the cave wall and the western side is under the stratigraphic profile of the Cabin area.) There is no clear evidence that the hearth had been built in a prepared pit, although one of the elongated rocks in its center is vertical, which hints at the existence of a hole.

A summary of the above shows that there are evident hearths in the early Magdalenian deposits, found on or in very densely packed cultural surfaces. These surfaces represent both "living floors" and "middens". People were clearly living in the midst of their trash. In other words, they discarded much of their debris at the locus of abandonment, in the areas at or near where the objects had been created and/or used. Such debris runs the gamut from microdebitage to cores, from long-bone splinters to antlers and horn cores, from fire-cracked rocks to bone needles, from hearth sweepings such as ash and charcoal to engraved scapulae. The hearths in a few clear cases were prepared in the sense that basins were dug, fires built and cobbles added. In other cases it appears more likely that the fires were simply built on the surface, but cobbles were also often involved to transmit and conserve heat, probably for cooking (by roasting, boiling or steaming). Many of the cobbles were (also) used as anvils, perhaps (among other things) for breaking marrow bones. In some cases, there are concentrations of fire-cracked (blackened, reddened) rocks, but without associated ash or massive amounts of charcoal (or charcoal-stained sediments, since, in fact, large chunks of charcoal are rare in the Magdalenian deposits). These may have been the result of hearth-cleanings, or (less likely) of simple surface hearths where all the charcoal had been washed away by running or percolating water. Analyses of the spatial distributions of not only fire-cracked cobbles, but also of crazed and pot-lidded lithics and burnt (charred or–far more rarely calcined) bones, will hopefully reveal the location of latent hearths, even where obvious charcoal and ash patches are not present. Indeed, the early Magdalenian levels must have had many hearths, given the great amounts of ash, charcoal-stained silt and burnt materials that characterize all these intensive occupation deposits. The remarkable aspects of these levels, we reiterate, are the frequency and intensity of reoccupation of Mirón Cave and the extensive nature of the use of space (even into the ramp and inner cave areas).

These indications and the sheer abundance and diversity of faunal and cultural remains from the banal to the artistic (both portable and rupestral art of proven early Magdalenian age) all point to El Mirón having been a long-term major nexus of the human settlement of the montane zone of eastern Cantabria. The novelty of this site, especially at this time, is that it was not simply being used as a specialized ibex-hunting locus like other sites in the Cordillera (e.g., Collubil in the Picos de Europa of Asturias, El Rascaño in Cantabria, Bolinkoba in Vizcaya, or Erralla in Guipúzcoa). On at least many occasions, this was a large-scale, multi-purpose site, probably relatively long-term in nature, although this remains to be demonstrated by seasonality analyses of faunal remains. Some levels are rich in remains of red deer, as well as ibex, and many have yielded large numbers of fish bones, despite the fragility of such items.

EL MIRÓN IN THE CONTEXT OF THE EARLY MAGDALENIAN OF CANTABRIA

In addition to sites with more limited early Magdalenian deposits (e.g., Rascaño), this period is well-known in the territory of what is today Cantabria Province at the classic sites of Altamira and El Juyo in the coastal zone and El Castillo on the edge of the Cordilleran foothills. The Lower Cantabrian Magdalenian of Juyo (Freeman and González Echegaray 1984; Freeman et al. 1988) is famous for its diversity of structures, including pits and arrangements of stones, as well as patches of variously colored sediments reminiscent of lenses of red ochre, greenish-grey silt, grey clay, yellowish beige silty clay uncovered recently in Mirón. The full Lower Magdalenian sequence of levels in Juyo totals about 1.5 m thick, a fact which, combined with the density of the midden-like cultural residues of the levels, indicates the repetitive, intensive nature of the human occupations during a relatively short span of time. The similarly rich Lower Magdalenian deposit in Altamira (up to 90 cm thick) has also revealed the existence of stone-lined pits during a recent excavation of limited extent (Freeman and González Echegaray 2001). Currently under excavation, La Garma Cave in the coastal zone of central Cantabria is also yielding spectacular evidence of major Middle Magdalenian occupations with very clear stone structures (Ontañón 2003; Arias et al. 2005). One should recall that Obermaier's Magdalenian Beta (Cabrera's Level 8) in Castillo was a massive deposit (1.2-2 m-thick), composed of a series of hearths and sub-dividable into at least three layers. Generally black in color, this deposit which filled the vast Castillo vestibule had lenses of charcoal, ash, fire-reddened sediments, ochre, and grey silts. First encountered by Alcalde del Río in 1903, this early Magdalenian horizon was dug out by Obermaier in 1911-12 (Cabrera 1984). Among the many finds, it is worth noting the presence of archaic-looking macroliths and anvils, associated with a variety of sagaies including quadrangular types with geometric engravings, all reminiscent of artifacts from Mirón.

To the East, in Guipúzcoa, other Lower Magdalenian sites have also yielded evidence of stone structures and definite hearths with associated bones, antlers and artifacts (interpreted by their excavators as evidence of ritual activity, as is the case at Juyo): Erralla (Altuna et al. 1985), Praile Aitz (Peñalver and Mujika 2005). Other major Basque Country early Magdalenian sites include Ekain, Ermittia, Urtiaga, and Aitzbitarte in steep terrain near the coast, Bolinkoba in the Cordillera, and Abauntz in Navarra, with well-described activity areas (Utrilla and Mazo 1992).

The record from the West, in Asturias, also includes numerous major early Magdalenian sites in the coastal zone (Cueto de la Mina, La Riera, El Cierro, La Lloseta, Cova Rosa, La Paloma) or at the edge of the Cordilleran foothills (Las Caldas, La Viña), with one high mountain site likely to be of this age (Collubil). Constructed hearths are cited in the literature: a particularly complex example has been extensively described from the Middle Magdalenian of Caldas (Corchón 1982), and a stone-lined pit has been described from Riera (Straus and Clark 1986).

The total number of sites reasonably attributable to the early Magdalenian in the entire Vasco-Cantabrian region is about 60. This high number is indicative of a relatively dense human population, especially considering the facts that these were foraging people and that we totally lack knowledge of their open-air sites. Virtually every major river valley along the whole coast, from the Urumea in the East to the Nalón in the West, has at least one early Magdalenian site–and more usually a string of sites distributed between the coast and the mountains. Undoubtedly, some strictly littoral sites of this period were lost at the time of the early Holocene marine transgression. Major concentrations of early Magdalenian sites seem to exist near the mouth of the Deva River in Guipúzcoa, in the area around the present day Bay of Santander (lower courses of the Ríos Miera, Pas and Saja), in the eastern coastal strip of Asturias (lower courses of the Ríos Bedón and Sella), and along the middle course of the Río Nalón in central Asturias. There exists the possibility that lowland sites complementary to Mirón might exist along the middle and lower courses of the Río Asón in such caves as Valle (with a recently obtained ^{14}C date of 13,820 from an otherwise poorly defined level in a small test pit [García-Gelabert 2005]) and Otero. The general picture for human settlement of the Vasco-Cantabrian region during Dryas I is that there were major, repeatedly and intensively used hub sites (multi-purpose base camps)– usually in the coastal zone (the prototypes being Altamira, Juyo and Garma, together with Castillo) and now, with the excavation of Mirón, also known from the montane zone. The difference between these sites, all with thick, organically-rich deposits containing abundant, diverse cultural materials and structures such as hearths, pits and possible walls is that, while the near-coastal sites are dominated by remains of red deer and often shellfish such as limpets, Mirón has very abundant ibex remains, sometimes together with red deer and salmon. Other high mountain sites that are smaller caves (Bolinkoba, Rascaño, Collubil) contain faunal assemblages almost completely composed of ibex remains, suggesting that they were specialized hunting camps, not multi-purpose residential bases. The landscape is dotted with other, lesser sites of early Magdalenian age–probably a variety of logistical sites, short-terms camps, and other localities that were "satellites" of the major hub base camps (Straus 1986; Butzer 1986; Utrilla 1994). It is worth reiterating that in the Lower Cantabrian Magdalenian Mirón was linked to Altamira, Castillo, Juyo, Rascaño, Pendo and Cierro by the presence of red deer scapulae engraved with images of red deer hinds done with fine multiple striations. These distinctive objects seem to mark a real socio-cultural territory (perhaps that of a regional band) that extended from the Río Sella in the West to the Asón in the East, essentially covering the area of the modern province of Cantabria and the eastern sector of Asturias (González Morales, Straus and Marín n.d.). Mirón thus further enriches the varied record of early Magdalenian sites types as a major, but montane base camp. As at other cave sites in Cantabrian Spain, just as at rockshelter and open-air sites in France, Switzerland and Germany, human modification of living space was both common and extensive.

Acknowledgments. Excavations in El Mirón Cave since 1996 have been funded by the Fundación M. Botín, National Geographic Society, Gobierno de Cantabria, L.S.B. Leakey Foundation, Spanish Ministry of Education and Science, U.S. National Science Foundation and University of New Mexico. Material support has been provided by the Town of Ramales de la Victoria and Universidad de Cantabria. Scores of students from Spain, the United States and many other countries of the Americas and Europe have participated. The figures, drawn by Straus, R. Schwendler and Y. Nakazawa, were redrafted by R. Stauber. Our sincere thanks to all and to "Pencho" Eguizábal and Jean Auel!

REFERENCES

ALTUNA, J.; BALDEON, A.; MARIEZKURRENA, K. (1985) - *Cazadores Magdalenienses en la Cueva de Erralla*. San Sebastián: Munibe 37.

ARIAS, P., et al. (2005) - La estructura magdaleniense de La Garma A. Aproximación a la organización espacial de un hábitat paleolítico. In BICHO, N., ed. - *O Paleolítico: Actas do IV Congresso de Arqueologia Peninsular*. Faro: Promontoria Monográfica 2, p. 123-141.

BUTZER, K.W. (1986) - Paleolithic adaptations and settlement in Cantabrian Spain. *Advances in World Archaeology* 5: 201-252.

CABRERA, V. (1984) - *El Yacimiento de la Cueva de "El Castillo."* Madrid: Biblioteca Prehistórica Hispana 22.

CORCHON, M.S. (1982) - Estructuras de combustión en el Paleolítico: a propósito de un hogar de doble cubeta de la Cueva de Las Caldas. *Zephyrus* 34/35:27-46.

FREEMAN, L.G.; GONZALEZ ECHEGARAY, J. (1984) - Magdalenian structures and sanctuary from the Cave of El Juyo. In BERKE, H.; HAHN, J.; KIND, C.-J.KIND, eds. - *Jungpaläolithische Siedlungsstrukturen in Europa,* Tübingen: Institut für Urgeschichte, p. 39-49.

FREEMAN, L.G. and GONZALEZ ECHEGARAY, J. (2001) - *La Grotte d'Altamira*. Paris: Seuil.

GARCIA-GELABERT, M.P. (2005) - El trabajo sobre hueso en el Magdaleniense superior final del grupo humana de la Cueva del Valle, Rasines, Cantabria. *Zephyrus* 58:111-134.

GONZALEZ MORALES, M. and STRAUS, L.G. (2005) - The Magdalenian sequence of El Mirón Cave (Cantabria, Spain): an approach to the problems of definition of the Lower Magdalenian in Cantabrian Spain. In DUJARDIN, V., ed. - *Industrie Osseuse et Parures du Solutréen au Magdalénien en Europe*. Paris: Mémoires de la Société Préhistorique Française 39, p. 209-219.

GONZALEZ MORALES, M., STRAUS, L.G.; MARIN, A.B. (n.d.) - Los omóplatos decorados magdalenienses de Cueva del Mirón (Ramales de la

Victoria, Cantabria) y su relación con las Cuevas del Castillo, Altamira y El Juyo. In BAQUEDANO, E.; MAILLO, J.M., eds. - *Homenaje a Victoria Cabrera*. Alcalá de Henares, Zona Arqueológica (in press).

ONTANON, R. (2003) - Sols et structures d'habitat du Paléolithique supérieur, nouvelles données depuis les Cantabres: la Galerie Inférieure de La Garma. *L'Anthropologie* 107:333-363.

PENALVER, X.; MUJIKA, J.A. (2005) - Praile Aize I (Deba, Gipuzcoa): evidencias arqueológicas y organización espacial en un suela magdaleniense. In BICHO, N., ed. - *O Paleolítico: Actas do IV Congresso de Arqueologia Peninsular*. Faro: Promontoria Monográfica 2, p.143-156.

STRAUS, L.G.(1979) - Caves: a paleoanthropological resource. *World Archaeology* 10:331-339.

STRAUS, L.G. (1986) - Late Würm adaptive systems in Cantabrian Spain. *Journal of Anthropological Archaeology* 5: 330-368.

STRAUS, L.G. (1990) - Underground archaeology. In SCHIFFER, M., ed. - *Archaeological Method and Theory*. Tucson: University of Arizona Press, Tucson, vol.2, p.255-304.

STRAUS, L.G. (1997) - Convenient cavities: some human uses of caves and rockshelters. In BONSALL, C.; TOLAN-SMITH, C., eds. - *The Human Use of Caves*. Oxford: British Archaeological Reports S-668, p. 1-8.

STRAUS, L.G.; CLARK, G.A. (1986) - *La Riera Cave*. Tempe, Anthropological Research *Papers* 36.

STRAUS, L.G.; GONZALEZ MORALES, M. (2003a) - El Mirón Cave and the 14C chronology of Cantabria Spain. *Radiocarbon* 45: 41-58.

STRAUS, L.G.; GONZALEZ MORALES, M. (2003b) - Early-Mid Magdalenian excavations in El Mirón Cave. *Eurasian Prehistory* 1(2):117-137.

STRAUS, L.G.; GONZALEZ MORALES; M., FARRAND, W.; HUBBARD, W. (2001) - Sedimentological and stratigraphic observations in the Cantabrian Cordillera, northern Spain. *Geoarchaeology* 16: 603-630.

STRAUS, L.G.; GONZALEZ MORALES; M.; FANO, M.; GARCIA-GELABERT, M.P. (2002a) - Last Glacial human settlement in eastern Cantabria. *Journal of Archaeological Science* 29:1403-1414.

STRAUS, L.G.; GONZALEZ MORALES, M.; GARCIA-GELABERT, M.P. and FANO, M. (2002b) - The Late Quaternary human uses of a natural territory: the case of the Río Asón drainage. *Journal of Iberian Archaeology* 4:21-61.

UTRILLA, P. (1994) - Campamentos-base, cazadores y santuarios. Algunos ejemplos del Paleolítico peninsular. In LASHERAS, J.A., ed. - *Homenaje al Dr. Joaquín González Echegaray*, Madrid: Monografías del Museo y Centro de Investigación de Altamira 17, p.97-113.

UTRILLA, P.; MAZO, C. (1992) – L'occupation de l'espace dans la grotte d'Abauntz. In RIGAUD, J.-P.; LAVILLE, H.; VANDERMEERSCH, B., eds. - *Le Peuplement Magdalénien*, Paris: CTHS, p.365-376.

ANSWER TO THE PROBLEM OF THE DIACHRONIC AND SYNCHRONIC RELATIONSHIP OF ARQUEOPALEONTOLOGICAL ELEMENTS IN SITES WITH HOMOGENEOUS SEDIMENTS IN THE MIDDLE-PLEISTOCENE: THE EXAMPLE OF GRAN DOLINA, SIERRA DE ATAPUERCA

RÉSOLUTION DU PROBLÉME DE LA RELATION DIACHRONIQUE ET SYNCHRONIQUE D'ÉLEMENTS ARCHÉOPALEONTHOLOGIQUES EN GISEMENTS À SÉDIMENT HOMOGÈNE DU PLEISTOCÈNE MOYEN: L'EXAMPLE DE GRAN DOLINA, SIERRA DE ATAPUERCA

Rosana OBREGÓN[1] and Antoni CANALS[2]

[1] Rosana Obregón: Centro Nacional de Investigación sobre la Evolución Humana. Fundación Atapuerca. Avda. de la Paz 28 entreplanta. 09005 BURGOS. Rosana.obregon@cenieh.es

[2] Antoni Canals: Universitat Rovira i Virgili. Àrea de Prehistòria. Pl. Imperial Tarraco, 1. 43005 TARRAGONA. antoni.canals@prehistoria.urv.cat

Abstract. In a study of lower Palaeolithic sites with homogeneous sediments, we are forced to conclude that the remains are accumulations that took place over enormous periods of time. Interpretations made under these conditions are problematic. The answer to this problem is computerised processing of the data and the application of the archaeo-stratigraphic method. In this way we can establish more accurate limits between the archaeopaleontological assemblages, because the empty spaces are identified that separate the assemblages with this method. We can establish depositional patterns and therefore define diachronic relationships.

Keywords: Archaeostratigraphy, profile, archaeological assemblages, gap, palaeosurface.

Résumé. L'étude de gisements de le Paléolithique Inférieure, ou le sédiment est très homogène et nous ne pouvons pas avoir des datations absolues précises, nous trouvons que au moment de commencer l'étude de ces éléments, nous rattachons des restes du produit d'accumulations arrivées au long d'énormes périodes de temps. Quelconque intérprétation que nous faisions dans cette conditions ne serait pas correcte. La solution à ce problème-ci se présente pendant l'ultérieur travail de laboratoire, où nous pouvons, grâce au traitement informatique des données, apliquer la méthode archéostratigraphique. De cette façon, nous arrivons à établir des limites beaucoup plus précis entre les ensembles archéopaleonthologiques, étant donné que nous les individualisons à travers des espaces vides moyenant entre eux. C'est pour ça qu'on peut instaurer des modèles de déposition, et des relations diachroniques certaines.

Mots-clés: Archaeostratigraphy, profil, assemblages archéologiques, espace, palaeosurface

Homogeneous sedimentary units can be problematic when studying the spatial distribution of the archaeological remains that they contain. These remains end up being associated with the unit as a whole and treated as a single depositional stratum. A possible solution is to establish more precise, thinner archaeological divisions within each sedimentary stratum. This is usually not possible during excavation and becomes a task for digital data processing. That is, we can analytically group items that appear clustered and separate them from other clusters with gaps. Such a task can be accomplished if the three dimensional coordinates for each item have been accurately recorded.

Our methodology follows the nomenclatures established by geological studies. The archaeological record corresponds spatially with pre-established geological units with regard to specific sedimentary phenomena (Meléndez and Fuster 1978:242-244). Hence, we do not override these pre-established nomenclatures. Our analysis is framed within homogeneous geological units or layers. We consider a homogeneous geological unit that which does not exhibit sufficient differences that can be observed or followed in the field. Homogeneous geological units may include limited horizontal or vertical variations or facies. However, these are usually diffuse and localized, and for our purpose considered as part of the larger geological unit.

PROBLEMS IN THE STUDY OF ARCHAEOLOGICAL ASSMEBLAGES IN HOMOGENEOUS SEDIMENT

The study of lower and middle Pleistocene sites is characterized by a set of methodological steps from excavation process to the data recording to interpretation. We face a number of problems at each step:

1. In deep time we do not have precise absolute dates because the error ranges are in thousands of years. Likewise, relative dating does not provide accurate chronological control because technological and faunal changes in old Pleistocene chronologies are not as rapid as to characterize paleontological and much less archaeological assemblages (Isaac 1972:13-34)
2. On numerous occasions, a geologically and stratigraphically defined sedimentary deposit is more than one meter thick; its lithological features being homogeneous and lacking smaller subdivisions. As a consequence, the archaeological remains contained in this deposit need be considered as a single assemblage unless intra-deposit gaps are visible, and therefore we are forced to assume that such remains are a product of hominid activity taking place over enormous periods of time. Any other interpretation derived under these conditions risks being erroneous (Butzer 1982: 96).
3. During the process of excavation, all the archaeo-paleontological items found in a stratum are classified and filed as belonging to a single deposit, when in fact we know that they are a product of different hominin occupations of possibly different intensity, duration and function. In this way, we accumulate errors throughout the field recording, in data entering and in the subsequent archaeological analysis. The activities carried out by hominins might have been very diverse, and the times at which they were carried out might have been punctuated, or continuous in time. At times archaeological assemblages may approach palimpsests, (Meignen 1994:81) which do not allow for differentiation between occupation events. In other cases we might encounter a mixed situation with a palimpsest and a well differentiated occupation event overlapping each other. (For example, we might encounter an undifferentiable concentration of remains from an intensive hominin occupation - a palimpsest, diachronically bound by archaeologically sterile sediments that separate it from a different set of remains representing a more punctuated hominin occupation within which specific activity areas can be ascertained).

ARCHAEOSTRATIGRAPHIC METHOD: A THEORETICAL INTRODUCTION

The databases from successive excavation seasons are the method's working tools. From these, we first extract three-dimensional coordinates, which can differ between seasons, projects, teams, etc. according to the excavation methods they employed (Renfrew and Bahn 1993). Therefore, the first thing we must do is build a new database in which all the data are standardized to a single correlative/grid system, such as would be obtained with a total station (McPherron 2005: 2, 3, 11).

To apply our method, it is necessary to have appropriate recording of the objects during the process of excavation. These have to be recorded three-dimensionally in order to know their relative position. The more accurate the recording, the more precise the stratigraphic contacts will be (McPherron et al. 2005b:248). However, we think one should not become obsessed about millimetric errors. These do not distort the assignment of each object to its corresponding group, because the latter is normally separated by a sterile gap of several centimetres, which will allow the interpretation of the spatial clustering to be correct.

With the described database, we digitally generate a block whose sides or planes are the axes of the stratigraphic plots. Then, the archaeological items are visualized as points in a two-dimensional surface. For any given item, its corresponding point represents its lowermost "z" measurement. For the configuration of the plots it is necessary to make a grid, which we normally base on the main axes of the excavation. Subsequently, the items contained in no less than ten centimetre thickness and no more than twenty (depending on the overall density of materials and the slope of the deposit). This helps avoid the distortion that the slope might add along the "z" axis (distortion that increases with thickness). Once the items are converted to points in the two dimensional grid, we manually or digitally trace lines that delimit archaeological assemblages according to visible gaps between clusters of points.

Once all the visible archaeological layers are identified in the entirety of the plots, we verify longitudinal and transverse readings (i.e., plots) by superimposing both at their contact zone. This crossing of dividing lines is thus a verification process. If both the gaps and the spaces containing archaeological items coincide longitudinally and transversally, it is considered that the divisions established are correct.

For an appropriate reading of the projections it is necessary to keep in mind the following three general criteria that are always present (see Canals 1993).

The slope

Archaeological objects are always deposited on a surface and follow the topography. Whether it be a primary anthropogenic deposit or secondary reworked material in a mudflow, they both adjust to the surface exposed at time of deposition. Hence, our imaginary line should correspond to a real slope. From this basic principle, we can deduce that in the moment that we produce a profile, the materials that we have encoded as points, can unite in an imaginary line that shows the slope at the time of deposition of the archaeological materials.

This means that the spatial distribution of objects in an empty space is a way to reconstruct different paleosurfaces within the site, the evolution in the formation of the deposit, and the slope that it acquired after each new sedimentation episode. In a cave context, it is on these paleosurfaces that the paleontological materials and limestone blocks derived from roof spall were deposited.

We understand paleosurfaces "as the result of a static process formed in situ". "They often make up archaeological deposits, which also form the substrate upon which occupation takes place" (Goldberg and Macphail 2006:42, 46)

Object clusters

We can define as a significant cluster those objects which displays the nearest distances among the points in a profile. They can be considered density clusters, as they represent different density than the surrounding sediment. As a general rule, these clusters are observed on the horizontal plane. They can be found within a stratum or in isolation. In the latter case, one detects large concentrations that decrease gradually. Vertical density changes allow us to draw density maps of objects with clusters that might delimit areas of specific hominin activity.

Gaps

Gaps can be defined as the absence of archaeological objects within a stratum. Gaps allow us to establish sharp boundaries between separate archaeostratigraphic events. For this, the absence of objects must be continuous. A lack of remains is interpreted as the absence of anthropogenic input; and the sedimentary deposit was formed solely by natural processes. Nevertheless, postdepositional processes may filter anthropogenic items into a sterile stratum through the action of various factors, such as the fall of big blocks or trampling (Fiorillo 1989:61-71). A final point regarding gaps, is that the more lines we draw on the profile (separating clusters), the more risk of error there is, because boundaries become less clear, with less clearly visible gaps. However, such a blurry situation is close to the reality of what the original hominin paleosurfaces must have looked like.

RESULTS OF APPLYING THE ARCHAEOSTRATIGRAPHIC METHOD

Previous studies of Lower and Middle Pleistocene sites in homogeneous sedimentary units have been approached from a functional perspective. In other words, the significance of non-anthropogenic space has not been realized and such space has been incorporated within the analytical framework (Vaquero 1997:73). As a result, due to the bulkiness of the available data, interpretations regarding changes in the use of space or activities has not been possible. In other words, there is no control of diachronic change and no conception of the diachronic aspects of hominin occupations. Instead, interpreting a whole archaeological assemblage as belonging to a single geological unit gives us the impression that all the activities carried out by hominins were contemporaneous.

In the past several years, there has been some awareness of the limitations mentioned above. Under this perspective studies have developed methods with which to alleviate the problem. For instance, in the cave of Lazaret, France, an entire monograph has been devoted to a single exceptional archaeological layer that could be isolated right from the time of excavation. Its boundaries were later corroborated by means of an archaeostratigraphic method (De Lumley et al. 2005). Thanks to the possibility of isolating clusters of materials belonging to different hominin occupations, it was possible at Lazaret to determine that the occupation took place during the first half of the month of November and that its main objective was the procurement and storage of meat for winter subsistence (DeLumley et al. 2005:389).

In the Sierra de Atapuerca, this method has also been applied. Specifically in the site of Gran Dolina, in layer TD6, where remains of Homo antecessor were found (Carbonell et al. 1995; Bermúdez de Castro et al. 2004). Through this archaeostratigraphic study, it has been possible to determine that not all of the hominin remains belong to the same archaeological stratum, known as "Aurora stratum". One human remain appeared above that level, demonstrating that cannibalism (Canals et al. 2003:501) was not a single occurrence but rather a recurrent event.

HOW TO APPLY THIS MEHOD TO SITES WITH COMPLEX SEDIMENTATION

When applying this method to the study of sites with a complex genesis, it is necessary to reconstruct some aspects of formation first. By "complex genesis" we understand a group of sedimentary units formed from input of different processes, whether they be multiple entries, or a single entry yielding deposits going in different directions. If the successive flows that form a stratum do not have the same depositional intensity or amount of water or mud, this will produce paleosurfaces formed by beds of sediment of different lengths and thicknesses. Some of these might not occupy the entire space of a cave. In such contexts, there is also the possibility of puddle zones, and other phenomena caused by topographic irregularities. Nevertheless, all of these processes do not affect the homogeneity of the sediment. As long as climatic and atmospheric conditions determining sedimentation remain the same, the deposit will retain its homogeneous lithological features all throughout the space in a cave.

Up to now, we have briefly described some of the possible consequences at a geological level. The implications of the processes described are also important at an archaeological level. If the sedimentary flows that enter a cave are able to carry with them anthropogenic items that were previously laying on a surface at the entrance of the cave, these will be deposited differentially on the surface where the flow is finally deposited. The heaviest materials will be deposited near the entrance area, and the smallest and lightest materials will reach more distal portions of the flow, which will indicate to us the extent of the latter. We believe that the latter case is less probable since high-intensity flows have high probabilities of occupying the whole surface within a given cave.

In case the flows of water or mud are of low intensity and hence unable to drag along archaeopaleontological material, such material will be deposited on the pre-existing paleosurfaces (in situ), and the sediments will simply bury them. If the flows do not occupy the entire surface, some items will be buried while others will be left exposed and ultimately merge with newly deposited ones on the same paleosurface. This exposure results in higher chances of postdepositional disturbance from anthropogenic, biogenic or pedogenic agents.

HOW TO DETECT THESE PHENOMENA IN OUR READING OF THE PROJECTIONS?

From our perspective, object clusters may contribute new data. Frequently we encounter object clusters that abruptly step downwards or upwards (about 6-10 cm steps). If it is unlikely that such steps existed originally on the paleosurface and had material accumulated on top of them (except in the case where the natural topography has shown to yield such step-like formations) (Carbonell et al. 2002:58), then we can interpret these steps as boundaries between two different sediment flows. Sometimes these phenomena are accompanied by abrupt changes in density. If we observe a dense object cluster in an overall empty space, and it has very clear limits in one end that do not fit well with what appears along its sides, we can interpret such an abrupt end as the lateral section of a paleochannel. This can be later corroborated, as the whole section of such a paleochannel will be visible on a perpendicular projection. In the same way, we can identify clearly delimited depressions filled with material. In such pools of material, abrupt density changes would again point to us the possible presence of a vertical chimney or cut and fill sequence. All such incidents deform a surface that is hypothetically continuous, forming object clusters.

Archaeostratigraphic gaps may be horizontal and continuous (Canals 1993), but also vertical, and if they are not attributable to the presence of big blocks, vertical gaps can be interpreted as contact zones between different sediment flows. These flows divide archaeological assemblages not only diachronically but

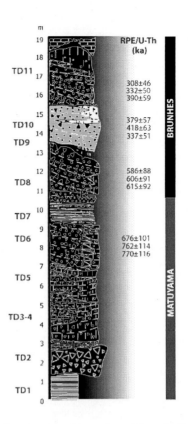

13.1 Stratigraphic section of Gran Dolina.

also possibly synchronically. Their identification allows us to identify precise episodes within the formation of the archaeological record. We often encounter such phenomena (of synchronicity) in the analysis of lithic assemblages. There can be very complete reduction chains represented, and yet few specimens refit. In sum, the archaeological record is more spatially fragmented than it seems.

In the case of Gran Dolina site, in Sierra de Atapuerca, we have focused on layer TD10-1 **(Figure 13.1)**. This stratum exhibits a silty clayey sediment, deposited close to the original cave entrance (Mallol 2004:170-172). From bottom to top, the sediment consists of an increase in centimetre size angular pebbles. Some huge rocks and medium-size blocks alternate with angular pebbles that have fallen to the ground from the cavity roof and its walls (Hoyos y Aguirre 1995:43). It is an opening scenario with two sediment entrance directions: to the north and to the south of the front wall of the excavation (Pérez González et al. 2001:40). Despite these changes TD10-1 does not present clear visible smaller scale sedimentation, so that we can consider it a homogeneous deposit and thus appropriate to apply the archaeostratigraphic method. This deposit was cross sectioned vertically by a railway trench. Up to the present, between 70 cm and one meter of TD10-1 have been excavated in an approximately 90 m² surface area. Here, we have been able to apply the archaeoestratigraphic method, as the use of strictly horizontal stratigraphic divisions was not possible due to the complex genesis of the deposit. We also had to

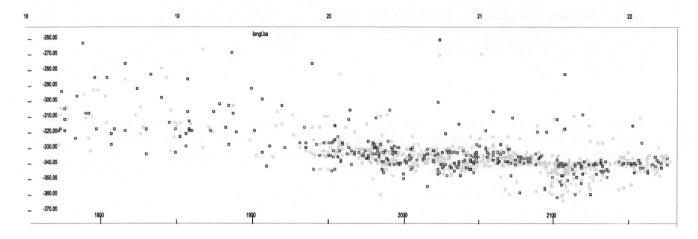

13.2 Backplot showing a high concentration of material and lateral change in density.

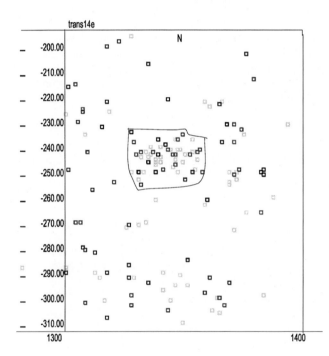

13.3 Backplot showing a possible paleochannel.

modify the foundations of the method and adapt it to the complex sedimentation that this unit comprises. **Figure 13.2** shows a backplot with a high concentration of materials and lateral change of density, while **Figure 13.3** shows a backplot with possible paleochannel.

CONCLUSION

We have reviewed the solution to the problem of archaeological layers within homogeneous sediments that are not separable during the excavation process. It is later, in the lab, where we can carry out computer analysis of the data and apply the archaeostratigraphic method (Canals 1993). In this way, we establish much more precise limits among archaeological units, by individualizing them through the empty spaces or gaps between them. We can then make certain observations regarding modes of deposition, and establish diachronic relationships. Also, when characterizing the different assemblages, we can establish synchronic relationships among the objects of a single archaeological unit and in this way delimit even smaller units into which the record can be subdivided. Although we can never affirm that we are separating cumulative objects belonging to a single occupation, we are effectively grouping together items that were isolated under a single layer of sterile sediment. In the case of units with very abundant materials, we must assume that their accumulation is the result of a palimpsest.

On the other hand, if objects accumulated repeatedly and intermittently, our theoretical framework allows us to appreciate the evolution of the identified paleosurfaces, as well as slope variations that came into place throughout the formation of the deposit.

The method can apply to cave as well as open-air sites, provided that they have stratigraphies with homogeneous sedimentary units (this is the only pre-requisite). If the thickness of the layer exceeds 20 cm we can expect materials from more than a single human occupation event.

The archaeostratigraphic method is valid for primary sites as well as secondary ones, that is, from sites representing accumulations of objects resulting from transport and reworking following their original abandonment by humans. In the former, our method shows the number of times the site was used and abandoned. In the latter, it shows the number of times that the "immediate surroundings of the site" were used and abandoned. Open air sites are especially interesting, as they represent a specific spot chosen for human activity (of the plethora of possibilities), and this might have been due to culturally relevant reason that we might be able to infer.

The method can also contribute to paleontological sites, since we can establish sequences and count the number of occupation and abandonment episodes within a given space. Especially interesting is the application of the method to sites with alternating human and animal occupations, as it may alert us to the establishment of seasonal occupation patterns.

In sum, we can expect a series of accomplishments, listed here from concrete to abstract:
1- We separate items coming from genetically different pathways.
2- We obtain a sense of synchrony and diachrony among different groups of objects.
3- We delimit activities carried out in different episodes.
4- We can help to define the type of occupation that has occurred in each case.
5- We can see if there is diachronic variation in the use of a single space.
6- We would see, ultimately, if the significance or the use of a single space changed over time and in relation with the environment.

We do not seek to present the archaeostratigraphic method as a default, nor do we think that it should be followed to the exclusion of other methods, but rather that it should be combine with other disciplines (e.g., sedimentology, technological, or zooarchaeological analyses), and especially with refitting studies, in order to corroborate or reject hypothetical relationships between adjacent items. It entails a difficult interdisciplinary effort, which will accomplish our ultimate research goal: the characterization of the different kinds of hominin activity that took place during the time in which a particular site was occupied.

REFERENCES

BERMÚDEZ DE CASTRO, J.M. MARTINÓN-TORRES, M. CARBONELL, E. SARMIENTO, S. ROSAS, A. VAN DER MADE, J. LOZANO, M. (2004). The Atapuerca sites and their contribution to the knowledge of human evolution in Europe. *Evolutionary Anthropology* 13:25-41

BUTZER KARL W. (1982). Archaeology as human ecology. Cambridge University press, Cambridge.

CANALS A. (1993). *Méthode et techniques archéo-stratigraphiques pour l'étude des gisements archéologiques en sédiment homogène: application au complexe CIII de la frotte du Lazaret, Nice (Alpes Maritimes)*. Unpublished doctoral dissertation. Museum national d'histoire Naturelle. Paris (France).

CANALS, A. VALLVERDÚ, P. CARBONELL, E. (2003). New archaeo-stratigraphic data for the TD6 level in relation to *Homo antecessor* (Lower Pleistocene) at the Site of Atapuerca, north-central Spain. *Geoarchaeology* vol 18 n°5:481-504.

CARBONELL, E. BERMÚDEZ DE CASTRO, J.M. ARSUAGA,J.L. DÍEZ,J.C. ROSAS,A. CUENCA-BESCOS,G. SALA,R. MOSQUERA,M. RODRÍGUEZ,X.P. (1995). Lower Pleistocene hominids and artifacts from Atapuerca- TD6 (Spain) *Science* 269:826-830.

CARBONELL, E. (2002) *Coordinator: Abric Romaní nivell i Models d'ocupació de curta durada de fa 46.000 anys a la Cinglera del Capelló* (Capellades, Anoia, Barcelona. Universitat Rovira y Virgili.

FIORILLO, A.R. (1989). *An experimental study of trampling: implication for the fossil record in Bone modifications* Bonnichsen and Song eds. Centre for the study of the Americans.

GOLDBERG, P. MACPHAIL, R. (2006). *Practical and theoretical geoarchaeology*. Blackwell Science Ltd., a Blackwell Publishing company U.K.

HOYOS, M. AGUIRRE, E. (1995). El registró paleoclimático pleistoceno en la evolución del karst de Atapuerca (Burgos): el corte de Gran Dolina. *Trabajos de prehistoria* 52 n°2: 31-45

MALLOL, C. (2004). *Micromorphological observations from the archaeological sediments of Ubeidilla (Israel) Dmanisi (Georgia) and Gran Dolina-TD10 for the reconstruction of hominid occupation context*. Unpublished Doctoral thesis. Harvard University.

McPHERRON Shanon J.L. (2005). Artifact orientations and site formation processes from total station proveniences. *Journal of Archaeological Science*.

McPHERRON, J.P. DIBBLE, L. GOLBERG, P. (2005b). Z *Geoarchaeology* vol 20 n° 3:

MEIGNEN, L. (1994). L'analyse de l'organisation spatiale dans les sites du paléolithique moyen: structures évidentes, structures latentes. *Anthropologie Méditerranées* 3:7-23.

MELENDEZ,B. FUSTER, JM. (1978). *Geología* ed. Paraninfo Madrid.

PÉREZ-GONZÁLEZ, A. PRATS, J.M. CARBONELL, E. ALEIXANDRE, T. ORTEGA, A.I. BENITO, A. MARTÍN, M.A. (2001). Geologie de la Sierra de Atapuerca et stratigraphie des remplissages karstiques de Galeria et Dolina (Burgos, Espagne) *L'ánthropologie* 105:27-43.

RENFREW, C. BAHN, P. (1993). *Arqueología: teorías, métodos y práctica* Akal Ed.

VAQUERO, M. (1997). *Tecnología lítica y comportamiento humano. Organización de las actividades técnicas y cambio diacrónico en el Paleolítico Medio del Abric Romaní, Capellades, Barcelona*. Doctoral thesis Universitat Rovira i Virgili.

STRATIGRAPHIE ET CHRONOLOGIE EN ARCHÉOLOGIE PRÉHISTORIQUE

STRATIGRAPHY AND CHRONOLOGY IN PREHISTORIC ARCHAEOLOGY

Françoise DELPECH

Institut de Préhistoire et Géologie du Quaternaire, PACEA UMR 5199 du CNRS, Avenue des Facultés, Bâtiment B18, Université Bordeaux 1, 33405 Talence Cedex, France; delpech@ipgq.u-bordeaux1.fr

Abstract. Much of what we know about prehistoric human life is drawn from materials preserved in the rockshelters and caves in which people once lived. Because these materials are routinely found in stratified deposits, the principals and methods of stratigraphy are fundamental to our efforts to establish spatial and temporal relationships among archaeological entities. Failing to apply, or incorrectly applying, these principles and methods leads to the formulation of invalid hypotheses concerning the chronology of past events. We use examples drawn from five sites in southwestern France to illustrate this fact and to present a series of alternative hypotheses concerning the chronology of these sites derived from a stratigraphic reanalysis of them.

Keywords: Biostratigraphy, relative chronology, prehistoric archaeology, southwestern France

Résumé. Ce que l'on connaît de la vie des hommes préhistoriques est pour une large part issu des restes conservés dans les abris ou entrées de grottes qui leur ont servi d'habitat. Ces vestiges proviennent de strates superposées de façon plus ou moins régulières. La stratigraphie, avec ses principes et ses méthodes, est donc à la base de tous les travaux visant à établir des relations spatiales et temporelles entre des entités archéologiques. La non application ou l'application incorrecte de ces principes et méthodes conduit à établir des hypothèses non valides sur la chronologie d'événements passés. Sur la base de quelques exemples tirés de cinq gisements du sud-ouest de la France, nous justifierons ces assertions et présenterons des hypothèses alternatives issues d'une revisite stratigraphique de ces gisements.

Mots-clés: Biostratigraphie, chronologie relative, archéologie préhistorique, sud-ouest de la France

Au cours de la Préhistoire, les hommes vivant dans les régions karstiques ont habité les entrées de grottes et les abris sous roche. Ils y ont abandonné des vestiges d'origine et d'âge divers. Les processus d'enfouissement et de fossilisation ont conduit (ou non) à leur préservation tout en (ré)organisant leur répartition spatiale dans des strates plus ou moins bien superposées. En un même lieu, ce sont souvent les traces de plusieurs « tranches de vie humaine » qui reposent les unes sur les autres.

Les strates et leur contenu constituent une bonne part des archives des périodes de la Préhistoire. Pour lire ces archives, lors des fouilles et des études qui suivent, on réunit dans un même ensemble les données représentant une même tranche de vie et on classe les ensembles dans un ordre chronologique en suivant le principe selon lequel l'ordre de succession des strates non perturbées est aussi l'ordre de leur dépôt (Hedberg, 1979). L'application de ce principe permet le classement chronologique immédiat des ensembles provenant des strates non perturbées d'un même gisement mais il y faut rechercher d'autres moyens pour classer dans un ordre chronologique les ensembles issus des strates de plusieurs gisements. Or, ceux offerts par l'une des branches de la stratigraphie, la biostratigraphie, apportent des résultats tout à fait performants dans le domaine de la chronologie relative (cf., notamment Mein 1975; Hedberg 1979;

Guérin 1982; Cordy 1982; Delpech 2005). Rappelons que l'unité biostratigraphique (ou biozone) n'est pas la couche qui est individualisée lors de la fouille mais c'est l'ensemble des couches présentant le(s) même(s) caractère(s) paléontologique(s)[1]. La biostratigraphie n'est cependant pas suffisamment prise en compte en Archéologie préhistorique ; on peut encore trouver des exemples de gisements du sud-ouest de la France, livrant des technocomplexes paléolithiques, qui, pour diverses raisons, ne tiennent pas compte de la biostratigraphie et nombreux sont les gisements pour lesquels la chronologie relative des ensembles archéologiques a été établie en considérant des caractères dont l'évolution au cours du temps n'est pas connue. En outre, pour certains gisements, la chronologie relative des ensembles est plus admise que démontrée : l'exploitation des sites préhistoriques en Aquitaine et particulièrement en Périgord remonte à près de deux siècles et il est difficile de nier des idées qui sont depuis longtemps passées dans la mémoire collective et livresque (Delpech 2005, 2007).

Dans cet article sont présentés quelques exemples d'études et révisions biostratigraphiques qui ont conduit à

---
[1] Dans ce travail, ainsi que l'ont préconisé Mein, Guérin et Cordy, trois caractères paléontologiques ont été considérés, à savoir : 1) le degré d'évolution des taxons caractéristiques des lignées-guides, 2) l'association des taxons, 3) la présence de formes caractéristiques.

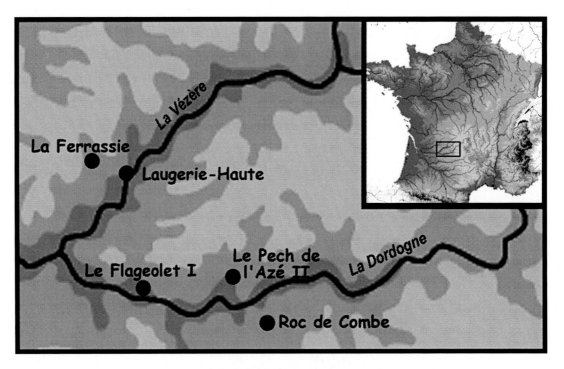

14.1 Situation géographique des gisments cites dans le texte

revoir la chronologie, jusque là classiquement retenue, de certains événements de la Préhistoire. Ces révisions ont été menées à deux niveaux : 1) au sein d'un même gisement où la biostratigraphie (notamment) a pu montrer la présence de strates perturbées ce qui conduit à nier le classement chronologique des ensembles archéologiques qui en proviennent et à faire quelques propositions pour la pertinence des recherches futures ; 2) dans plusieurs gisements où les performances de la biostratigraphie en matière de chronologie relative permettent d'avancer des hypothèses concernant les « relations » entre technocomplexes du Paléolithique supérieur.

Ce travail s'appuie essentiellement sur les données de cinq gisements du sud-ouest de la France ; l'un : le gisement de Roc de Combe, commune de Payrignac, est situé dans le Lot ; les quatre autres sont situés dans le département de la Dordogne : l'abri de La Ferrassie à Savignac de Miremont, l'abri du Flageolet I à Bèzenac, le grand abri de Laugerie-Haute aux Eyzies de Tayac et le gisement du Pech de l'Azé II à Carsac (**Figure 14.1**). Ce travail s'intéresse aux strates, aux couches de terrain, et à leurs relations chronologiques ; peu importe, ici, le contenu archéologique et l'âge « absolu » de chacune. Rappelons toutefois que le gisement du Pech de l'Azé II date du Pléistocène moyen et supérieur et livre des technocomplexes du Paléolithique moyen ; les sites de La Ferrassie et du Roc de Combe datent du Pléistocène supérieur et livrent des technocomplexes du Paléolithique moyen et du Paléolithique supérieur tandis que les sites du Flageolet I et de Laugerie-Haute qui datent, eux-aussi, du Pléistocène supérieur livrent du Paléolithique supérieur.

CHRONOLOGIE RELATIVE DES STRATES D'UN MÊME GISEMENT

« L'ordre de succession des strates non perturbées est l'ordre de leur dépôt » ; c'est le premier principe de la stratigraphie. Dans un gisement donné, l'ordonnancement des strates devrait donc suffire pour les classer dans un ordre chronologique et donc dater de façon relative les éléments que chacune d'entre elles contiennent. Toutefois, certains processus de formation peuvent perturber cet ordonnancement et rendre le principe non applicable. On va voir que, dans certains cas, la biostratigraphie aide à déceler ces anomalies et conduit à définir l'unité (la tranche de terrain) sur laquelle on doit se fonder quant on mène des recherches concernant l'évolution ou le sens de changement au cours du temps d'un phénomène quel qu'il soit. Les exemples sont tirés des trois gisements périgourdins : La Ferrassie, Laugerie-Haute et le Pech de l'Azé II ; ce sont les associations d'ongulés et leurs variations, d'une couche à l'autre, qui ont été analysées.

La Ferrassie (Savignac de Miremont, Dordogne)

Le cas de La Ferrassie a été traité dans un article sous presse (Delpech 2007); on y trouvera tous les détails qui peuvent manquer ici.

C'est dans les années 1980 qu'ont été publiés les résultats des travaux menés sur les vestiges récoltés par Henri Delporte lors des fouilles qu'il avait entreprises en 1968. La stratigraphie synthétique du site avait alors été établie sur la base des seuls travaux de sédimentologie (Laville et Tuffreau 1984). Cette stratigraphie soulevait nombre de

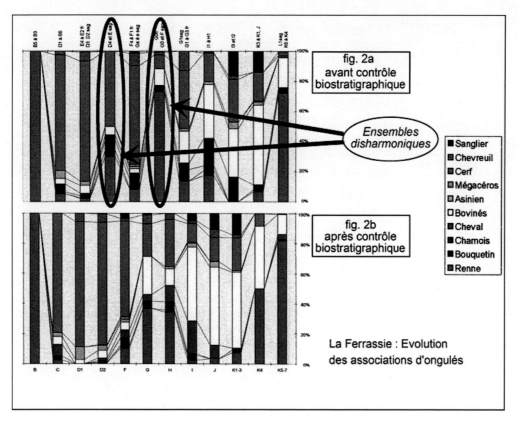

14.2 La Ferrassie. Evlution schématique des associations d'ongulés prenant en compte a) la totalité des données provenant des coupes frontale et sagittale; b) les données de la seule coupe frontale.

problèmes en particulier au physicien chargé des datations (Delibrias 1984) et au paléozoologue chargé de l'étude des grands mammifères. La succession temporelle des associations d'ongulés, établie d'après l'étude des restes fauniques, ne permettait d'envisager aucune évolution logique (Delpech 1984); deux ensembles, en particulier, paraissaient en totale disharmonie : D4 + E_{sag} et G_0 + F_{sag} (**Figure 14.2a**).

Près de 10 ans plus tard, une révision stratigraphique du gisement fut entreprise dans le cadre d'un projet collectif de recherche (PCR) du service régional de l'Archéologie d'Aquitaine qui intéressait plusieurs sites de référence périgourdins (projet mené sur trois ans, dirigé par Jean-Philippe Rigaud puis par Jean-Pierre Texier (Texier 1998)). Le contrôle biostratigraphique effectué a conduit au rejet d'une partie du matériel, celui provenant essentiellement de la « coupe sagittale » dans laquelle l'homogénéité et la succession chronologique des strates n'étaient apparemment pas respectées (Delpech 2007; Texier, 1998). Ce matériel rejeté, la succession des associations d'ongulés paraît beaucoup mieux ordonnée : une logique environnementale peut lui être trouvée (**Figure 14.2b**). Les travaux de géologie menés dans le cadre du PCR sur les processus de formation du gisement de La Ferrassie conduisent aussi à nier les conclusions d'ordre chronologique des travaux antérieurs (Texier 1998). Remarquons, en outre, que les datations radiométriques conduisent à soutenir ce point de vue :

elles montrent en effet que les strates de la coupe sagittale ne sont pas chronologiquement ordonnées tandis que celles de la coupe frontale paraissent mieux classées dans le temps (cf., Delpech et Rigaud 2001 fig. 2; Texier 2001 fig. 11).

Aussi, pour tous travaux s'appuyant sur la chronologie des strates, le matériel provenant de la « coupe sagittale » des fouilles Henri Delporte ne peut être pris en compte. Le gisement de La Ferrassie est un gisement de référence dans beaucoup de domaines. En ce qui concerne la séquence aurignacienne notamment, il est nécessaire d'envisager une révision des ensembles lithiques, après rejet du matériel provenant de ces zones perturbées.

Laugerie-Haute (Les Eyzies, Dordogne)

On considérera ici les niveaux avec solutréen. Les vestiges fauniques examinés (cf., Prat *in* Laville 1964 p. 48; Delpech 1983) sont issus des fouilles François Bordes à Laugerie Haute Est et des fouilles François Bordes et Philip Smith à Laugerie Haute Ouest (Bordes 1958; Smith 1966). Ainsi que Denis et Elie Peyrony (1938) l'avaient indiqué suite aux renseignements fournis par Stehlin, dans tous les niveaux solutréens le Renne domine fortement suivi par le Cheval, beaucoup plus rare (**Tableau 14.1**). Tous les autres taxons (Cerf, Bovinés, Bouquetin, Chamois, Mammouth, Ours, Loup, Renard, Lièvre) ne sont représentés que par de rares vestiges et de

Couches	Solutréen final			Solutréen supérieur				Solutréen moyen				Sol. inf ?	Solutréen inférieur			
	1	2	3	4	5	6	7	8	9	10	11	11A	12a	12b	12c	12d
Bouquetin								1	1			1		2		
Bovinés		1													1	
Cerf		3												1		
Chamois										1						
Renne	3	236	139	194	432	190	198	199	87	146	5	121	576	509	382	155
Cheval		3	1	3	2	2	1	4		16	2	12	12	15	6	4
Mammouth		20			1			1	21	24		11	5	1	2	5

Tableau 14.1 Laugerie-Haute-Ouest, niveaux Solutréens.
Répartition taxonomique par couche archéologique des restes d'ongulés déterminés

façon épisodique: la composition des associations fauniques (essentiellement des associations d'ongulés) varie peu d'un ensemble stratigraphique à l'autre.

Aussi toutes les strates solutréennes de Laugerie-Haute se rangent dans une même biozone. Les datations radiométriques obtenues pour diverses couches solutréennes justifient cette position (Delpech et Rigaud 2001 fig. 4; Roque *et al.* 2001 fig. 6). En conséquence, d'un point de vue chronologique, les « niveaux solutréens » de Laugerie-Haute doivent être traités comme un seul ensemble; on ne peut donc utiliser les données lithiques qui en proviennent pour justifier des successions temporelles comme celle qui était retenue dans les décennies passées, à savoir la succession: « Solutréen inférieur à pointes à face plane » -« Solutréen moyen à feuilles de laurier » - « Solutréen supérieur à feuille de saule » (cf. Bordes 1984:262).

Le Pech de l'Azé II (Carsac, Dordogne)

Le site du Pech de l'Azé II fut découvert en 1949 par François Bordes et Maurice Bourgon, fouillé par ceux-ci en 1950 puis par François Bordes à partir de 1951 (Bordes et Bourgon 1951; Bordes 1954, 1955, 1972). Les restes fauniques ont fait l'objet de travaux divers, chacun intéressant un domaine paléontologique ou archéozoologique particulier (Bordes et Prat 1965; Delpech et Prat 1980; Guadelli 1987; Huertas-Viciana 1989; Prat 1968; Suire 1969). On n'examinera ici que les associations d'ongulés, celles représentées dans les couches inférieures 9 à 6 et celles présentes dans les couches 4 à 2 qui les recouvrent (**Figure 14.3**). Les associations des couches inférieures 9 à 6 sont quasi-identiques ; toutes pourraient être issues d'une même population et appartiennent donc à une même biozone qui se serait formée alors que l'environnement faunique ne présentait pas de variation notoire. Au contraire les couches 4 à 2 livrent des associations fauniques non seulement très différentes mais dont les variations d'une couche à l'autre sont tout à fait désordonnées. Chacune d'elles représente une population initiale particulière ; en outre leur évolution est disharmonique d'une couche à l'autre ce qui suggère l'existence de fortes perturbations stratigraphiques. Les datations ESR ne contredisent pas ces points de vue (Grün, *et al.* 1991). La nouvelle lecture géologique du site les conforte (Texier 2006).

Ainsi pour tous travaux nécessitant des bases de chronostratigraphie, on doit grouper d'une part les données des couches 9 à 6 et d'autre part celles des couches 4 à 2 tout en sachant que l'ensemble 9 à 6, biostratigraphiquement homogène, s'est sans doute formé au cours d'une période stable d'un point de vue environnemental à l'opposé de l'ensemble 4 à 2 qui révèle des environnements différents mais dont on ne peut rétablir la succession.

CORRÉLATIONS STRATIGRAPHIQUES ENTRE SITES ET CHRONOLOGIE RELATIVE DES STRATES

Nous ne donnerons qu'un seul exemple qui porte plus particulièrement sur trois gisements du Périgord : la Ferrassie (fouilles H. Delporte), Le Flageolet I (fouilles J.-Ph. Rigaud) et Roc de Combe (fouilles F. Bordes et J. Labrot). Les corrélations biostratigraphiques proposées concernent des strates formées lors d'une période particulièrement importante dans l'histoire de l'homme d'Europe de l'Ouest et de ses « cultures » à savoir la fin du Paléolithique moyen et le début du Paléolithique supérieur ainsi que la disparition des néandertaliens et l'arrivée de l'homme anatomiquement moderne. Nombre de travaux concernent cette période et ces gisements en particulier. Nos références, cependant, se limiteront à trois articles traitant essentiellement de paléontologie et de stratigraphie dans lesquels le lecteur trouvera une bibliographie beaucoup plus complète (Delpech *et al.* 2000; Grayson et Delpech 2002; Delpech et Texier 2007). Le premier de ces travaux a pour thème le gisement du Flageolet I. Il a notamment conduit à définir huit biozones qui livrent, chacune, des associations d'ongulés caractéristiques ou, plus précisément, parmi lesquelles ont pu être identifiées des unités repères comme celle qui, à La Ferrassie et au Flageolet I, se caractérise par une faune de grands mammifères de forêt tempéré (couches VII du Flageolet I et D2 à F de La Ferrassie) ou encore celle qui, dans les trois gisements, indique un environnement particulièrement rude avec petit Renne et Renard polaire, (couches 7 du Roc de Combe, XI du

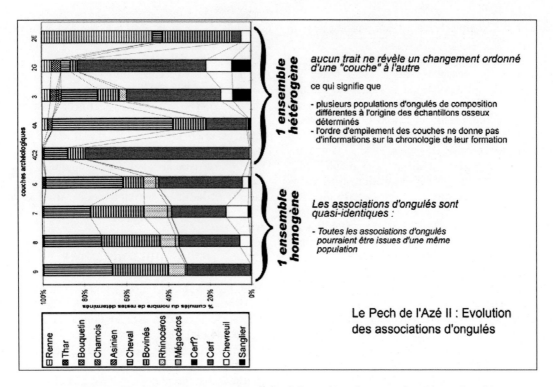

14.3 Le Pech de l'Azé II. Evolution des associationd d'ongulés et interpretations biostratigraphiques

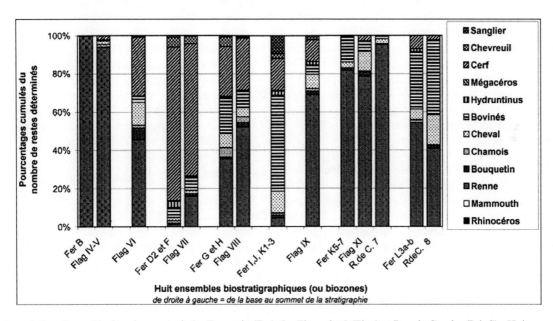

14.4 Association d'ongulés des gisements de La Ferrassie (Fer), Le Flageolet I (Flag) et Roc de Combe (RdeC). Huit ensembles biostratigraphiques ont été identifies. (Les letters ou chiffres qui suivent l'indication du nom du gisement font reference à la couche ou au niveau archéologique).

Flageolet I et K5-7 de La Ferrassie) succédant à des conditions plutôt douces (**Figure 14.4**) (Delpech et al. 2000). Le second de ces travaux ne concerne pas le thème développé aujourd'hui : il s'intéresse au comportement de prédation des hommes du début du paléolithique supérieur[2] (Grayson et Delpech 2002) tandis que le troisième développe de nouveau les questions de stratigraphie en essayant de les appliquer à l'échelle du continent européen (Delpech et Texier 2007).

Au cours de la période correspondant à la formation des huit biozones, trois grands technocomplexes ont été

---

[2] Dans cet article, l'hypothèse selon laquelle les hommes du Paléolithique supérieur auraient pratiqué une chasse plus spécialisée que les hommes du Paléolithique moyen a été testée. En développant une étude quantitative utilisant un échantillon de 133 ensembles d'ongulés représentés dans des sites avec Moustérien, Castelperronien et Aurignacien, il a été notamment montré que cette hypothèse ne pouvait plus être retenue.

identifiés : le Castelperronien, l'Aurignacien et le Gravettien. Les recherches biostratigraphiques ont montré que, dans la région concernée, le Périgord, ces trois technocomplexes se succèdent dans le temps : rien n'a permis de retenir une éventuelle contemporanéité de l'une ou l'autre de ces entités lithiques. Quant à l'étude menée à l'échelle du continent européen qui concerne principalement le technocomplexe Gravettien, elle a permis de situer précisément cette entité tant dans l'espace que dans le temps. Elle a aussi montré qu'il n'était pas pertinent d'établir des corrélations fines sur la base de datations radiocarbones, qu'il s'agisse de dates calibrées ou non, pour des périodes qui se situent près des limites de la méthode. Elle conduit aussi à rejeter certaines hypothèses comme celle selon laquelle, entre 27800 BP et 17000 BP, les hommes auraient évité la Moravie pendant les interstadiaires (cf., Jöris et Weninger 2004). Le degré de résolution temporelle que permet d'atteindre les datations radiométriques est beaucoup trop bas pour accréditer une hypothèse reposant sur ces seules datations.

CONCLUSION

La richesse du Périgord en sites préhistoriques, notamment paléolithiques a attiré depuis bientôt près de deux siècles amateurs et professionnels intéressés par l'histoire de l'homme. Au même titre que la région qui les contient, ces sites pour nombre d'entre eux sont devenus des sites « classiques » ce qui confère un caractère d'« authenticité » et, par là, de « vérité » aux écrits les concernant. De ce fait, les critiques et les discussions de ces travaux ne sont pas systématiques et l'édifice de la connaissance s'est souvent érigé sans que l'on ne doute de ses fondements eux-mêmes.

En Archéologie préhistorique, il a sans doute été longtemps difficile de remettre en cause des idées ancrées dans le temps presque aussi profondément que la science archéologique elle-même mais, dans la majorité des domaines, ce n'est plus le cas aujourd'hui. Nombre de disciplines sont apparues ou se sont développées et contribuent à l'avancée des connaissances. La biostratigraphie, longtemps restée à l'écart, est cependant toujours beaucoup trop en retrait. Rappelons : 1) la biostratigraphie aide à repérer des perturbations dans les sédiments ; 2) elle offre un moyen performant pour l'établissement des chronologies relatives ; 3) elle permet de définir l'unité stratigraphique c'est à dire la tranche de terrain qui contient les éléments chronologiquement homogènes ; 4) la révision biostratigraphique de gisements archéologiques classiques peut conduire ou non à une remise en cause de traits culturels considérés jusque là comme évolutifs.

Quant aux implications archéologiques des révisions biostratigraphiques effectuées, elles sont encore considérées avec trop de circonspection. Dans ce travail cependant, nous avons noté en particulier : 1) la nécessité de rejeter une partie du matériel provenant de la coupe sagittale du gisement de La Ferrassie; 2) la nécessité de regrouper les données provenant de plusieurs entités stratigraphiques traitées jusqu'à aujourd'hui séparément : les données solutréennes du gisement de Laugerie-Haute comme celles des couches 4 à 2 d'une part, 9 à 6 d'autre part, du Pech de l'Azé II forment chacune un seul ensemble. Afin de tester les hypothèses archéologiques retenues à ce jour, ces gisements doivent donc faire l'objet de nouvelles recherches qui prennent en compte les ensembles chronologiques définis biostratigraphiquement ; pour tous travaux intéressant les aspects évolutifs et chronologiques des événements du passé, les travaux de biostratigraphie ont une importance primordiale.

BIBLIOGRAPHIE

BORDES, F. (1954) Les gisements du Pech de l'Azé (Dordogne). *L'Anthropologie*, 58 : 56, p. 401-432.

BORDES, F. (1955) Les gisements du Pech de l'Azé (Dordogne) avec note paléontologique par J. Bouchud. *L'Anthropologie*, 59 : 1-2, p. 1-38.

BORDES, F. (1958) Nouvelles fouilles à Laugerie-Haute-Est. Premiers résultats. *L'Anthropologie*, 62, p. 205-244.

BORDES, F. (1972) *A tale of two caves*. Harper and Row publishers, New York, Evanston, San Francisco, London.

BORDES, F. (1984) Le Paléolithique en Europe. In CNRS éd. : *Leçons sur le Paléolithique* tome.2, Cahiers du Quaternaire, 7.

BORDES, F. et BOURGON, M. (1951) Le gisement du Pech de l'Azé-Nord. Campagnes 1950-1951. Les couches inférieures à Rhinoceros mercki. *Bulletin de la Société Préhistorique Française*, 48 :11-12, p. 520-538.

BORDES, F. et PRAT, F. (1965) Observations sur les faunes du Riss et du Würm I en Dordogne. *L'Anthropologie*, 69 : 1-2, p. 31-46.

CORDY, J.-M. (1982) Biozonation du Quaternaire post villafranchien continental d'Europe occidentale à partir des grands mammifères. *Annales de la Société géologique de Belgique*, Liège, 105, p. 303-314.

DELIBRIAS, G. (1984) La datation par le carbone 14 des ossements de La Ferrassie. In DELPORTE, H, éd. *Le grand abri de La Ferrassie. Fouilles 1968-1973*. Etudes Quaternaires, 7, p. 105-107.

DELPECH, F. (1983) *Les faunes du Paléolithique supérieur dans le Sud-ouest de la France*. Cahiers du Quaternaire, 6, Editions du CNRS : 453 p.

DELPECH, F. (1984) La Ferrassie : Carnivores, Artiodactyles et Périssodactyles. In DELPORTE, H., éd. *Le grand abri de La Ferrassie. Fouilles 1968-1973*. Etudes Quaternaires, 7, p. 61-89.

DELPECH, F. (2005) Utilité et utilisation de la biostratigraphie en archéologie préhistorique. *Bulletin de la Société préhistorique française*, 102, 4, p. 749-755.

DELPECH, F. (2007) Le grand abri de La Ferrassie, source de réflexion sur la biostratigraphie d'un court moment du Pléistocène. In *Ouvrage en hommage à H. DELPORTE*. Editions du CTHS, sous presse.

DELPECH, F.; GRAYSON, D. K.; et RIGAUD, J.-Ph. (2000) Biostratigraphie et paléoenvironnements du début du Würm récent d'après les grands mammifères de l'abri du Flageolet I (Dordogne, France). *Paléo*, 12, p. 97-126, 18 fig., 32 tabl.

DELPECH, F. et PRAT, F. (1980) Les grands mammifères pléistocènes du Sud-ouest de la rance. In CHALINE, J., éd. *Problèmes de stratigraphie quaternaire en France et dans les pays limitrophes*. Supplément au Bulletin de l'Association Française pour l'Etude du Quaternaire, N.S. : 1, p. 268-297.

DELPECH, F. et RIGAUD J.Ph., (2001) Quelques exemples sur l'apport des datations en archéologie préhistorique. In BARRANDON, J.-N. ; GUIBERT, P. ; MICHEL, V., éds. *Datation*. Actes des XXI° Rencontres Internationales d'Archéologie et d'Histoire d'Antibes (2000), APDCA, p. 315-332.

DELPECH, F. et TEXIER, J.-P., (2007) Stratigraphies, chronologie et corrélations : rappel de quelques principes de base. In RIGAUD, J.-Ph., éd. : « *Le Gravettien : entités régionales d'une paléoculture européenne* ». Paléo, , sous presse.

GRAYSON, D. K. et DELPECH, F. (2002) Specialized Early Upper Paleolithic Hunters in Southwestern France? *Journal of Archaeological Science*, 29, p. 1439-1449.

GRÜN, R.; MELLARS, P.; et LAVILLE, H. (1991) ESR chronology of a 100,000-year archaeological sequence at Pech de l'Azé II, France. *Antiquity*, 65, p. 544-551.

GUADELLI, J.-L. (1987) *Contribution à l'étude des zoocoenoses préhistoriques en Aquitaine (Würm ancien et interstade würmien)*. Thèse de Doctorat de l'Université Bordeaux I, n° 148, 548 p.

GUÉRIN, C. (1982) Première biozonation du Pléistocène européen, principal résultat biostratigraphique de l'étude des Rhinocerotidae (Mammalia, Perissodactyla) du Miocène terminal au Pléistocène supérieur d'Europe occidentale, *Géobios*, 15, 4, Lyon, p. 593-598.

HEDBERG, H. (1979) *Guide stratigraphique international. Classification, terminologie et règles de procédure*. Doin éd., Paris, 233 p.

HUERTAS-VICIANA, I. (1989) *Carnivores rissiens du Pech de l'Azé II (couche 7)*. Mémoire de DEA, Institut du Quaternaire, Université Bordeaux I.

JÖRIS O.; WENINGER, B., (2004) Coping with the cold: on the climatic context of the Moravian Mid Upper Palaeolithic. In Jiři A. Svoboda and Lenka Sedláčková eds: The Gravettian along the Danube. *The Dolni Věstonice Studies*, 11, p. 57-70.

LAVILLE, H., (1964) Recherches sédimentologiques sur la paléoclimatologie du Würmien récent en Périgord. *L'Anthropologie* (Paris), 68, p. 1-48 et 219-252

LAVILLE, H et TUFFREAU, A., (1984) Les dépôts du grand abri de La Ferrassie : stratigraphie, signification climatique et chronologie. In DELPORTE, H., éd. *Le grand abri de La Ferrassie. Fouilles 1968-1973*. Etudes Quaternaires, mémoire 7, p. 25-50.

MEIN, P. (1975) *Résultats du groupe de travail des Vertébrés. Report on activity on the RCMNS working groups (1971-1975), IUGS ; regional committee on Mediterranean Neogene stratigraphy*, Bratislava, p. 78-81.

PEYRONY, D. et PEYRONY, E., (1938) *Laugerie-Haute près des Eyzies (Dordogne)*. Archives de l'Institut de Paléontologie Humaine, 19, 84 p.

PRAT, F. (1968) *Recherches sur les Equidés pléistocènes en France*. Thèse de Doctorat d'Etat ès Sciences naturelles, Faculté des Sciences de l'Université de Bordeaux, n° 226, 696 p.

ROQUE, C. ; GUIBERT, P. ; VARTANIAN, E. ; BECHTEL, F. ; SCHVOERER, M. ; OBERLIN, C.; EVIN, J.; MERCIER, N.; VALLADAS, H.; TEXIER, J.-P.; RIGAUD, J.Ph. ; DELPECH, F. ; CLEYET MERLE, J.-J. ; ET TURQ, A., (2001) Une expérience de croisement de datations TL/C14 pour la séquence solutréenne de Laugerie-Haute, Dordogne." *Datation, Actes des XXI° Rencontres Internationales d'Archéologie et d'Histoire d'Antibes, 19-21 Octobre 2000*, éd. APDCA, p.217-232.

SUIRE, C. (1969) *Contribution à l'étude du genre Canis d'après des vestiges recueillis dans quelques gisements pléistocènes du sud-ouest de la France*. Thèse de 3ème cycle. Faculté des Sciences de l'Université de Bordeaux, n° 638, 179 p.

TEXIER, J.-P. (1998) *Rapport final de synthèse du PCR «Révision litho- et biostratigraphique de quelques sites de référence périgourdins"*. Document du SRA Aquitaine.

TEXIER, J.-P. (2001) Sédimentogénèse des sites préhistoriques et représentativité des datations numériques. In BARRANDON, J.-N. ; GUIBERT, P. ; MICHEL, V., éds. *Datation. Actes des XXI° Rencontres Internationales d'Archéologie et d'Histoire d'Antibes (2000)*, APDCA, p. 159-175

TEXIER, J.-P. (2006) – Nouvelle lecture géologique du site paléolithique du Pech de l'Azé II (Dordogne, France). *Paléo*, 18, sous presse.

CAVES AND ROCKSHELTERS OF THE TRIESTE KARST (NORTHEASTERN ITALY) IN LATE PREHISTORY

Manuela MONTAGNARI KOKELJ

Department of Antiquity Sciences "Leonardo Ferrero", University of Trieste, Via Lazzaretto Vecchio, 6 – 34123 Trieste, Italy; e-mail: montagna@units.it

Abstract. The C.R.I.G.A. project (the acronym stands for Informatic Cadastre of Archaeological Caves) is an interdisciplinary study aimed at the analysis of the geo-environmental factors that might have conditioned the use of the Trieste Karst caves in prehistory. The project started in the 1990s and is still in progress. A dedicated geo-referenced database, containing the data of 165 cavities of archaeological interest, is almost completely implemented at present. A GIS analyses carried out recently give interesting results in predictive terms, which, combined with more traditional studies, allow the formulation of new hypotheses. The main steps of the project and the most recent results are presented in this article.

Keywords: Trieste Karst; C.R.I.G.A. project; dedicated database; GIS analyses

Résumé. Le C.R.I.G.A. est un projet interdisciplinaire dont le but est l'analyse des facteurs géo-environnementaux qui auraient conditionnés l'utilisation des cavernes du Karst de la région de Trieste durant la préhistoire. Le projet a débuté dans les années 90, il est encore en progression. A ce jour une base de données dédiée et géo-référencée, contenant les informations de 165 cavités archéologiques, est quasiment achevée; les analyses GIS récemment ont procuré des résultats intéressants en termes prédictifs, qui, associés à d'autres études plus traditionnelles, permettent de formuler de nouvelles hypothèses. Les principales étapes ainsi que les résultats plus récents du projet sont exposés dans cet article.

Mots-clés: Karst de Trieste; Projet C.R.I.G.A.; base de données dédiée; analyses GIS

This paper presents the result of an interdisciplinary collaboration with a number of colleagues of the Department of Geological, Environmental and Marine Sciences of the University of Trieste, in particular Franco Cucchi, Susanna Erti, Alessio Mereu, Chiara Piano, Anna Rossi and Luca Zini. Our collaboration started in the late 1990s and focused on three main projects (all inclusive of database and GIS): ArcheoGIS, the archaeological map of the Isonzo River valley (north of the Trieste Karst); C.R.I.G.A., the cadastre of the archaeological caves of the Trieste Karst; the study of lithic raw materials, from geological formations to archaeological assemblages in Friuli Venezia Giulia and nearby areas. The first ended in 2001 with the publication of the monograph *Gorizia e la valle dell'Isonzo: dalla preistoria al medioevo*; while the others are still in progress.[1]

THE TRIESTE KARST

Before discussing the C.R.I.G.A. project, few parameters about the area under examination are necessary. In terms of the geographical setting and physiographic characters **(Figure 15.1)**, the Trieste Karst occupies the southwestern area of the Classical Karst, a plateau of low rounded hills and low mountains ranging from 100-200 m to 800-900 m above sea level. A few major peaks in this region reach the maximum height, that cover the easternmost part of northern Italy and the southwestern part of Slovenia. The outcropping rocks are chiefly limestones, crossed by two flysch (marl and sandstone) belts, a dozen kilometres wide. The first runs through central Istria (southeast of the Karst) in ESE-WNW direction and borders the Gulf of Trieste as a thin rim, while the second lies parallel to the first, about 20 km to the north, and then bends northwards bordering the Selva di Tarnova (Slovenia).

The limestone area is typically karstic, very poor of water, with common dolines randomly scattered over the plateau, wide rock outcrops and heavy clayish soils *(terra rossa)*.

Due to these characteristics and the general scarcity of water, this territory is, and presumably was in the past, basically unsuitable for cultivation and more appropriate for animal grazing, as a long tradition shows. But this activity would not offer sufficient means of subsistence throughout the year. From the purely geomorphological point of view, the alluvial plains lying to the west (the Friuli Plain) and to the southeast (parts of the Istria Peninsula) represent the ideal complementary areas for an integrated year-round subsistence.

The physiographic characters just outlined and, in particular, the high number of caves present in the limestone belt of the Trieste Karst conditioned not only

---

[1] All the three projects were presented in 2001 at the UISPP XIV Congress in Lièges (see MONTAGNARI KOKELJ *et al.* 2003a, 2003b, 2003c); the volume edited by MONTAGNARI KOKELJ 2001 contains the results of ArcheoGIS; updated info on the more recent developments of the C.R.I.G.A. and lithics projects can be found in MEREU *et al.* 2003, MONTAGNARI KOKELJ *et al.* 2006, D'AMICO *et al.* 2006.

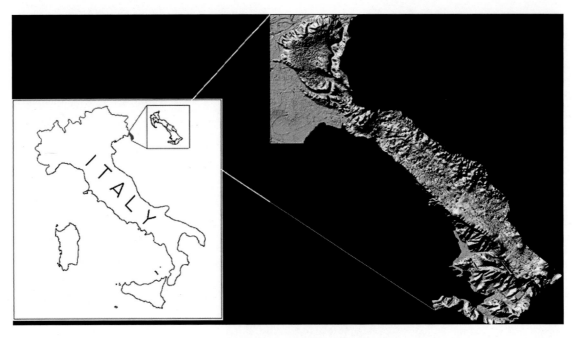

15.1 Location map of the Trieste Karst (Northeastern Italy)

the use of the territory in prehistory, but also the archaeological research since about the 1870s. Both professionals and amateurs were in fact attracted by caves, and the consequences are that, on the one hand, practically no open-air sites are known until the Middle Bronze Age, when massive stone structures, called *castellieri*, were built on top of many hills in the Karst and Istria. On the other hand, the quality of the data from caves is variable, and the number of systematic investigations is relatively low (see below; see also Knavs, this volume).

THE C.R.I.G.A. PROJECT

The origins of the project go back to the early 1990s, when a control of the materials datable to the late prehistory kept in the deposits of the local *Soprintendenza Archeologica* revealed that our knowledge was strongly biased by the discrepancy between the high quantity of artefacts collected in previous cave investigations and the minimum number of published ones. Consequently, systematic revisions were planned to try to fill the gap. Since then the Neolithic-to-Iron-Age materials of about 30 cavities have been studied and almost completely edited, while all the Roman and Medieval artefacts were the object of two dedicated studies.[2]

In the late 1990s the project assumed an interdisciplinary character. A purely archaeological approach was in fact insufficient to try to discover to what extent the physical environment might have influenced the human use of Karst caves in antiquity, which was one of the most interesting questions emerging from the previous analyses. Nevertheless these had concentrated on the materials of excavated caves, which are about 25% of the 165 cavities where archaeological finds are reported according to the data of the Cave Regional Cadastre, based on the historical records of the *Commissione Grotte "E. Boegan"*, *Società Alpina delle Giulie*, the oldest speleological society in the world, active since 1893. A thorough control of these historical records, aimed at assessing the reliability of the information and the possibility of recovering the materials mentioned, as well as systematic field surveys to check the geo-environmental characteristics of the 165 caves and their surroundings, were consequently added to the ongoing critical revisions of existing archaeological collections.

The necessity of storing and managing the increasing quantity of data resulted in the creation of a dedicated geo-referenced database, that later was adapted for GIS applications (see below). The structure of the database is the result of constructive theoretical discussions between archaeologists and geologists (who are also experts in informatics), combined with subsequent tests on the data. Testing has actually continued also after 2001, when the essentials of the database structure were set up,[3] because the different subsections of the project had different rates of implementation that sometimes imply modifications of single parameters. In particular, minor changes of specific fields were introduced with the implementation of the

---

[2] The revisions and the relative publications rest on the principle of reproducing graphically all the typologically identifiable artefacts and giving them a chrono-cultural contextualization. The references for the studies of prehistoric collections already concluded are indicated in MONTAGNARI KOKELJ et al. 2002, note 3, while the publication of the materials from the caves investigated in the first half of the 20th century by Raffaello Battaglia is in preparation; for Roman and Medieval materials see respectively DURIGON 1999 and BIN 2001-02.

[3] For details see MONTAGNARI KOKELJ et al. 2003a.

physiographic part of the database, after field surveys and relative studies had been completed between 2001 and 2003,[4] while modifications of the relationships among tables were necessary for the following GIS analyses (modifications that, obviously, did not alter the contents).

THE DATABASE

The main characteristics of the original database are the following: Microsoft Access 97 was the software chosen to construct the database, because it is easy to enter data and is commonly used by scientific institutions as well as by most people of medium informatic culture. The original structure of the database was quite complex, due to the quantity and variety of elements included. The elements correspond in fact to about 100 fields, grouped into 53 tables coherently organized in 5 main sections. The pivot of the structure is the section called topography, which contains one table with the data essential to locate the cave. In this database the field ID is used as primary key to link this basic table to all the others, by means of both one-to-one and one-to-many relations. The other four sections are dedicated to archaeology, documentation found in archives, physiographic aspects and general notes.

The archaeological area is similar to the topographical one, in so far as there is one invariable table, called investigations, containing the elements that identify a specific excavation, and this is linked with others of variable content and number. The latter include the essential data on each natural layer or group of layers, identified as a chrono-cultural phase: as different phases are usually documented in any investigated deposit, each of them will be described separately in order not to mix up all data.

As to the other main sections of the database, the archives contain tables recording graphic, photographic, audio-visual and written documentation, plus a specific table devoted to archaeological information. The last table is often, especially when the source is unknown, the only evidence of a possible human use of the cave in a usually undeterminable past period.

The physiographic data are recorded in three distinct tables, relative to the physiographic characters of the area outside the cave (morphology, geomorphology, lithology, hydrology, present context, etc.), the inside characters (plan of the cavity, walkable surface, recent collapses, presence of water, etc.) and those of the entrance (exposure, width, height, daylit surface, etc.). In our opinion, some of the parameters introduced into the last two tables might be sensitive indicators to identify the conditions that favoured or limited the human use of a cave.

As far as the database layout is concerned, various forms, corresponding to the subdivisions just indicated, were created to facilitate the implementation and visualization of the records. In particular, the basic topographic data of the cave are always visible on the left side of the main form, while the other sections and subsections will appear on the right side simply by pressing the corresponding button **(Figure 15.2)**.

As indicated above, after the implementation of the physiographic part of the database and in view of the following GIS applications it was necessary to simplify the database structure, that was too heavily weighed towards its original design, the cadastral use, rather than for spatial and statistical analyses. The simplification consisted of eliminating the dictionary tables and the combination of all the one-to-one-related tables. This operation reduced the original number of tables from 53 to 9 **(Figure 15.3)**.

At this point it was decided to evaluate which of three different GIS software systems – *MapInfo*, *GeoMedia* and *ArcGIS*, would be the best for the goals of the C.R.I.G.A. project. On comparative grounds, *ArcGIS* was chosen because it allows an easy, immediate and bidirectional conversion from raster to vector data as well as a friendly use of 3D models with relative conversion into raster data.[5]

GIS ANALYSIS

The GIS analyses carried out so far has necessarily taken into consideration the basic difference in quality between the data present in the Cave Regional Cadastre and those obtained from systematic investigations. Moreover, for the purpose of GIS application, investigations have been considered as systematic also when the stratigraphic data are only partially valid, but the recovered materials can be assigned to a known chrono-cultural phase.[6]

Notwithstanding these limitations, the first results of these studies are highly promising, though it must be underlined that GIS analyses are meant to be heuristic devices that should be used to orient future research developments and desirable new excavations.

The first analyses concerned the physiographic characters of the immediate surroundings of the caves and their

---

[4] Among field activities there was also the control of the exact position of all caves, because only c. 20% of the data present in the Regional Cadastre had been obtained by using PDA with integrated GPS. The degree of reliability of the coordinates registered in the updated cartography is now appreciable as *Roma40* and *ED50* geographical coordinates, *Gauss-Boaga* metric coordinates and those obtained with GPS are differentiated.

[5] The evaluation was made by Chiara Piano, who is also the author of the most promising GIS analyses (PIANO 2003-04).
[6] The chrono-cultural attributions are based partly on direct control of the preserved materials and partly on the literature; a recent overview made by me when I studied the contemporary use of caves and *castellieri* in the Bronze and Iron Ages was used for GIS analyses (see ERTI 2002-03, 72).

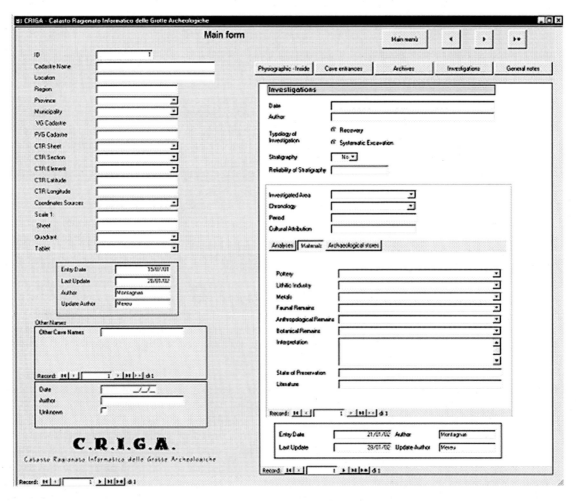

15.2 The C.R.I.G.A. project: example of database forms, showing topographic (left) and archaeological data (right)

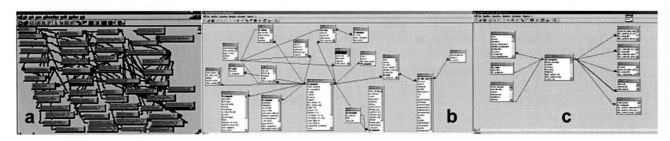

15.3 The C.R.I.G.A. project: simplification of the database structure

inside characteristics.[7] All the archaeological cavities included in the Regional Cadastre were considered first, and then compared with the 45 which have a chrono-cultural attribution ranging from the Mesolithic to the Iron Age. The 5 caves datable to the Lower and Middle Palaeolithic were excluded from the analyses mainly due to their extremely low number, and to the difficulty of reconstructing the past geomorphological aspects, certainly different from the present ones. On the contrary, all the other caves entered the study because, in spite of the changes in sea level and coastline position from the Mesolithic to Neolithic, one can assume that the general conditions of the territory have not changed significantly from the Mesolithic onwards.

The comparison among all the archaeological caves listed in the Regional Cadastre and those datable from the Mesolithic to the Iron Age do not show any observable difference as to the external physiographic characters. The result was expected, in consideration of the location of the high majority of all caves (about 90%) on the Karst plateau, while the remaining are randomly scattered in the valley of the Rosandra drainage, in the southwest part of the area under examination.

The cave entrances and the internal aspects do show some patterns. The percentage of shaft entrances, already

---

[7] These analyses were carried out mostly by Susanna Erti (ERTI 2002-03).

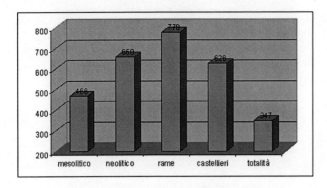

15.4 The C.R.I.G.A. project, GIS analyses: values of the walkable surface inside caves of different chronology versus the total

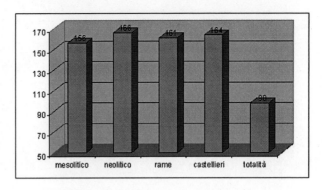

15.5 The C.R.I.G.A. project, GIS analyses: values of the sunlight surface inside caves of different chronology versus the total

relatively low when all caves are considered (30%), decreases even more in the case of those with datable evidence (about 15% of the total). In the latter the entrance is usually level or only slightly inclined, while the degree of steepness is generally higher in the former. Also the inner surface is normally significantly more horizontal in caves where datable episodes of occupation have been recognized. In the same sites two other elements that might have favoured their choice show higher values, i.e., the size of the inner surface where walking is made easier by the absence of fallen rocks and other debris **(Figure 15.4)** and the size of the area better lit during the day **(Figure 15.5)**.

The results of this first series of GIS analyses suggested a test to check the methodology used and its predictive potential. All the 165 archaeological caves included in the Regional Cadastre were re-examined to see which of them would fit the conditions best. First caves with shaft entrances, medium-high steepness at the opening and inclination of the interior higher than 30% were eliminated from the test, together with those with walkable surface less than 100 m^2 and daylit surface less than 30 m^2. The result was a new list of sites that included some of those not yet explored and almost all those already systematically investigated. In the case of the former, only a direct control of the deposits could validate the predictive value of the analysis. In the case of the latter, the exclusion of certain cavities with unquestionable evidence of human occupation is likely to be due to an error in our choice of parameters. One probable error is the acritical use of the indication "shaft entrance" when found in the Cave Regional Cadastre files, without distinguishing true shafts from cave openings located in collapsed dolines. More attention in the future to possible misinterpretations of this kind will certainly improve the quality of both data entry and consequent analyses.

A second set of GIS analyses focused again on the identification of the caves most suitable for occupation, but this time on the basis of their accessibility to primary resources.[8] On the basis of parallel and related traditional archaeological studies, springs (fresh water), the sea (fish, molluscs and salt) and the so called *terre rosse* (i.e., soils formed before the Holocene, usually located at the bottom of dolines, particularly suitable for agriculture and grazing) were identified as the main resources in prehistory. Their value is largely determined by their accessibility, measurable in terms of the time necessary to reach them from a specific site, which is function of both the distance and the geomorphology of the area (steepness and natural barriers).

The methodology chosen to relate all these factors was the multivariate analysis of accessibility maps. The first step was the creation of a DTM based on the regional cartography (orthophotos of the Friuli Venezia Giulia Autonomous Region), realized by using the 3D Analyst extension of *ArcGIS*. Primary resources and natural barriers were positioned on the DTM. Then the slope map was obtained by using the Spatial Analyst extension of the software. It is a raster image with 50 m grid size, subdivided into 10 intervals depending on the steepness degree (< 2%, 5%, 7,5%, 10%, 15%, 20%, 25%, 30%, > 40%), later reclassified into 10 classes (values from 1 to 10) where the first represents the lowest steepness, that is, the best accessibility. The following step was the analysis of the Euclidean distance (operationalized again with the Spatial Analyst extension), that is, the calculation of the straight-line from each cell to the source (= the primary resource) present in the surrounding area. The analysis was applied to springs **(Figure 15.6)**, *terre rosse* and datable archaeological caves (the actual choice of a specific site would in fact confirm its suitability for occupation), and produced the relative equidistance maps, that is, raster images with 50 m grid size, subdivided into regular metric intervals (< 100, 500, 1000, 1500, 2000, 2500, 5000, 7500, 10000, > 10000), again reclassified into 10 classes (values from 1 to 10) where the first represents the minor distance, consequently the most

---

[8] This second set of analyses was carried out by Chiara Piano, who had to limit her study to the NW part of the Trieste Karst, because at that time the revision of the Geological-Technical Map of the Autonomous Region Friuli Venezia Giulia (scale 1:5.000) was incomplete (PIANO 2003-04).

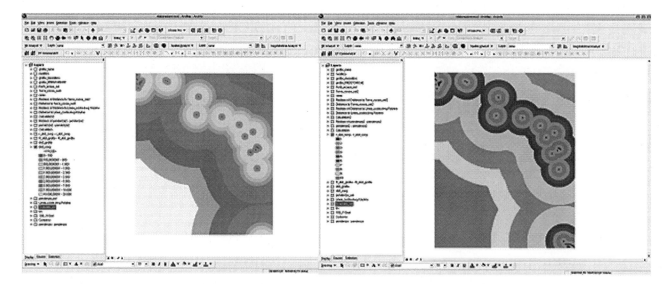

15.6 The C.R.I.G.A. project, GIS analyses: map of the site-resource Euclidean distance calculated for springs (left), and further reclassification (right)

advantageous path. The map of the distance from the sea was created with the same procedure. Unfortunately, the almost complete absence of reliable data on the evolution of the coastline in the period under examination had compelled the use of the present position of the sea as basic parameter. In order to minimize the problem, the first distance interval was enlarged to 500 m, a range that would be sufficient to comprehend the past variations of the coastline.

Eventually, all the maps were assembled and weighted. Each map was assigned a normalized value calculated on the basis of the relative importance of the specific resource. For example, in the case of the Neolithic and Copper Age the values assigned in the raster calculator were respectively 0.3 to steepness, 0.2 to springs, distance from the sea and *terre rosse*, and 0.1 to the presence of caves actually occupied in those periods. The result is a raster map for each period, with the area subdivided into 10 classes representing the accessibility of the analysed resources, that is, the probability, highest in class 1, lowest in class 10, for the area to be chosen by humans in consideration of its potential.

The final step of this set of GIS analyses was the application of the results to a "backward study" of predictive character, carried out on all the 2560 natural caves recorded in the Informatic Cave Regional Cadastre, and aimed at identifying the most suitable for occupation on the basis of both their physiographic characters and accessibility to primary resources.

The first operation, the selection of all the cavities without shaft entrance, reduced the total number to 466. A simple query was used to create a table with the basic data, Regional Cadastre number and name and coordinates, of these 466 caves. The table was later imported in *ArcGIS* to obtain the shape of the relative points. At the same time the raster of the most suitable areas was transformed into polygons (where each polygon corresponds to a class) in order to attribute (by means of spatial query) a value of likely accessibility to the primary resources to all the mapped caves in each period. If only the sites present in the first two classes of accessibility are taken into consideration, the result of the analysis indicates that only 2 and 18 caves enter class 1 and 2 respectively in the Mesolithic, 10 and 19 in the Neolithic, and so on. But if only the caves included in class 1 and 2 in all periods are considered, that is, if we try to identify the optimal locations in a diachronic perspective, their total number drops to 11 **(Figure 15.7)**.

TRADITIONAL STUDIES[9]

These results must be considered preliminary, as input data are still to be augmented and GIS analyses have a predictive value that requires field controls. Nevertheless, the reciprocal influence and integration between such analyses and traditional studies must be acknowledged, and it has been so since the beginning of the interdisciplinary collaboration in the C.R.I.G.A. Project.

At the time when the database was almost definitely structured and its implementation started, studies based on the results of the systematic revisions of old collections were already in progress, and have continued afterwards. The revisions gave a statistical basis to what previously was only an impression. From the Neolithic to the Early Bronze Age in the caves of the Trieste Karst there is a relatively high number of artefacts that are exotic to the local region, when this can be identified, like in the case of the Early-Middle Neolithic Vlaška Group, and would even represent the only evidence in periods of

---

[9] This section summarizes the main results of studies of the last decade: references can be found in one of my last works (MONTAGNARI KOKELJ 2005). For precise quotations in the text see the following notes.

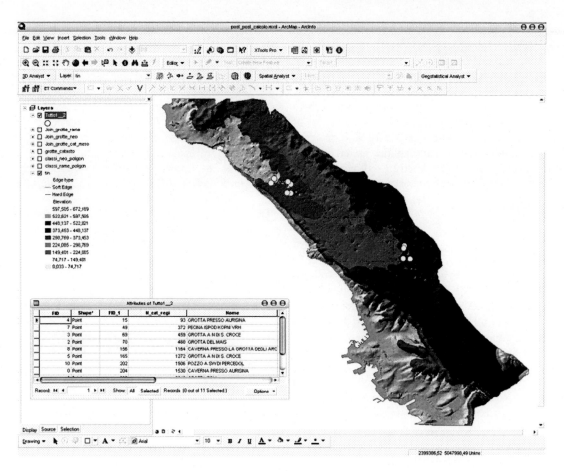

15.7 The C.R.I.G.A. project, GIS analyses, maps of accessibility to primary resources:
the 11 caves included in class 1 and 2 in all periods

highly discontinuous, or almost null occupation of the area.

The recognition of the importance of exotic finds throughout the late prehistory stimulated new archaeometrical analyses: flint, *greenstone* (HP metaophiolites) and clay artefacts were processed, with results of different reliability, because the natural features (i.e., the raw material), of the different classes of artefacts are suitable to characterization in variable degree, and the cultural features (i.e., typology, style) are often of limited support, but, in general, quite promising.

In addition to the specific analytical problems, the difficulty of disentangling the various mechanisms through which an object is introduced into a given area can never be underestimated. Once an object is identified as exotic, besides a "problem ... of equifinality: that a number of different processes can lead to the same resulting pattern" (Scarre 1993:2), one has in fact to consider that its presence may be connected with "1) the movement of objects alone (trade and gift exchange), 2) objects moving with individuals (traders, craftspeople, bride exchange, etc.), 3) objects moving with groups of people (colonization, war and foraging), and 4) the movement of ideas, not objects" (Olausson 1988:18).

The analysis of the context is crucial for this purpose, and must be applied to "the spatial and/or temporal context in which an artefact was lost or discarded ... the circumstances relevant to that loss/discard event ... [i.e.] the specific behavioural context ... the social, economic and political conditions ... - the link, in other words, between material culture and culture generally" as well as the "academic and philosophical context in which items are studied" (Schofield 1995:4). Our research started from the analysis of single sites to move towards the identification of the cultural systems that might have operated in the Karst under specific physiographic and environmental constraints.

Sedimentological and soil micro-morphological analyses carried out recently in some important deposits of the Trieste Karst gave relevant results. They showed in fact that certain areas inside the caves, characterized by layered heaps of ash and charcoal with high quantities of spherulites and phytoliths, were used in the past for stabling animals, in particular sheep and goats. A control of stratigraphic profiles described and reproduced graphically in previous publications and new field analyses confirmed and extended the evidence for this site model, that now is recognizable in most other Italian Karst cavities as well as in Slovene and Croatian ones (Boschian, Montagnari Kokelj 2000; Boschian 2006).

According to the French scholars who first identified, in the early 1970s, such specialized corraling sites in the French Midi, usually located on plateaux and visited seasonally by shepherds moving from complementary open air settlements in lowlands or valleys. Among the cultural correlates of the model of *grottes bergeries* there is usually an observable decrease in the number of artefacts, often associated with a relatively high presence of exotic materials. The revision of the archaeological data of the Karst deposits submitted to sedimentological and soil micro-morphological analyses actually revealed a similar situation, compatible also with the more general picture emerging from the other re-examinations.

What might have interesting implications is the fact that the presence of exotic objects of different origin is considered to be one of the archaeological correlates also of salt production and exchange. Recently I have used this and other indirect indicators[10] to put forward the hypothesis that salt might have been a key element in the exploitation of an area with no other attractive resources, such as the Karst. The hypothesis assumes that shepherds knew about the economic potential of this territory already in the Neolithic and came repeatedly to collect salt formed by solar evaporation in natural ponds along the coast, while systematic production of salt would have started only in the Bronze Age.

CONCLUSION

The results achieved so far by the C.R.I.G.A. project indicate the importance of the combination of studies relative to archaeology, archaeometry and geomatics: the interdisciplinary approach will continue to test the interpretative hypotheses generated by the most recent studies.

Acknowledgements. The project is financially supported by the Italian Ministry of Education, University and Research – MIUR, project PRIN 2004, 2004109799_003.

REFERENCES

BIN, M. (2001-02) – *Materiali medievali da grotta e Progetto C.R.I.G.A.* Trieste [unpublished dissertation]. 98 p.

BOSCHIAN, G. (2006) – Geoarchaeology of Pupičina Cave. In MIRACLE, P.; FORENBAHER, S., eds. – *Prehistoric herders of northern Istria. The archaeology of Pupičina Cave. Volume 1.* Monografije I katalozi, 14. Pula: Arheol. Muz. Istre, p. 123-162.

BOSCHIAN, G.; MONTAGNARI KOKELJ, E. (2000) – Prehistoric shepherds and caves in the Trieste Karst (Northeastern Italy). *Geoarchaeology: An International Journal*. 15: 4, p. 331-371.

D'AMICO, C. et al. (2006) – Greenstone shaft-hole axes of north-eastern Italy, Slovenia, Croatia: archaeometrical data and cultural contexts. In *Materie prime e scambi nella preistoria italiana, XXXIX Riunione Scientifica dell'Istituto Italiano di Preistoria e Protostoria, Firenze, 25-27 novembre 2004.*

DURIGON, M. (1999) – La frequentazione delle grotte carsiche in età romana. *Archeografo Triestino*. Trieste. S. 4, 59, p. 29-157.

ERTI, S. (2002-03) – *Geologia e geomorfologia delle grotte archeologiche del Carso triestino*. Trieste [unpublished dissertation]. 92 p.

MEREU, A. et al. (2003) – Informatics and archaeological caves of the Trieste Karst (north-eastern Italy), in *Enter the past: the E-way into the four dimensions of cultural heritage*, CAA 2003, Workshop 8 – Archäologie und Computer, Vienna (Austria), 8-12 April 2003 [poster]

MONTAGNARI KOKELJ, E., ed. (2001) – *Gorizia e la valle dell'Isonzo: dalla preistoria al medioevo*. Gorizia: Comune di Gorizia. 156 p. 1 CD-ROM.

MONTAGNARI KOKELJ, E. (2005) – Some considerations on salt exploitation at Trieste Karst in prehistory. *Godišnjak*. Sarajevo. 34, p. 47-81.

MONTAGNARI KOKELJ, E. et al. (2002) – La grotta Cotariova nel Carso triestino (Italia nord-orientale):materiali ceramici degli scavi 1950-1970. *Aquileia Nostra*. Trieste. 73, cc. 37-190.

MONTAGNARI KOKELJ, E. et al. (2003a) – GIS and caves: an example from the Trieste Karst (north-eastern Italy). In *Section 1: Théories et méthodes, sessions générales et posters, Actes du 14ème Congrès UISPP, Université de Liège, Belgique, 2-8 septembre 2001.* Liège: Le Secrétariat du Congrès, p. 63-71. (BAR International Series; 1145).

MONTAGNARI KOKELJ, E. et al. (2003b) – ArcheoGIS of the Isonzo Valley (north-eastern Italy) from prehistory to medieval times. In *Section 1: Théories et méthodes, sessions générales et posters, Actes du 14ème Congrès UISPP, Université de Liège, Belgique, 2-8 septembre 2001.* Liège: Le Secrétariat du Congrès, p. 73-77. (BAR International Series; 1145).

MONTAGNARI KOKELJ, E. et al. (2003c) – Surface lithic scatters: interpreting a north-eastern Italian site. In *Section 1: Théories et méthodes, sessions générales et posters, Actes du 14ème Congrès UISPP, Université de Liège, Belgique, 2-8 septembre 2001.* Liège: Le Secrétariat du Congrès, p. 79-85. (BAR International Series; 1145).

MONTAGNARI KOKELJ, E. et al. (2006) – Dallo studio di complessi litici di superficie alla carta geo-litologica del Friuli Venezia Giulia (Italia nord orientale). In *Materie prime e scambi nella preistoria*

---

[10] The other main indicators that I used are the geo-morphological characters of the coastal area, suitable to the formation of evaporation basins; the vicinity of the caves to the salt pans, and their usually long, though discontinuous, but often specialized pastoral use; the ethno-historical evidence of saline. Indirect indicators, though less conclusive than the presence of artefacts directly connected with the production of salt, are normally admitted and used by scholars who study an evanescent substance such as this.

italiana, XXXIX Riunione Scientifica dell'Istituto Italiano di Preistoria e Protostoria, Firenze, 25-27 novembre 2004.

OLAUSSON, D. (1988) – Dots on a map - thoughts about the way archaeologists study prehistoric trade and exchange. In HÅRDTH, B. *et al.*, eds. – *Trade and exchange in prehistory: studies in honour of Berta Stjernquis*. Lunds: Lunds University, p. 15-24 (Acta archaeologica lundensia: 16).

PIANO, C. (2003-04) – *I sistemi informativi territoriali come punto d'incontro tra geologia e archeologia: alcuni progetti e software a confronto*. Trieste [unpublished PhD dissertation]. 118 p.

SCARRE, C. (1993) – Introduction. In SCARRE, C.; HEALY, F., eds. – *Trade and exchange in prehistoric Europe*. Oxford: Oxbow Monograph 33, p. 1-4.

SCHOFIELD, A.J. (1995) – Artefacts mean nothing. In SCHOFIELD, A.J., ed. – *Lithics in context: suggestions for the future direction of lithic studies*. Oxford: Lithic Studies Society Occasional Paper 5, p. 3-8.

THE SECRET CAVE CITY HIDDEN IN THE CLIFFS
(LOVRANSKA DRAGA CANYON, ISTRIA, CROATIA)

Darko KOMŠO* and Martina BLEČIĆ**

*Archaeological Museum of Istria, Pula, Croatia; e-mail: darko.komso@pu.htnet.hr
**Ministry of Culture, Croatia, e-mail: martyna7@yahoo.com

Abstract. Lovranska Canyon is very important archaeological area, with more than 30 caves. Project Oraj with the main goal of understanding the use of caves in the microregion started in 2006. Fourteen caves were recorded and mapped, and one cave was excavated. The most important record is the understanding of complex structural organization of the whole network of caves during the Late Roman period. The preliminary results are excellent, and confirm that projects whose main goal is to study entire network of sites in the region can obtain more structured and sophisticated results then the ones that focus on the single sites.
Keywords: Croatia, cave, burial, site structure, site network

Résumé. Le canyon de Lovranska est une zone archéologique très importante, qui compte plus de 30 grottes. Le projet « Oraj » a débuté en 2006 avec pour objectif de connaître l'utilisation des grottes dans cette microrégion. Quatorze grottes ont été répertoriées et cartographiées; une a été fouillée. Le résultat le plus important a été la compréhension de l'organisation structurale complexe de tout un réseau des grottes pendant l'Antiquité tardive. Les résultats préliminaires sont excellents et confirment qu'un projet ayant pour but la connaissance d'un important réseau de sites dans ladite région peut obtenir des résultats encore plus concrets et affinés que les projets qui se focalisent sur un seul site.

Mots-clés: Croatie, grotte, sépulture, site structuré, réseau de site.

Caves represent an important source of archeological information worldwide. Their importance lies in the fact that they are spatially defined entities, natural shelters, places that attract both people and animals with their favorable conditions, and also provide excellent storehouses of human artifacts and activities. Human communities rarely inhabited only one cave for a long period of time. They rather used a network of caves in a region, moving between them depending on the season of the year, the different resources to be exploited and activities. Caves were used in numerous ways: as living space, livestock sheds, refugee camps, cult places, burial grounds, and so on. Caves are more important for prehistoric sequences although interesting results are achieved also for the Roman and the medieval periods.

The Istria peninsula has 227 recorded caves, 75 of which contained archeological finds. Most of the caves are located in the eastern and northern Istria, on the slopes of the Ćićarija and Učka Mountains (Komšo 2003). In this region, in the Lovranska Draga Canyon, during spring 2006, a multidisciplinary project Oraj was initiated. The main goal of this project was to understand the changes in patterns of cave use in the micro region from prehistory until today.

LOCATION AND HISTORY OF EXCAVATION

Lovranska Draga is situated in the eastern part of Istria coast, on the slopes of the Učka Mountain. It is about 4 km long, 1 km wide, starts at the sea and ends at about 800 m above the sea level. The canyon is closed and compact in its microclimatic conditions, with one side open towards the sea at the Medveja cove. In geological sense, the bedrock formations are dolomites and homogenous gray dolomites that turn into limestone of Cretaceous and Paleogene origin. The cliffs are carbonate in composition and in some areas there are compound overlaps of carbonate and flisch complex (Klepač 1987).

The numerous caves and rock shelters in the area have attracted the attention of many researchers ever since the beginning of the 20th century (Komšo 2003). The first recorded excavation in the Oporovina cave was carried out by Belario de Lengyel in 1929 when he registered the first prehistoric sequence in a cave site on the coast of the Kvarner Bay (Lengyel 1933; Malez 1986). Further research was conducted by Mirko Malez during 1953 (Malez 1960, 1974, 1986), with the discovery of a human burial that he dated to the Mesolithic period. At the end of the 1980's and the beginning of the 1990's, Ranko Starac conducted small-scale test excavations in the caves of Lovranska Draga, Oporovina and Vrtaška Cave. In Oporovina cave he recorded finds from Copper, Bronze and Iron Ages, as well as burials and numerous finds from the Late Roman period. In Vrtaška Cave, finds from the Bronze Age and Late Roman period were recorded (Starac 1987, 1994, 2000).

RESEARCH 2006

This paper presents the preliminary results from this year's initial season of research, lasting from 8th till 15th of April 2006. Previous investigations were concentrated on particular caves of the Lovranska Draga, while the Oraj project focused its attention on the Canyon as a whole.

Parts of northern and southern side of the Canyon have been surveyed. During this stage 14 caves were recorded

16.1 The eastern cliffs of the Lovranska Draga Canyon. 1. Oporovina Cave; 2. Abri Cisterna; 3. Abri Uho

and mapped: Oporovina, Zemunica, Abri Kosača 1 - 6, Abri Uho, Abri Cisterna on the east side (**Figure 16.1**) and Vrtaška caves 1 - 4 on the west side of the Canyon. In most of the caves numerous traces of human activities have been recorded (Oporovina, Abri Kosača 1, 3 - 6, Abri Cisterna, Abri Uho, Zemunica, Vrtaška cave 1). Two small test trenches were opened in the largest cave of the canyon, Oporovina.

OPOROVINA

Oporovina cave is located on the northern cliffs of the Lovranska Draga, at about 270 m above the sea level. The entrance has a triangular shape, and is 15 m wide and 14 m high. The main chamber is composed of two chambers: the main one is 63 m long and the smaller one is 20 m long. In front of the entrance there is a small terrace. The terrace and the entrance-space are especially interesting because of a great number of stairs, and semicircular and circular grooves and recesses, hewn into the bedrock during the 5th and 6th century AD. These grooves used to support a multi-storied wooden construction (6 levels of grooves were recorded), which was placed in the terrace and the entrance part of the cave (**Figure 16.2**). Well protected and illuminated by the sun all day, it was convenient for longer stays of large number of people.

16.2 Oporovina, remains of the grooves in the cave wall

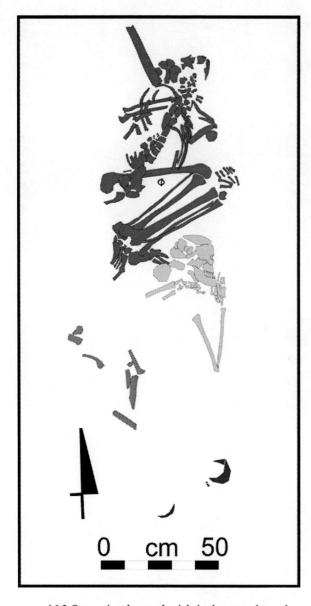

16.3 Oporovina, human burials in the second trench

The excavation was carried out in the main chamber, where two trenches were opened. The first trench was placed in the front part of the chamber close to the cave wall. The second one was placed in the back of the cave underneath the cascade-shaped layers of dripstones, where Malez recorded the human burial. The first trench, with a 4 m^2 surface area, was excavated to a depth of 65 cm, during which excavation various cultural layers with human occupation and artificial pits used for waste disposal were recorded. Pottery, objects made of bone and antler, flint artifacts, as well as metal fragments of attire and jewelry, were found. Together with these objects there were numerous animal bones and an extraordinary abundance of sea shells testifying of diverse nutritional choice and habits of the communities that occupied the cave. After preliminary analysis of the finds, we can confirm that the chamber space has been used during the Late Neolithic and Copper age, as well as Late Roman period. The most interesting finds came from the second trench, where five human burials were uncovered (**Figure 16.3**). Differently from other contemporaneous burials in the Istria peninsula, the bodies where laid down in contorted position and according to the types of costume of the deceased, they can be dated to the 5th and 6th century AD. Malez's mesolithic date for the burial he excavated could therefore be rejected. These burials are probably contemporaneous with the above mentioned modification of the terrace in front of the cave. This is the time of the first penetrations of invading enemies into Istria and Italic peninsula. It is possible that this cave was initially used by the local population to avoid contact with the intruders. Further modification to the cave environment, the building of complex structures, and the presence of burials testify to the long duration of what was perhaps supposed to be just a temporary shelter.

ABRI CISTERNA

Abri Cisterna is located high on a vertical cliff, several dozens meters to the south, at the same altitude as Oporovina cave. A path hewn into the cliff itself leads from Oporovina cave to Abri Cisterna. This rockshelter cotains visible remains of a Late Roman water reservoir, grooves and carved stairs on the bedrock. It is evident that this cave was used as a water reservoir for inhabitants of Oporovina.

ABRI UHO

Abri Uho is situated at about 50 m north from Oporovina cave, 228 m above the sea level. It is fascinating that, except for the hewn grooves and stairs that lead to a higher level, the rockshelter also contains 15 carved crosses (**Figure 16.4**). They appear in different compositions and are of different styles. Therefore it is possible that they are not all of the same origin. Their set up and way of depiction is interesting, most intriguing being the ones with co-directional arms placed in circles. This rockshelter might have been used as a hermitic monastic housing or a place dedicated to meditation and religious practices by the Oporovina inhabitants.

VRTAŠKA CAVE 1

On a very inaccessible place at the opposite side of the Canyon, 256 m above the sea level are Vrtaške Caves. Among the local population they are also known as the Greek caves. Stone constructions with preserved height of about 4 m, probably a fortress wall, as well as grooves and remnants of carved stairs in the bedrock, were still preserved. Former excavations dated the use of cave to the Late Roman period identical to the one in Oporovina cave, as well as to the Bronze Age.

CONCLUSION

Based on preliminary results, we can propose how the pattern of cave use changed over time in Lovranska Draga. During the Late Neolithic and Copper Age only

16.4 Abri Uho, carved and hewn crosses

Oporovina and Abri Kosača 1, located on the eastern side of the Canyon were more or less intensively used as seasonal habitation. Vrtaška cave 1, situated on the western and less accessible side of the Canyon was occasionally used during the Bronze Age. During this period also Oporovina is still in use.

Lovranska Draga underwent dramatic change and gained true significance during the 5th and 6th century, in the Late Roman period, when, threatened by the invading enemies coming from the east, the whole network of caves was transformed into a functional rescue settlement. Every rockshelter has its own special quality that distinguishes it from others, and still they are all connected with the largest and most important Oporovina cave. The Oporovina cave is the central point of this network, where a multistoried wooden construction has been built and where the largest part of the community lived and buried their dead. Numerous rock shelters were linked to the main locale and they were probably used by smaller, more specialized groups. At a close distance to Oporovina cave, a water reservoir was built in the Abri Cisterna, while Abri Uho seems to have been a place where religious practices were the main focus. Vrtaška cave was transformed into a fortress and it was probably the last resort for the population to seek shelter from invading enemies. We can see how the local population erected the secret cave city hidden in the cliffs of Lovranska Draga under such a threat. The caves were then abandoned and went forgotten until recently.

Our first research season returned interesting results and confirmed some of our initial expectations. Also, it has confirmed that projects whose main goal is to study entire network of sites in the region can obtain more structured and sophisticated results then the ones that focus on the single sites.

REFERENCES

KLEPAČ, K. (1987). Geološka podloga Lovrana i okolice. *Liburnijske teme* 6, Opatija, p. 17-24

KOMŠO, D. (2003). Pećine Istre - mjesta življenja od prapovijesti do srednjega vijeka. *Histria Antiqua* 11, p. 41-54

LENGYEL, B. (1933). Scoperta dell'uomo preistorico nelle caverne della riviera liburnica. Atti del 1. Congresso Speleologico Nazionale, Trieste

MALEZ, M. (1960). Pećine Ćićarije i Učke u Istri. *Acta Geologica* II, Zagreb, p. 163-264

MALEZ, M. (1974). Istraživanje paleolitika i mezolitika na području Liburnije. Liburnijske teme 1, Opatija, p. 17-50

MALEZ, M. (1986). Pregled paleolitičkih i mezolitičkih kultura na području Istre. *Izdanja HAD-a* 11, Zagreb, p. 3-47

STARAC, R. (1987). Stanje istraženosti arheoloških lokaliteta na području Lovranštine u vremenskom razdoblju od neolitika do antike. *Liburnijske teme* 6, Opatija, p. 47-52

STARAC, R. (1994). Rezultati novijih arheoloških istraživanja obavljenih na području Lovranštine,

Moščeništine i Brseštine. *Liburnijske teme* 8, Opatija, p. 9-30

STARAC, R. (2000). *Pregled povijesti naseljavanja Učke s gledišta arheološke topografije. Učka - živjeti s planinom i od planine,* Bilten Općine Lovran

REFLECTIONS ON THE TAKARKORI ROCKSHELTER (FEZZAN, LIBYAN SAHARA)

Stefano BIAGETTI[1] and Savino DI LERNIA[1,2]

[1]The Italian-Libyan Archaeological Mission in the Acacus and Messak. *Sapienza*, University of Rome, Italy.

[1,2]Dipartimento di Scienze dell'Antichità. *Sapienza*, University of Rome, Italy. dilernia@uniroma1.it; www.acacus.it; To whom correspondence should be addressed.

Abstract. Rockshelters have hosted ancient Saharans since the Late Pleistocene. After a long hiatus, human occupation of such sites was remarkable during the Holocene, when hunter-gatherers and pastoralists exploited rockshelters in different manners. Italian scholars have been excavating these locations since the 1950s, shedding light on various aspects of the archaeology of the region. Fresh data have emerged from the excavation of the Takarkori rockshelter, located in the Acacus Mts. (southwestern Fezzan, Libya), which is presently under excavation. Shifts in the shelter's use throughout the Holocene will be discussed and weighed against our current knowledge of the whole region, accumulated over decades of research.

Keywords: Sahara, Holocene, Tadrart Acacus, rockshelter.

Résumé. Les abris sous roche ont logé anciens Sahariens du Récent Pléistocène. Après un long hiatus, l'occupation humaine de ces sites s'est agrandi beaucoup pendant l'Holocène, lorsque les chasseurs-cueilleurs et les bergers ont utilisé les abris de façons différentes. Les chercheurs italiens ont creusé ces sites dès années cinquante, et ont mis en lumière beaucoup d'aspects de l'archéologie de la région.
Nouvelles données vont émerger de l'abri de Takarkori situé dans les montagnes de Acacous (Libye sud occidental) que aujourd'hui est en train d'être creusé. Les changements dans l'utilisation de l'abri seront objet de discussion et de comparaison avec les connaissances accumulées pendant les années sur la région.

Mots-clés. Sahara, Holocène, Tadrart Acacous, abris sous roche.

THE SAHARA DURING THE HOLOCENE

Africanists and especially Saharan archaeologists have been concerned with the role of rockshelters in Holocene prehistory for many years. The history of the study of these locations is deeply rooted in the archaeology of the area and began with the first 19th century explorers who discovered rock carvings decorating the walls of shelters. That tradition continued throughout the subsequent 150 years, accurately representing the interests, vogues, methods, aims and development of African archaeology itself.

Saharan rockshelters and caves are unevenly distributed in today's "desert" and are to be found along the valleys crossing the high massifs rising in the central regions. Consequently, these are protected sites, where thick archaeological deposits may be preserved. Generally speaking, Saharan rockshelters may feature palimpsests dating from the Aterian (up to 100,000 yrs bp[1] and roughly corresponding to the European Middle Palaeolithic) to the present day.

The central Saharan massifs are surrounded by vast regions of sand dunes (*erg*) and by "stone deserts," consisting of plateaus of bare rock (*hammada*) or areas of coarse gravel (*reg*). These areas are densely punctuated by several thousand open-air sites, whose date ranges from 10,000 years BP to the historical period. Nevertheless, Acheulean, never yet identified in cave or shelter deposits of the central Sahara, and Aterian work sites can also be found. Conversely to what has been stressed about the shelters, Holocene open-air sites are normally easily recognizable but often severely affected by erosional processes so that only rarely can their stratigraphy be recorded.

In this context, where sheltered and open-air sites formed part of diverse settlement systems in the Holocene, modern research (for recent surveys see Phillipson 2005; Smith 2005) tends to emphasize the role of interaction between the climate and humans throughout this period, shedding light on the timing and impact of environmental changes. It is known that much of the Sahara had very little, if any human settlement during the hyper-arid period comparable to the last Ice Age in the northern hemisphere. Shortly after 12,000 years ago, the monsoon rains and humans returned. Increased rainfall and lower evaporation made larger water resources available. The Holocene was a period of rapid climatic fluctuations, where short and abrupt dry spells occurred several times (e.g., Gasse and Van Campo 1994; Cremaschi 2001; Hassan 2002). These arid episodes were of diverse duration and did not take place simultaneously throughout the Sahara. In any case, the magnitude of some of these events was dramatic and possibly

---

[1] Throughout the text, the notion BP refers to as uncalibrated years before present, based on Libby's half life.

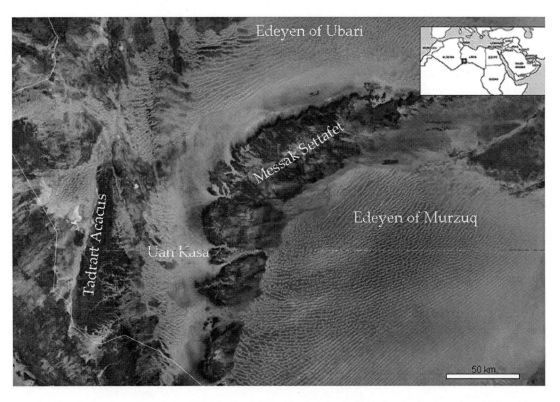

17.1 The area under concession to 'The Italian-Libyan Archaeological Mission in the Acacus and Messak' of the Sapienza, University of Rome.

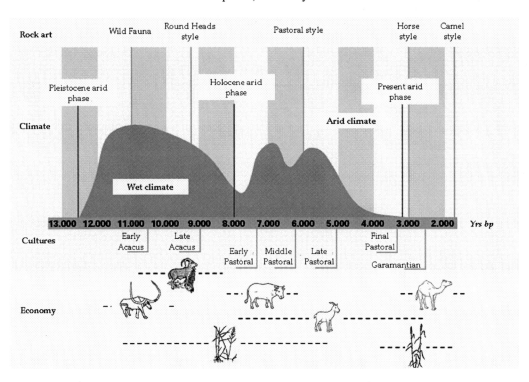

17.2 The Holocene cultural sequence in the Acacus Mountains and its surroundings.

perceptible within a few generations. Notwithstanding the incompleteness of research on this topic, which remains patchy, it is clear that climate oscillations affected human occupation in several areas of the desert, resulting in migratory drifts, back and forth movements, the intermingling of different populations, and the adoption of different lifestyles. The Holocene sequence in the Acacus Mts. (**Figure 17.1**), as outlined by the fifty years of work by the 'The Italian-Libyan Archaeological Mission in the Acacus and Messak', is particularly informative (**Figure 17.2**).

SOUTHWESTERN LIBYA: FIFTY YEARS OF RESEARCH

Since the 1930s, during the Italian colonial occupation of Libya, a series of scientific expeditions have been organized, especially to the southwestern edges of the country, the Fezzan. These missions attempted to outline the historical and archaeological heritage of the region, but did not cover the Tadrart Acacus, unknown until the 1950s.

Fabrizio Mori ideally founded the Italian research endeavors in the Acacus Mountains, starting in 1955. His studies focused on rock art (**Figure 17.3**), the paintings and carvings located in the mountain range, where he also carried out the first excavations, digging trenches in some shelters located in the central Acacus (Mori 1965, 1998). In the 1970s – 80s other shelters such as Ti-n-Torha (Barich 1974), Wadi Athal (Barich and Mori 1970) and Uan Muhuggiag (Barich 1987) underwent archaeological testing, significantly contributing to our knowledge of the Holocene occupation of the region. This testing was a reflection of its time, offering some crucial insights into the prehistory of this area of the Sahara, previously unknown. However, it should be stressed that these studies were basically site-oriented, and that excavations were limited in extent. As a consequence, the archaeological landscape in its entirety was practically unknown until the 1990s, when a landscape approach was adopted. Since 1990, systematic extensive surveys (Cremaschi and di Lernia 1998) and selected excavations (di Lernia 1999; Garcea 2001; di Lernia and Manzi 2002) have made it possible to gather a large set of data about landscape exploitation, settlement patterns, the ^{14}C dating of rock art, genetics and burial customs within Holocene cultural trajectories (di Lernia 2006). Theoretical and

17.3 Painting in the Middle Pastoral style at Uan Amil (central Acacus).

17.4 The Takarkori rockshelter during excavation.

methodological advances, alongside the more accurate knowledge of the archaeology of the region achieved during recent years form the basis for a new view of the use of rockshelters, as seen from a different perspective.

WHY THE TAKARKORI ROCKSHELTER?

The Takarkori project aims to improve the picture drawn by past excavations of shelters, where small trenches stressed the *vertical* rather than the *horizontal* dimension. It thus involves the excavation of large areas and emphasizes the use of electronic devices. Additionally, this research focuses on the site's vicinity, aiming to place it within a wider regional framework.

The Takarkori rockshelter lies around 100 meters above a dried river valley. It is fairly large, especially when compared to other excavated shelters in the area, some 80 m long and about 10 m wide. The site was chosen for the exceptional state of preservation of the archaeological deposit (apparently about 2 meters thick) over a large area (Biagetti et al. 2004).

After three fieldwork campaigns the bottom of the sequence has not yet been reached. However, an area of more than 100 m² has already been exposed (**Figure 17.4**), the largest in the region. Layers dating to more than four thousand years of human occupation have been unearthed. Evidence of Late Acacus (Mesolithic in European terminology) occupation, along with Early, Middle, and Late Pastoral Neolithic remains, has been discovered, dating, according to radiocarbon measurements, from about 8300 to 4000 years BP.

As expected, such a large excavation required special efforts to tackle the loose sandy sediments which systematically characterize the archaeological deposits of the region. The nature of these soils represents a serious challenge for archaeologists, affecting the reading of the layers, whose contours are often blurred. Furthermore, the degree of seasonal and repeated use of the shelter contributes to the formation of complex palimpsests. Finally, syn- and post-depositional processes, in particular animal/human trampling and erosion, have been remarkably effective.

METHODS

A Geographic Informative System (GIS) was adopted in conjunction with the excavation and surveys. At the intrasite level, the stratigraphic units were recorded in 3D, with each being vectorially documented *in situ*. When fully operational, this method makes it possible to perform a volumetric evaluation of each stratum, and to improve our interpretation of the degree and development of depositional and post-depositional processes. This information is complemented by the plotting of artefacts. All archaeological materials were recorded using a total

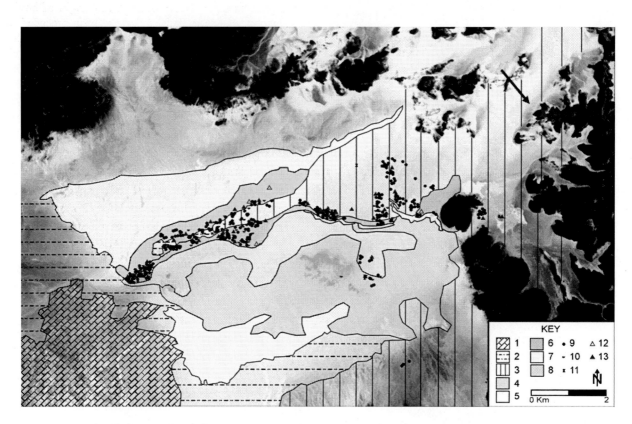

17.5 Open-air sites in the vicinity of Takarkori. Key: 1 - bedrock; 2 - old gravel; 3 - slope deposits; 4 - red sand; 5 - yellow dunes; 6 - shore deposits; 7 - swamp deposits; 8 - basin bottom; 9 - fireplaces; 10 - grinding stones; 11 - tethering stones; 12 - Palaeolithic sites; 13 - Pastoral sites (map by A. Zerboni and M. Cremaschi).

station with the goal of shedding light on latent or hidden which require statistical and distributional processing in order to be identified. One of the most significant aspects of our work lies in the attention devoted to fireplaces and stone structures. Among these apparent structures, the more than 100 excavated fireplaces have revealed variability in position, shape, size, volume and associated materials throughout the ages. Traditional multidisciplinary analyses, palaeobotanical and micromorphological in particular, were performed by specialists whose prolonged on-site assistance and cooperation was of primary importance. Stone structures underwent similar treatment, and when possible, were also recorded in three dimensions. Visual modelling was used to obtain an easy-to-consult, dynamic documentation. This will serve for future applications, acting as a support for dioramas and the reconstruction of particularly interesting segments of the excavation, the latter having been requested by the Libyan authorities.

On a larger scale, systematic surveys were carried out around the site. More than 100 rock art sites were studied and a dozen burial tumuli counted. Finally, some open air sites were discovered, mainly along a dry river system, a few kilometres to the west (**Figure 17.5**).

RESULTS AND FEEDBACK

The ongoing Takarkori excavation is refining our perception of Holocene cultures. In particular, we would like to emphasise our increased awareness of the true complexity of human trajectories during the Holocene, oversimplified in past and recent research, including our own. The earliest occupation of the shelter dates back to the these periods. At Takarkori, Late Acacus hunter-gatherer-fishers of the ninth millennium BP settled for much of their seasonal cycle. Small groups also exploited the shelter to corral wild animals. As suggested for Uan Afuda in the central ranges of the Acacus Mountains. (di Lernia 1999), a few Barbary sheep (*Ammotragus lervia*) were tamed and corralled on site. Faunal analyses have not yet been completed, but Barbary sheep coprolites were identified inside a stone pen where fodder was accumulated (**Figure 17.6**). This unique feature is found alongside a stone hut that currently has few parallels in the region. Moreover, a large hearth with its associated discard area was discovered and is thought to be roughly contemporary with the aforementioned features. Wild plants and seeds were collected and processed, as humans had already done during the previous Early Acacus period, and as they still do today in the region. Of major interest are several concentrations of seeds located close to fireplaces, probably heated to prevent their germination (Mercuri *pers. comm.*). Finally, some human remains were found along the shelter wall. As a whole, the Late Acacus evidence recorded at Takarkori strongly supports the premise of some degree of permanency in settlement and the beginning of the delayed use of vegetable and animal resources.

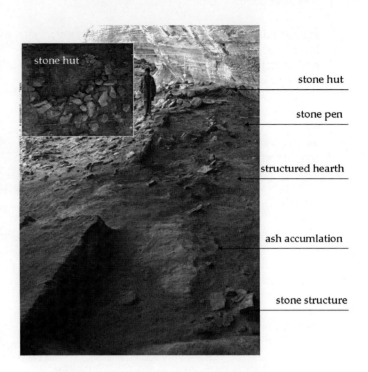

17.6 Late Acacus layers in the Takarkori rockshelter.

The advent of food production, in the Sahara corresponding with pastoralism, as agriculture will emerge only four millennia later, is well attested at Takarkori. The introduction of domesticates to the region, ovicaprids from the Near East and the origin of bovids still uncertain, does not seem to have been preceded by abrupt climatic change, and people continued to exploit the mountain ranges rather than the lowlands (di Lernia 1999). The Early Pastoral stage, dated from the very end of the eighth millennium up to the mid-seventh, had not been clearly detected before. Despite the strong wind erosion affecting the Early Pastoral layers, as well as other excavated sites in the area, several fireplaces were discovered, generally not dissimilar to those of Late Acacus date. However, Takarkori yielded further important information about the less recognized Pastoral Neolithic period. Early Pastoral burials are particularly important, as a few inhumations were unearthed along the shelter wall. This is among the first funerary evidence for this period.

Two burials were found in the Middle Pastoral layers, whose peculiarity lies in the fact that they underwent an impressive process of natural mummification (**Figure 17.7**). After a short arid spell at the end of the seventh millennium BP, during this period Takarkori, like other mountain sites, became a site for seasonal transhumance, where herdsmen spent the dry months with their ovicaprids. The short-term and repeated use of the shelter is attested by the large number of fireplaces of decreasing size, whilst in the lowlands hundreds of more permanent open air sites flourished along the lake shores.

17.7 Naturally mummified burial of the Middle Pastoral period at the Takarkori rockshelter.

	Early Acacus 10.000 – 9.000 uncal. yrs bp	Late Acacus 8.900 – 7.000 uncal. yrs bp	Early Pastoral 7.000 – 6.300 uncal. yrs bp	Middle Pastoral 6.100 – 5.000 uncal. yrs bp	Late Pastoral 5.000 – 3.500 uncal. yrs bp
N. of seasons spent in the shelter	(* *) *not reached yet at Takarkorit*	* * *	* * *	* *	*
Role of the shelter in the settlement pattern	base camp (?)	residential site	main site	transhumance site	transient site
Location of burials (*inside/outside of the shelter*)	(?)	inside	inside	inside	outside

17.8 Uses of rockshelters in the Acacus Mountains throughout the Holocene.

A few preserved layers belonging to the Late Pastoral horizon (5000 – 3500 years BP) shed light on the radical shift that occurred after the onset of arid conditions 5000 years ago. Late Pastoral *strata* consist of hardened dung surfaces with a few very small fireplaces and no structures. This likely indicates that the shelter was mainly used by specialized herders to corral small livestock. The Late Pastoral system involves long-distance movements, probably related to the emergence of the first social stratifications. The fifth millennium is marked by the appearance of burial tumuli throughout the Sahara, including the Takarkori rockshelter.

To conclude, and to illustrate how the Takarkori project is shedding new light on the prehistory of the region, we should note the striking shift through time of the role played by this rockshelter, along with others, in the Acacus Mountains (**Figure 17.8**). During the Late Acacus period, humans exploited the Takarkori rockshelter for living in a semi-permanent manner, along with their tamed prey. The first small groups of pastoralists of the Early Pastoral did the same with their fully domesticated animals. Conversely, at the apex of the pastoral civilization in the 6th millennium, the Takarkori rockshelter became part of a transhumance system, where men, women and animals thus shared the same site. Longer distance nomadism involved Late Pastoral people, who used the shelter as an enclosure for ovicaprids.

From the Late Acacus to the Middle Pastoral periods, rockshelters represented the milieu for different types of burials. The Takarkori evidence is particularly significant as all the adult burials, with seven out of 14 being inhumations of women, together with juveniles or infants

17.9 Burial tumuli in the Acacus Mountains.

(Tafuri et al. 2006). The custom of burying people in rockshelters ended abruptly at the beginning of the Late Pastoral period. Only with the onset of desert conditions, around 5000 years ago, was there a spatial separation between life and death. Whilst living areas and rock art remained inside the rockshelter, death was moved outside, giving origin to the well-known megalithic architecture of the region (**Figure 17.9**). The formation of the desert paralleled this spatial separation between life and death, a separation which still exists among the Kel Tadrart Tuaregs (di Lernia 2006), who never use the shelters for disposing of their dead. The Tuaregs perceive rockshelters as a place for humans and animals, and therefore for life.

REFERENCES

BARICH, B.E. (1974) La serie stratigrafica dell'Uadi Ti-n-Torha (Acacus, Libia). *Origini* VIII: p. 7-157.

BARICH, B.E. (1987), (ed.) *Archaeology and environment in the Libyan Sahara. The excavations in the Tadrart Acacus, 1978-1983*, BAR International Series 368. Oxford.

BARICH, B.E. and F. Mori (1970) Missione paletnologica italiana nel Sahara Libico. Risultati della campagna 1969. *Origini*, IV: p. 79-144.

BIAGETTI, S.; MERIGHI, F. and di LERNIA, S. (2004) Decoding an Early Holocene Saharan stratified site. Ceramic dispersion and site formation processes in the Takarkori rock-shelter, Acacus Mountains, Libya. *Journal of African Archaeology* 2 (1): p. 11-36.

CREMASCHI, M. (2001) Holocene climatic changes in an archaeological landscape: the case study of Wadi Tanezzuft and its drainage basin (SW Fezzan, Libyan Sahara). *Libyan studies* 32: p. 5-28.

CREMASCHI, M. AND di LERNIA, S. (eds.), (1998) *Wadi Teshuinat. Palaeoenvironment and Prehistory in south-western Fezzan (Libyan Sahara)*. CNR (1998), Milano.

di LERNIA, S. (1999) Discussing pastoralism. The case of the Acacus and surroundings (Libyan Sahara). *Sahara* 11: p. 7-20.

di LERNIA, S. (ed.), (1999) *The Uan Afuda Cave (Tadrart Acacus, Libyan Sahara). Hunter-gatherer societies of Central Sahara*. Arid Zone Archaeology Monographs 1. All'Insegna del Giglio, Firenze.

di LERNIA S. (2006). Cultural landscape and local knowledge: a new vision of Saharan archaeology. Libyan Studies 37.

di LERNIA S. and MANZI G. (eds.), (2002) *Sand, Stones, and Bones. The archaeology of death in the Wadi Taenzzuft Valley (5000-2000 BP)*. AZA Monographs 3, All'Insegna del Giglio: Firenze.

GARCEA, E.A.A. (ed.), (2001) *Uan Tabu in the settlement history of Libyan Sahara*. Arid Zone Archaeology Monographs 2. All'Insegna del Giglio, Firenze.

GASSE, F. and VAN CAMPO, E. (1994) Abrupt post-glacial events in West Asia and African monsoon domains. *Earth and Planetary Science Letters* 1256: p. 435–456.

HASSAN, F. (ed.), (2002) – *Droughts, Food and Culture: Ecological Change and Food Security in Africa's Later Prehistory*. Kluwer Academic/Plenum Publishers, New York.

MORI, F. (1965) *Tadrart Acacus. Arte rupestre e culture del Sahara preistorico*. Einaudi, Torino.

MORI, F. (1998) *The Great Civilizations of Ancient Sahara*. L'Erma di Bretsheider, Roma.

PHILLIPSON, D.W. (2005) *African Archaeology*. Cambridge University Press, Cambridge.

SMITH, A.B. (2005) *African Herders. Emergence of Pastoral Traditions*. Altamira Press, New York.

TAFURI, M.A.; BENTLEY, R. A.; MANZI, G. and di LERNIA, S. (2006) – Mobility and kinship in the prehistoric Sahara: Strontium isotope analysis of Holocene human skeletons from the Acacus Mts. (southwestern Libya). *Journal of Anthropological Archaeology* 25. 390–402.

PART III – CURRENT RESEARCH IN AMERICAS

COLLAPSED ROCKSHELTERS IN PATAGONIA

Luis A. BORRERO[1], R. BARBERENA[1], F.M. MARTIN[2] and K. BORRAZZO[1]

[1]DIPA-CONICET, Saavedra 15, Buenos Aires, Argentina; Borrero - laborrero@hotmail.com; Barberena - ramidus28@fibertel.com.ar; Borrazzo – kborrazzo@yahoo.com.ar
[2]CEQUA, Punta Arenas, Chile; Martin-fabs10@hotmail.com

Abstract. Rockshelters may have been used only for short occupation episodes, but their sequences still produced the basic archaeological framework for several American regions. Patagonia is one of those cases. Sequential changes occurred at many rockshelters, including the reduction of sheltered space produced by sedimentation and roof fall. These changes were important determinants of the human selection of those locales. Places that offered protection against wind and rain in the past ended as accumulations of rocks that were no longer attractive for human habitation. A review of cases in Southern Patagonia is presented, as well as some implications for our understanding of the archaeological record.

Keywords: Rockshelters, Geoarchaeology, collapse

Résumé. Les abris pourrait avoir été des endroits utilisés seulement pendant les périodes courtes, mais leurs séquences calmes ont produit le cadre archéologique fondamental pour plusieurs régions américaines. La Patagonie est un de ces cas. Les changements séquentiels sont arrivés à beaucoup des abris, y compris la réduction d'espace abrité produit par la sédimentation et le toit tombe. Ces changements étaient des déterminants importants de la sélection humaine de ces endroits. Les endroits que la protection offerte contre le vent et la pluie dans le passé terminé comme accumulations de rochers qui étaient non plus long attrayant pour l'habitation humaine. Une revue de cas dans Patagonie Méridionale est présentée, de même que quelques implications pour notre compréhension des archéologique.

Mots-clés. Abri sous-roche, geoarchéologie, collapse

"*Rockshelters form, mature, degrade, and ultimately cease to exist*" (Collins 1991:158).

Caves and rockshelters constitute the main repositories of archaeological findings used to reconstruct the history of human exploration and colonization of Patagonia. Their visibility is many times taken for granted. However, many more rockshelters may have existed in the past which are no longer recognizable. In that sense, collapsed rockshelters constitute one of the main elements of what may be called "the hidden component of the archaeological record". The situation with dolomite caves, as in South Africa, is one of the best known examples where roof erosion exposed *breccia* deposits (i.e., Sterkfontein, Brain 1981). Collapsed rockshelters are important within a regional framework of archaeological inquiry, and not to consider them in the initial phases of research may produce systematic error. We stress the necessity to include the potential existence of collapsed rockshelters in research designs. There is certainly a role for geoarchaeological studies in this search, since sometimes there are geological clues that may help in their detection. Geoarchaeology can also contribute to the study of the evolution of these sheltered spaces through stratigraphic and morphological studies. Several sedimentological proxies may inform about different processes producing bedrock degradation, like thermoclastism or chemical weathering (Laville *et al.* 1980; Farrand 2001).

EVOLUTION OF PATAGONIAN ROCKSHELTERS

There is a great lithological variability in Patagonian rockshelters and caves, establishing different initial conditions for their evolution. For instance, all of the shelters in the Pali Aike Volcanic Field, where the caves Las Buitreras, Fell and Pali Aike are located, are of volcanic origin. Therefore, they have a more stable morphological history than shelters on sedimentary rocks, like many from the central Patagonian Plateau or the Baguales range. Volcanic shelters in Patagonia show a comparatively small endogenous sedimentation, whereas sedimentary shelters, like Cerro León 3, a sandstone rockshelter on which we comment further below, have a higher rate of endogenous sedimentation, intimately related to morphological variations.

Collins (1991) makes a distinction between persistent weathering and episodic collapse events. Certainly, the importance of both processes in Patagonia can be demonstrated. Nevertheless, all kind of shelters would not be equally sensitive to these processes. Sedimentary shelters are more prone to persistent chemical weathering, and therefore provide a large part of their fine-grained sedimentary infilling, whereas Patagonian volcanic shelters tend to produce mainly large blocks. Our geoarchaeological work in Pali Aike supports this, since almost all the fine-grained sediments in Cóndor and Orejas de Burro 1 caves have an exogenous source. Mineralogical analyses are being conducted on samples

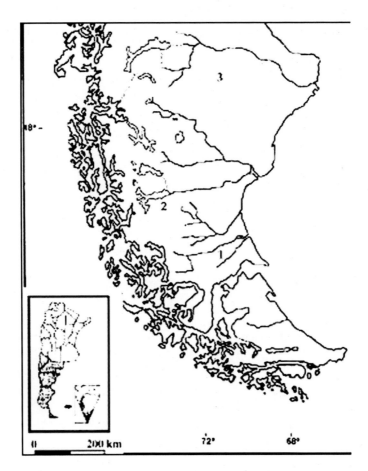

18.1 Location of areas mentioned in the text.

from Cerro León 3 site which will allow us to test this proposition further. Borrero *et al.* (1991) report several collapsed rockshelters at Ultima Esperanza. This is an area where sedimentary shelters predominate, and where rock weathering is important, as reported for Cueva del Mylodon, at which a layer of "rock meal" was formed on the surface (Wellman 1972; Saxon 1979; Favier, Dubois and Borrero 1997).

There are several agents causing episodic collapse events (Farrand 2001). Earthquakes are one of them, and have been proposed to explain the rockfalls at Las Buitreras and Fell Caves (Caviglia *et al.* 1986; Bird 1988). However, several reasons make it difficult to substantiate this interpretation (Borrero and Martin 2006). The presence of similar rockfall events well outside the region where these two caves are located, like Cueva Casa del Minero in the Central Plateau (Paunero *et al.* 2004), or most of the known archaeological sites in Patagonian caves have evidences of roof fall events, some of them very impressive (Borrero *et al.* 1991:104; Prieto 1991:77; Franco and Borrero 2003; Paunero 2003:134; Paunero *et al.* 2004:799; Aschero *et al.* 2005). This suggests that a wider supra-regional explanation is required. Thermoclastism near the end of the Pleistocene may be a mechanism applicable to the whole area.

In other South American regions, like the Pampas, similar processes are well recorded. Caves at Tandilia, in Buenos Aires province, were formed by dissolution of quartzite during wet periods of the Cenozoic. However, preserved sediments only began to accumulate during the Pleistocene-Holocene Transition. Mechanical weathering associated with the colder and drier conditions characteristic of that period were instrumental (Martínez and Osterrieth 2003). Rockfall events are also extensively documented at Arroyo Malo 3 (Diéguez and Neme 2003) and Agua de la Cueva (García *et al.* 1999), in the mountains of Mendoza, central Argentina. However, it is the importance of changes in morphology and collapse in Patagonia what we want to stress. Our research East of Baguales Range at site CL3 is relevant. We began our work at an open air site, located close to a rock formation. The site was found using a shovel test systematic study, and there were no indications of the existence of the site on the surface. However, as we excavated we found evidences of huge rocks that probably constituted part of a stone roof. This is not unusual as it was observed at many other sites. Site CL3 presents a stratigraphic sequence of 1.5 m in depth and archaeological evidence was recovered up to 1.33 m below the surface. We found that the horizontal distribution of cultural material can be used as a proxy for discussing the former existence of a rockshelter, as well as for the reconstruction of the past roofed space or the minimum estimated surface covered by the rockshelter. Effectively, at CL3 we found an abrupt density decrease in lithic artifacts as we proceed away from the sheltered space. These archaeological remains are strictly associated with the rocks, defining an occupation restricted to the protected space under the roof. The site offered important evidence for the distribution of blade assemblages in South Patagonia, a type of evidence that was not available from other sites in Baguales. This case shows the need to analyze the relationship between rockshelters and their respective talus in order to accurately assess the extension and characteristics of the inhabited space.

In conclusion, we observe great variability of shelter evolution in southern Patagonia during the Holocene, partly conditioned by lithological differences. Rockshelters seem to have different historical trajectories in the volcanic landscape of Pali Aike and the sedimentary contexts in Lago Argentino area, Ultima Esperanza and the Central Plateau (**Figure 18.1**).

BEHAVIORAL IMPLICATIONS

We begin our short survey of the implications of change in shelter morphology with a positive example for archaeological research. Probably most of the spalls with paintings recovered at sites like Cueva de las Manos are to be considered the result of thermoclastism, and as such they are part of a process of roof fall and cave disintegration. This process, which negatively affects the preservation of the paintings, provides us with one of the few instances of chronological control over wall paintings. Another important process is the infilling of sediments that cover engravings or paintings (Crivelli Montero et al. 1993).

These morphological changes also have behavioral implications for past human populations. Binford (1972) suggested that the analysis of archaeological sequences in caves and rockshelters implied a constantly changing behavioral habitat. In other words, many sites were changing their size and proportions through time. The process of collapse clearly produces different contexts through time, with some places progressively becoming more open, like dolomitic caves, and other places becoming progressively restricted. Accordingly, human behavior should be responsive to these changes. An example of the latter is dramatically demonstrated at La Martita cave, Santa Cruz, Argentina, which at the time of excavations was a rockshelter almost completely filled with sediment (Aguerre 1987). Of course, most of the prehistoric occupations took place when the cave still provided room for humans, and perhaps it was the lack of habitable space that led to its abandonment. The site was no longer attractive.

Some of these behavioral implications of collapse were taken into account by Patagonian archaeologists since the beginnings of professional archaeology in the area. Junius Bird, for example, trying to organize his knowledge of Patagonian archaeology into a coherent framework, considered the possibility that the seven human skeletons recovered at Cerro Sota, not too far away from Fell Cave, were victims of the rooffall recorded at the latter site (Bird 1988:212). This interpretation was possible until those remains were dated some 6,000 years after the roof fall (Hedges et al. 1992). Beyond the resolution of this particular case study, evidence exists suggesting awareness by past inhabitants of the instability of rockshelters. Some of these morphological changes in rockshelters were of an abrupt nature, and may have posed threats for the safety of the occasional occupants. Occupation of different sectors of caves and rockshelters may have resulted from an assessment of their comparative stability. For example, it was suggested that among the reasons for selecting the rear of Picareiro Cave, Portugal, was "that in the central and front areas there is clear evidence for instability of the ceiling and high walls of the cave" (Bicho et al. 2006:5). Similar examples are found in Patagonia. Human responses involved the rearrangement of inner space or abandonment. For example, at Parque Nacional Perito Moreno, the Cerro Casa de Piedra (CCP) sites experienced morphological changes during the Holocene. At site CCP7, a roof fall dated about 3400 years BP truncated the archaeological sequence (Civalero and Aschero 2003:143). In the same locality, today there are two rockshelters that were probably part of a single larger rockshelter during the middle Holocene (CCP 6 and 9) (Aschero et al. 2005:104). Site CCP9 was filled with rocks from the fall and the access was restricted to what originally was the rear of the site. All these changes were occurring within the lapse of a few human generations at most.

Finally, the deposition of human remains may change the significance of a site within a network of sites, either attracting further occupations (i.e., Baño Nuevo, Mena et al. 2000), or limiting them (Gorecki 1991). In any case, this kind of occupations may have changed the subsequent perception of a locality in the short-term.

There are other consequences of the complete or partial collapse of rockshelters that are worth considering on a methodological level. Several cases of pseudoartifacts resulting from the process of cave collapse are known. Pseudoartifacts from Las Buitreras (Sanguinetti 1976) and Pikimachay (MacNeish et al. 1970) are such classic examples. Oliver (1989:74) remarked that: "A roof-fall block falling on bone in a cave ... is mechanically identical to a hominid hurling a hammerstone onto bone". Fallen rocks may even mimic cut-marks (Oliver 1989:91). Also, the role of fallen rocks in the vertical movement of materials is not well understood, but it is surely important and should be taken into account (Rowlett and Robbins 1982). Rockfall combined with infilling by wind carried sediments, water and gravity may produce an archaeologically ambiguous situation, whose understanding requires explicit geoarchaeological and taphonomic analysis.

CONCLUSIONS

Many implications can be derived from this short assessment of the importance of collapsed rockshelters for archaeological research.

(1) Many existing sites may be invisible today (Borrero *et al.* 1991; Mena and Jackson 1991: 172; Aschero *et al.* 2005:104), and this affects our perception of the archaeological record.

(2) The availability of light and protection from the wind and rain at the microscale of the site were variable through time, and this situation affects other larger scales of analysis as well.

(3) The lack of use of many caves may be related with their instability. Human perception of risk, specifically associated to rock instability, is potentially important for our understanding of the formation of the archaeological record. As mentioned with the example of Picareiro Cave, safety considerations were probably taken into account during the process of selection of places to inhabit.

(4) Both Borrero (1993) and Goñi (1995) highlighted the lack of representativness of the record from caves in the study of past human settlement in Patagonia. The critique to a shelter-centered sampling perspective was important, since it is becoming clearer that the information from open-air sites is changing our perception of the past. The case of collapsed rockshelters adds another dimension of variability, since it modifies our perception of the landscape of shelters available for human use in the past. Thus, we can now add the lack of representativeness of the rockshelters observed today, for the total system of rockshelters available at the time of prehistoric human occupation.

Certainly, there are ways to correct the biases introduced by changes in shelter morphology and collapse. The main step consists of evaluating the representativeness of these samples in relation to the wider systems that produced them (Binford 1978:325-336; Straus 1990; Gorecki 1991; Whalthall 1998). As previously argued (i.e., Collins 1991; Petraglia 1993), geoarchaeology is a very important tool for the detection of currently invisible rockshelters, and for the study of their morphological evolution.

REFERENCES

AGUERRE, A.M. (1987) Investigaciones arqueológicas en el "área de La Martita", Departamento Magallanes, Provincia de Santa Cruz. *Comunicaciones Primeras Jornadas de Arqueología de la Patagonia*, pp. 11-16, Rawson

ASCHERO, C.; C.T. BELLELLI; M.T. CIVALERO; S.L. ESPINOSA; R.A. GOÑI; G.A. GURÁIEB and R.L. MOLINARI (2005) Holocenic Park: arqueología del Parque Nacional Perito Moreno. *Anales de Parques Nacionales* 17: 71-119

BICHO, N.; J. HAWS and B. HOCKETT (2006) Two sides of the same coin –rocks, bones and site function of Picareiro Cave, Central Portugal. In press in: *Journal of Anthropological Archaeology*.

BINFORD, L.R. (1972) Directionality in Archaeological Sequences. *An Archaeological Perspective* (Ed. L.R. BINFORD), pp. 314-326, Seminar Press, New York

BINFORD, L.R. (1978) *Nunamiut Ethnoarchaeology*. New York, Academic Press.

BIRD, J. (1988) *Travels and Archaeology in South Chile*. University of Iowa Press, Iowa

BORRERO, L.A. (1993) Site Formation Processes in Patagonia: Depositional Rates and the Properties of the Archaeological Record. In LANATA J.L, ed. *Explotación de recursos faunísticos en sistemas adaptativos americanos*. pp. 107-121, (Arqueología Contemporánea 4), Edición Especial, Buenos Aires

BORRERO, L.A. and F. M. MARTIN (2006) Revisiting the association between megamammals and humans at Las Buitreras Cave, Santa Cruz, Argentina. MS

BORRERO, L.A.; J.L. LANATA and P. CÁRDENAS (1991) Reestudiando cuevas: nuevas excavaciones en Ultima Esperanza, Magallanes. *Anales del Instituto de la Patagonia* 20: 101-110

BRAIN, C. K. (1981) *The Hunters or the hunted? An introduction to african cave taphonomy*. Chicago, University of Chicago Press.

CAVIGLIA, S.E.; L.A. BORRERO and H.D. YACOBACCIO (1986) Las Buitreras: convivencia del hombre con fauna extinta en Patagonia meridional. In BRYAN, A. L. Ed. *New Evidences for the Pleistocene Peopling of the Americas*. pp. 295-317, Center for the Study of Early Man, Orono

CIVALERO, M.T. and C.A. ASCHERO (2003) Early Occupations at Cerro Casa de Piedra 7, Santa Cruz Province, Argentina. In MIOTTI, L., M. SALEMME and N. FLEGENHEIMER eds. *Where the South Wind Blows*. pp. 141-147, Center for the Study of the First Americans, Texas A&M University

COLLINS, M. (1991) Rockshelters and the Early Archaeological Record in the Americas. In MELTZER, D. and T. DILLEHAY, eds. *The First Americans. Search and Research*. pp. 157-182, CRC Press, Boca Raton

CRIVELLI MONTERO, E.; D. CURZIO and M. SILVEIRA (1993) La estratigrafía de la cueva Traful I (Provincia del Neuquén, República Argentina). *Praehistoria* 1: 9-160.

DIÉGUEZ, D. and G. NEME (2003) Geochronology of the Chorrillo Malo 3 Site and the First Human Occupations of North Patagonia in the Early Holocene. In MIOTTI, L., M. SALEMME and N. FLEGENHEIMER, eds. *Where the South Wind Blows*. pp. 87-92, Center for the Study of the First Americans, Texas A&M University

FARRAND, W. (2001) Archaeological Sediments in Rockshelters and Caves. In STEIN, J. K. and W. R. FARRAND, eds. *Sediments in Archaeological*

Context. pp. 29-66. Salt Lake City, The University of Utah Press.

FAVIER DUBOIS, C. and L. A. BORRERO (1997) Geoarchaeological Perspectives on Late Pleistocene Faunas from Ultima Esperanza Sound, Magallanes, Chile. *Anthropologie* XXXV/2: 207-213.

FRANCO, N.V. and L.A. BORRERO (2003) Chorrillo Malo 2: Initial Peopling of the Upper Santa Cruz Basin, Argentina. *Where the South Wind Blows.* In MIOTTI, L., M. SALEMME and N. FLEGENHEIMER, eds. *Where the South Wind Blows.* pp. 149-152, Center for the Study of the First Americans, Texas A&M University

GARCÍA, A.; M. ZARATE and M. PAEZ (1999) The Pleistocene-Holocene Transition and Human Occupation in the Central Andes of Argentina: Agua de la Cueva Locality. *Quaternary International* 53-54: 43-52

GOÑI, R. (1995) Aleros: uso actual e implicancias arqueológicas. *Cuadernos del Instituto Nacional de Antropología y Pensamiento Latinoamericano* 16: 329-341

GORECKI, P. (1991) Horticulturalists as hunter-gatherers: rock shelter usage in Papua New Guinea. In GAMBLE, C. and W. BOISMIER, eds. *Ethnoarchaeological Approaches to Mobile Campsites. Hunter-Gatherer and Pastoralist Case Studies.* pp. 237-262. Ann Arbor, Ethnoarchaeological Series 1, International Monographs in Prehistory.

HGES, R.; R. HOUSLEY; C. BRONK and G. VAN KLINKEN (1992) Radiocarbon Dates from the Oxford AMS System: Archaeometry Datelist 15. *Archaeometry* 34: 337-357.

LAVILLE, H.; J. P. RIGAUD and J. SACKETT (1980) *Rock Shelters of the Perigord. Geological Stratigraphy and Archaeological Succession.* New York, Academic Press.

MACNEISH, R.; A. NELKEN-TERNER and A. GARCÍA COOK (1970) Second annual report of the Ayacucho Archaeological-Botanical Project. Robert S. Peabody Foundation for Archaeology, Andover.

MARTÍNEZ, G.A. and M.L. OSTERRIETH (2003) The Pleistocene-Holocene Stratigraphic Record from Early Archaeological Sites in Caves and Rockshelters of Eastern Tandilia, Pampean Region, Argentina. In MIOTTI, L., M. SALEMME and N. FLEGENHEIMER, eds. *Where the South Wind Blows.* pp. 63-68, Center for the Study of the First Americans, Texas A&M University

MENA, F. and D. JACKSON (1991) Tecnología y subsistencia en Alero Entrada Baker (Región de Aisén, Chile). *Anales del Instituto de la Patagonia* 20: 169-103

MENA, F.; V. LUCERO; O. REYES; V. TREJO and H. VELÁSQUEZ (2000) Cazadores tempranos y tardíos en la cueva Baño Nuevo-1, margen occidental de la estepa centropatagónica (XI Región de Aisén, Chile). *Anales del Instituto de la Patagonia* (Serie Ciencias Humanas) 28: 173-196.

OLIVER, J.S. (1989) Analogues and Site Context: Bone Damages from Shield Trap Cave (24CB91), Carbon County, Montana, USA. In BONNICHSEN, R. and M. SORG, eds. *Bone Modification.* pp. 73-98, Center for the Study of Early Man, Orono.

PAUNERO, R. (2003) The Cerro Tres Tetas (C3T) Locality in the Central Plateau of Santa Cruz, Argentina. In MIOTTI, L., M. SALEMME and N. FLEGENHEIMER, eds. *Where the South Wind Blows.* pp. 133-140, Center for the Study of the First Americans, Texas A&M University

PAUNERO, R.; M. CUETO; A. FRANK; G. GHIDINI; G. ROSALES and F. SKARBUN (2004) Comunicación sobre campaña arqueológica 2002 en Localidad La María, Santa Cruz. In CIVALERO, M., P. FERNÁNDEZ and A. GURÁIEB, eds. *Contra Viento y Marea. Arqueología de Patagonia*, pp. 808-813. INAPL, Buenos Aires.

PETRAGLIA, M. (1993) The Genesis and Alteration of Archaeological Patterns at the Abri Dufaure: An Upper Paleolithic Rockshelter and Slope Site in Southwestern France. In GOLDBERG, P., D. NASH y M. PETRAGLIA, eds. *Formation Processes in Archaeological Context.* pp. 97-112, Monographs in World Archaeology 17. Madison, Wisconsin Prehistory Press.

PRIETO, A. (1991) Cazadores tempranos y tardíos en la cueva 1 del lago Sofía. *Anales del Instituto de la Patagonia* 20: 75-99

ROWLETT, R.M. and M.C. ROBBINS (1982) Estimating original assemblage content to adjust the post-depositional vertical artifact movement. *World Archaeology* 14: 73-83

SANGUINETTI de BÓRMIDA, A. C. (1976) Excavaciones prehistóricas en la Cueva de "Las Buitreras" (Provincia de Santa Cruz). *Relaciones de la Sociedad Argentina de Antropología* X: 271-292.

SAXON, E.C. (1979) Natural Prehistory: The Archaeology of Fuego-Patagonian Ecology. *Quaternaria* 21: 329-356

STRAUS, L. (1990) Underground Archaeology: Perspectives on Caves and Rockshelters. In SCHIFFER, M.B., ed. *Archaeological Metod and Theory.* pp. 255-304, University of Arizona Press, Tucson.

WALTHALL, J. A. (1998) Rockshelters and hunter-gatherer adaptation to the Pleistocene/Holocene transition. *American Antiquity* 63 (2): 223-238.

WELLMAN, R.W. (1972) Origen de la Cueva del Mylodon en Ultima Esperanza. *Anales del Instituto de la Patagonia* 3: 97-102

CHORRILLO MALO 2 (UPPER SANTA CRUZ BASIN, PATAGONIA, ARGENTINA): NEW DATA ON ITS STRATIGRAPHIC SEQUENCE

CHORRILLO MALO 2 (SUR LE BASSIN DE LA RIVIÈRE SANTA CRUZ, PATAGONIE, ARGENTINA): NOUVELLES DONNÉES SUR SA STRATIGRAPHIE

Nora FRANCO[1], Adriana MEHL[2], and Clara OTAOLA[3]

[1] CONICET (IMHICIHU-DIPA) and Universidad de Buenos Aires, Saavedra 15, 5to, Piso, C.P. 1083, Argentina; nvfranco@yahoo.com
[2] CONICET and Universidad de La Pampa, Argentina: adrianamehl@gmail.com
[3] Universidad de Buenos Aires and IMHICIHU (CONICET), Argentina; claraotaola@yahoo.com.ar

Abstract. Chorrillo Malo 2 is a rockshelter located in the Upper Santa Cruz Basin (Patagonia, Argentina), very close to the Andean range. Its occupation by hunter-gatherers begins about 9,700 years BP with evidence suggesting early human exploration phase of the area. Its stratigraphic sequence extends for about 8,000 years. New chronological data from recent excavations are presented. The new information suggests that the reason for the reoccupation of the rockshelter between about 6,200 and 5,500 years BP is the additional shelter that it provided from the prevailing winds. Data on the morphological change of the rockshelter, along with preliminary sedimentological, lithic and faunal data are presented.

Keywords: Patagonia, hunter-gatherers, rockshelter, human reoccupation

Résumé. Chorrillo Malo 2 est un abri qui se trouve sur le bassin de la rivière Santa Cruz (Patagonie Argentine), très près de la Cordillère des Andes. Son occupation par les chasseurs cueilleurs a commencé ca. 9700 ans A. P. et il y a des évidences qui suggèrent que sa correspond à une phase d'exploration humaine précoce dans la zone. Sa séquence stratigraphique se déroule pendant ca. 8000 ans. On présente de nouvelles données chronologiques provenantes de fouilles récentes. Cette nouvelle information suggère que la raison de la réoccupation de cette abris entre ca. 6.200 et 5.500 ans A.P. c'est le refuge additionnel fourni par les vents dominants. On présente des données sur le changement de la forme de cette abri avec les données sédimentologiques, lithiques et faunistiques préliminaires. On rapport la présence de structures différentes.

Mots-clés: Patagonia, chasseur-ramasseurs, rockshelter, réoccupation humaine

Chorrillo Malo 2 site is a rockshelter located in southern Patagonia, at the south of lakes Argentino and Roca, at about 200 m a.s.l. and at around 20 km from the Andean range (Franco and Borrero 2003) (**Figure 19.1**). Different proxy analysis has shown that the area was available for human occupation at least since 10,000 years BP (Mancini 2002; Mercer and Ager 1983), and that there have been changes in precipitation and temperature during the Holocene. Pollen spectra suggest the existence of dry conditions during the early Holocene, turning slightly more humid after 6,100 years BP (Franco *et al.* 2004). After this period, there is a decline trend in temperature (Mancini 2002), and cold pulses and glacial advances have been proposed for the area (e.g., Aniya 1995; 1996; Malagnino and Strelin 1992; Mercer 1968, 1970). Evidence shows that at least 5,000 years BP an ecotone forest-steppe environment existed in the area of Chorrillo Malo, and that dense forests were present at the Cerro Frías area, less than 12 km away from Chorrillo Malo (Franco *et al.* 2004).

Previous research at this area was carried out under projects directed by Luis Borrero. Tests were executed at different sites (e.g., Belardi et al. 1992; Borrero and Carballo; Marina 1998) and Chorrillo Malo 2 is the one that contained the longest human record in the area. The site has dates between about 9,740 to 1,950 years BP (Franco and Borrero 2003), although the latter date should be carefully evaluated because of the small carbon sample. Today, this rockshelter has a surface of 30 m^2, and is exposed to the prevailing westerly winds (**Figure 19.2**). During the last year, the excavation surface was extended. Preliminary results, new chronological data and an evaluation of the reason for the reoccupation of the rockshelter are presented.

GEOLOGICAL INFORMATION

Geological testing was carried out within and outside the rockshelter. Outside the shelter the deposits are composed by fine sand with low quantities of clasts, which hold a poorly developed edafic horizon with vegetation. Under

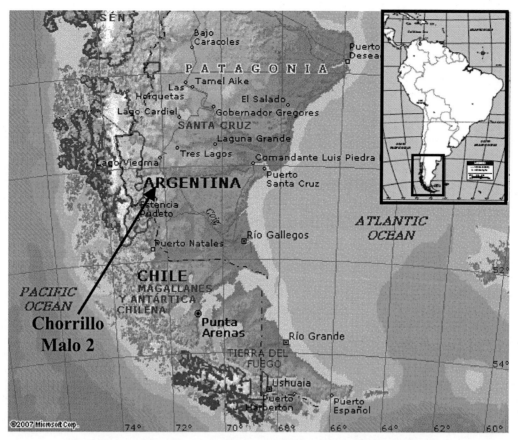

19.1 Location of Chorrillo Malo 2 site (modified from Microsoft Corp. 2007).

19.2 Chorrillo Malo 2 rockshelter at the beginning of the excavation

these, the deposits are composed of an upward fining sequence to the base of the test pit (fine sand to silt) with a progressive increase in the quantity of clasts to the base. Within the rockshelter three units were identified:

Unit I: dung. Three subunits were distinguished: a) dung; b) dung with medium to coarse sands; and c) dung with medium to coarse sands and clasts;

Unit II: composed of clasts and sand. The sand is coarser at the top of the deposit. A subunit with bigger laminar clasts, with medium sizes between 64 and 256 mm was distinguished (subunit IIa);

Unit III: composed of silt and clasts, with cobble and boulder sizes. They are tabular and prolate. Two subunits were distinguished: IIIa (pale brown, Munsell color 10YR 8/4), and III b (white, Munsell color 8/2 10YR)

The results of the analysis show that, although there is some external input, the coarse fraction of the sediment (coarse to medium sand and clasts) was generated by physical and chemical weathering of the surface of the rockshelter, which intensity fluctuated through time. Angular and laminar clasts of variable sizes composed the surface of the rockshelter. Physical alteration of the rock is related to thermal contraction and expansion of minerals, as well as freeze and thaw of water located within cracks in the rock. This fact is consistent with evidences of humidity in the profiles, and at the bottom of the rockshelter, and with the existence of low frequencies of manganese on archaeological bones. The presence of big rocks to the west and specially a big one, with a slope of 24° 10′ to the east, conditioned the early stages of sediment deposition within the shelter.

The geological information suggests that the dynamic processes in the interior of the rockshelter and the morphology of the shelter cavity controlled its evolution. The spalling of rocks, the main factor acting on the formation of the deposits, could be related to weather conditions, which were more intense during deposition of unit IIa. The cavity of the rockshelter was an efficient trap for sediments, and in a way protective from the erosive action of westerly winds.

ARCHAEOLOGICAL INFORMATION

New chronological data were gathered last season and results from the analysis of square 6 are presented below. The oldest deposits excavated during this field season (unit IIIa) were dated to 6,270 ± 45 years BP (Ua-32918), on a guanaco metapodial with lateral blows. At the moment, the big rock located at the west portion of the excavation area covers a part of the rockshelter surface. There were also big rocks to the south. This fact reduced the surface of the rockshelter, at least at the excavated area, but also provided additional protection from the prevailing winds. At the base of this unit, a rock structure with a semi-circular pattern was found. This structure had been covered by sediment and at some subsequent time a charcoal feature was superimposed over it. This implies human reutilization of the same place, probably related to the additional protection provided by the big rocks on the west and south. Evidence suggests that there is not a continuous human occupation at the rockshelter.

Side-scrapers (*sensu* Aschero 1975, 1983) are the most abundant tools. Most of the tools recovered are manufactured on flakes corresponding to the initial stages of manufacture. They are thick, with very little effort expended in their manufacture. They correspond to an expedient technology (Binford 1979) made on immediately available raw materials (dacite and basalt) and probably hand held. A cobble ground by use was also recovered in this deposit. The existence of a hammerstone and of core rejuvenation flakes suggests that cores were produced or used at the site.

There is good preservation of bones, since weathering 0 and 1 are the most frequent stages (Behrensmeyer 1978). *Lama guanicoe* (guanaco) was the main food resource. The most commonly occurring elements are from the appendicular skeleton. At least thirty percent of the guanaco bones of sizes of more than 4 cm were processed by humans. There are also tools on guanaco bone (such as pounders, *sensu* Hajduk and Lezcano 2005).

A big rock fell down some time after 5,395 ± 40 years BP (Ua-32920). The date was obtained on a guanaco tibia with butchering marks. Above it, the lowest part of unit IIa was dated at 2,860 ± 35 years B. P. (Ua-32917) on a guanaco humerus with cut marks. Possible rock structures were found. Expedient and curated artifacts were also recovered. Discoidal cores (*sensu* Aschero 1975, 1983) are frequent, although polyhedral ones and cores with isolated blows were also recovered. Side-scrapers and knives are the most abundant tools. Some of them are manufactured on *levallois* cores (*sensu* Nami 1992, 1997). Artifacts ground by use were also recovered. Ochre is abundant, as mineral pigment as well as on the surface of some tools, including ground and flintknapped ones. There is evidence of utilization of rocks probably from the Baguales range (sedimentary chalcedony and opal), located about 20 km from the site (Franco and Aragón 2002, 2004).

Faunal remains are mostly *Lama guanicoe*. The apendicular skeleton continues to be the most represented. More than thirty three percent of the specimens larger than 4 cm have anthropogenic marks. There are also impact marks on guanaco bones. The predominant weathering stages are 1 to 2. This deposit has abundant evidence of the action of roots, carnivores and rodents on the bones.

In the upper deposits there is evidence of regular utilization of raw materials coming from about 20 km for end-scrapers, mostly exhausted and fragmented. The evidence is consistent with the postulated existence of economic strategies for those raw materials probably coming from the Baguales range (Franco 2002), as can be seen in the presence of bipolar cores, in some cases made on tools, as well as in the discarded tool angles. Flakes made on pigments were also recovered. Side-scrapers and knives are abundant, although end-scrapers were also recovered. Most of the tools are fragmented. The cores and core rejuvenation flakes suggest that at least some of them were flintknapped at the site. A preform of a biface, probably a projectile point, and biface thinning flakes were also recovered in this deposit. Guanaco continues to be the main food resource and the apendicular skeleton is most frequently represented. Only 9% of bones larger than 4 cm have anthropogenic marks. Although the dominant weathering stages are 1 to 2, there are also bones in stages 3 and 4. We have to mention that, because this deposit has the highest weathering stages in the sample, butchering evidence may be underrepresented.

The most recent date obtained is 2,525 ± 35 years BP (Ua-32919) on guanaco metapodial with cultural marks.

CONCLUSIONS

In summary, evidence shows that, between about 6,270 and 5,400 years B. P., Chorrillo Malo 2 rockshelter provided shelter from the wind due to the presence of big rocks on the south and west sides. The better preservation of bones (the lowest weathering stages of the whole sequence) is consistent with this fact. Sedimentary data suggest some time between the building of the rock and charcoal structures mentioned above. Silt deposition between them suggests that there was no continuity in the human occupation of the rockshelter.

Between about 2,800 and 2,400 years BP, there is abundant evidence of use of Chorrillo Malo 2 rockshelter. There are higher depositional rates of artifacts, evidence of curated technology and an increase in the utilization of pigments. Some possible structures were also recovered.

There is also evidence for the utilization of the same rockshelter after that time, although it could not be dated. In general, we can say that changes in the shape of the rockshelter affected the way in which humans used it. The additional protection Chorrillo Mollo 2 provided between about 6,270 and 5,400 years BP was probably the reason why hunter-gatherers repeatedly chose this space at that time. However, additional excavations (inside and outside the rockshelter) as well as more analysis are needed to understand formation processes, changes in the shape of the rockshelter and in the way humans used it. This will help us understand its role within the area.

Acknowledgements. This research was supported by the Heinz Foundation. Additional support was obtained from University of Buenos Aires. We wish to thank Estancia Anita and especially señores Enrique Viel and Mayo Arredondo. We also want to thank Laura Miotti and Marcel Kornfeld for their invitation to participate in the Symposium, Luis Borrero for his useful comments, Laura Otaola, Bill Ross, the Asociación de Guías de Calafate and Andrés Gader. Important local support was provided by Martin Gray and Gerardo Povazsan. We also want to thank Virginia Mancini and all the members of the field team: Gabriela Armentano, Marcela Arredondo, Marcelo Cardillo, Valeria Ucedo, Florencia Ferrari, Erico Gaal, Ana Guarido, Brenda Gillio, Juan Maryañski, Mariana Ocampo, Melina Bednarz and Rodrigo Vecchi.

REFERENCES

ANIYA, M. (1996) Holocene variations of Ameghino Glacier, southern Patagonia. *The Holocene* 6:247-252.

ASCHERO, C. A. *Ensayo para una clasificaciòn morfológica de artefactos lìticos aplicada a estudios tipológicos comparativos. (Manuscrito)*. 1975. Report presented to the CONICET.

ASCHERO, C. A. *Ensayo para una clasificaciòn morfológica de artefactos lìticos aplicada a estudios tipológicos comparativos. (Manuscrito) 1983*. Report presented to the CONICET. Revision 1983.

BEHRENSMEYER, A. K. (1978) Taphonomic and ecologic information from bone weathering. *Paleobiology* 4 (2):150-162

BELARDI, J. B.; BORRERO, L. A.; CAMPAN, P. ; CARBALLO MARINA, F.; FRANCO, N. V.; GARCÍA, M. F.; HORWITZ, V. D.; LANATA, J. L. ; MARTIN, F. M.; MUÑOZ, F. E.; SAVANTI, F. (1992). Archaeological Research in the Upper Santa Cruz Basin, Patagonia. *Current Anthropology* 33(4):451-454.

BINFORD, L. R. (1979) Organization and formation processes: looking at curated technologies. *Journal of Anthropological Research* 35: 255-273.

BORRERO, L. A.; CARBALLO MARINA, F. (1998) Proyecto Magallania: La cuenca superior del río Santa Cruz. In *Arqueología de la Patagonia Meridional. Proyecto Magallania*. Compilado por: L. A. Borrero; pp. 11-27. Concepción del Uruguay, Ediciones Búsqueda de Ayllu.

FRANCO, N. V. *Estrategias de utilización de recursos líticos en la cuenca superior del río Santa Cruz*. (Manuscrito) 2002. Unpiblished PhD dissertation. Universidad de Buenos Aires.

FRANCO, N. V.; ARAGÓN, E. (2002) Muestreo de fuentes potenciales de aprovisionamiento lítico: un caso de estudio. In *Del Mar a los Salitrales. Diez mil Años de Historia Pampeana en el Umbral del Tercer Milenio*, ed. by D. Mazzanti, M.Berón and F.Oliva, F, pp. 243-250. Universidad Nacional de Mar del Plata and Sociedad Argentina de Antropología, Mar del Plata, Argentina.

FRANCO, N. V.; ARAGÓN, E. (2004) Variabilidad en fuentes secundarias de aprovisionamiento lítico: El caso del sur de Lago Argentino (Santa Cruz, Argentina). *Estudios Atacameños* 28:71-85.

FRANCO, N. V.; BORRERO, L. A. (2003) Chorrillo Malo 2: initial peopling of the Upper Santa Cruz

Basin. En *Where the South Winds Blow. Ancient Evidences of Paleo South Americans*. Ed. R. Bonnichsen, L. Miotti, M. Salemme y N. Flegenheimer, pp. 149-152. Center for the Studies of the First Americans (CSFA) and Texas A&M University Press, Texas, USA.

FRANCO, N. V.; BORRERO, L. A.; MANCINI, M. V. (2004) Environmental changes and hunter-gatherers in southern Patagonia: Lago Argentino and Cabo Vírgenes. *Before Farming* 3: 1-17.

HAJDUK, A.; LEZCANO, M. (2005) Un "nuevo-viejo integrante del elenco de instrumentos óseos de Patagonia: los machacadores óseos. *Magallania*, (Chile) Vol. 33 (1):63-80

MALAGNINO, E. C.; STRELIN, J. A. (1992) Variations of Upsala Glacier in Southern Patagonia since the late Holocene to the present. In *Glaciological Researches in Patagonia*, ed. R. Naruse and M. Aniya, pp.61-85.

MANCINI, M.V. (2002) Vegetation and climate during the Holocene in Southwest Patagonia, Argentina. *Review of Paleobotany and Palynology* 122:101-115.

MERCER, J. H. (1968) Variations of some Patagonian Glaciers since the late –glacial. *American Journal of Science* 266:91-109.

MERCER, J. H. (1970) Variations of some Patagonian Glaciers since the Late-Glacial: II. *American Journal of Science* 269: 1-25

MERCER, J. H. ; AGER, T. (1983) Glacial and Floral Changes in Southern Argentina since 14,000 years ago. *National Geographic Society Research* 15, 457-477.

NAMI, H. G. (1992) Noticia sobre la existencia de técnica "levallois" en Península Mitre, extremo sudoriental de Tierra del Fuego. *Anales del Instituto de la Patagonia (Serie Cs. Humanas)* 21:73-80.

NAMI, H. G. (1997) Más datos sobre la existencia de núcleos preparados y lascas predeterminadas en la Patagonia Austral. *Anales del Instituto de la Patagonia (Serie Cs. Humanas)* 25:223-227.

THE PALEOINDIAN OCCUPATIONS AT BONNEVILLE ESTATES ROCKSHELTER, DANGER CAVE, AND SMITH CREEK CAVE (EASTERN GREAT BASIN, U.S.A): INTERPRETING THEIR RADIOCARBON CHRONOLOGIES

Ted GOEBEL[1], Kelly GRAF[2], Bryan HOCKETT[3], and David RHODE[4]

[1]Center for the Study of the First Americans, Department of Anthropology, Texas A&M University, 4352 TAMU, College Station, TX 77843, U.S.A.; goebel@tamu.edu
[2]Department of Anthropology/096, University of Nevada Reno, Reno NV 89557, U.S.A.; kelichka7@yahoo.com
[3]Elko Field Office, U.S.D.I. Bureau of Land Management, Elko NV 89801, U.S.A; Bryan_Hockett@nv.blm.gov
[4]Desert Research Institute, Reno NV 89512, U.S.A.; Dave.Rhode@dri.edu

Abstract. Numerous caves and rockshelters in the Great Basin of western North America contain geological deposits chronicling human adaptive change through the terminal Pleistocene and early Holocene periods. This is especially the case in the western Bonneville basin of Nevada and Utah, where three caves in particular—Danger Cave, Smith Creek Cave, and Bonneville Estates Rockshelter—have yielded artifacts, faunal remains, floral remains, and hearth features in sealed, stratified contexts. Complex taphonomic histories, however, have led to much confusion in the radiocarbon dating of the caves' cultural deposits. In this paper we review the radiocarbon records of the three sites in detail, analyzing them in the context of site formation processes. Our results suggest that the main Paleoindian occupations of these caves date to between about 10,900 and 9500 radiocarbon years ago (12,850 and 10,650 calendar years ago), and that this corresponds to a cool, wet period in the eastern Great Basin that may relate to the Younger Dryas cooling event in the North Atlantic. In all three caves, Paleoindian occupations are followed by gaps in human occupations of at least 1500 years. These gaps appear to correlate to significant aridification of the Bonneville basin, starting at about 9000 radiocarbon years ago (10,200 calendar years ago).

Keywords: Paleoindian, radiocarbon dating, taphonomy, formation processes, Great Basin

Résumé. De nombreuses grottes et abris sous roche dans le Grand Bassin de l'ouest de l'Amérique du Nord contiennent des dépôts géologiques qui témoignent du changement d'adaptation humain pendant les périodes du Pléistocène final et du début de l'Holocène. Ceci est le surtout le cas dans le bassin de Bonneville de l'ouest au Nevada et dans l'Utah, où trois grottes en particulier—la grotte du Danger, la grotte de Smith Creek, et l'abri sous roche de Bonneville Estates—ont dévoilé des objets, des restes fauniques, des restes floraux et des restes de foyer dans des contextes scellés et stratifiés. Cependant, le contexte taphonomique complexe a compliqué la datation par le radiocarbone des dépôts culturels des grottes. Dans cet article nous réexaminons les données radiocarbone des trois sites en détail, les analysant dans le contexte des processus de formation de site. Nos résultats suggèrent que les occupations paléoindiennes principales de ces grottes se situent environ entre 10.900 et 9500 années radiocarbone (il y a 12.850 et 10.650 années calendaires), et que ceci correspond à une période froide et humide du Grand Bassin de l'est qui peut se corréler à l'événement du refroidissement du Dryas récent dans le Nord Atlantique. Dans chacune des trois grottes, les occupations paléoindiennes sont suivies par des intervalles sans occupations humaines d'au moins 1500 ans; ces intervalles semblent correspondre à une aridification significative du bassin de Bonneville, commençant il y a environ 9000 années radiocarbone (il y a 10.200 années calendaires).

Mots-clés: Paléoindien, datation radiocarbone, taphonomie, processus de formation, Grand Bassin

The Great Basin is an arid region of the western United States stretching from the Sierra Nevada Mountains in the west to the Rocky Mountains in the east. Like other deserts of the world, geological sedimentation rates in the Great Basin can be very slow to nonexistent, leaving ancient surfaces exposed for thousands of years. In addition, Middle Holocene aridity caused extensive erosion across much the Great Basin, exposing many previously buried early sites (Nials 1999). As a result, Great Basin archaeologists have had a difficult time finding well-buried prehistoric sites. This is especially the case for the period of the terminal Pleistocene and early Holocene (TPEH) (about 11,000-8000 ^{14}C BP). Very few sites ^{14}C-dated to the TPEH have been found in open-air (i.e., non-cave) geomorphic contexts that are buried, stratified, and sealed. For example, in a recent survey Beck and Jones (1997) reviewed ^{14}C age estimates from 35 Great Basin sites. Twenty-four of these are caves or rockshelters, and only 11 are from open sites. Of the open sites, virtually none have yielded expressive stone artifact assemblages, faunal remains, and features that can be shown to unequivocally predate 9000 ^{14}C BP. One exception is the extreme northwestern corner of the Great Basin in south-central Oregon, where sealed TPEH sites are relatively common (e.g., Jenkins et al. 2004).

Due to the difficulty in finding open-air sites in stratified, datable contexts, Paleoindian archaeologists in the Great Basin have for decades been drawn to the region's caves and rockshelters, many of which have been found to contain cultural deposits predating 8000 ^{14}C BP (Beck and Jones 1997; Willig and Aikens 1988). Although providing the bulk of information we currently have concerning Great Basin Paleoindian lifeways, this cave-centric perspective has come with a price. First, the cave/rockshelter record tells us only about what early prehistoric foragers were doing within and nearby these permanent features on the landscape, and little about what they did elsewhere. Since these Paleoindians likely practiced relatively high levels of mobility (Elston and Zeanah 2002; Jones et al. 2003), they probably spent much more time away from these natural shelters than they did near them. Second, humans were not the only inhabitants of these shelters; instead, they often shared them with other animals like rodents, carnivores, and raptors (Grayson 1983; Hockett 1991, 1994). The origin of animal and plant remains in these caves and rockshelters, therefore, may be from carnivore predation and gnawing, packrat (Neotoma sp.) nesting, or raptor activity, not necessarily from human activities.

This means, then, that archaeological research in the Great Basin's caves and rockshelters must be conducted through a taphonomic perspective that carefully analyzes natural as well as cultural site formation processes. This is the case when interpreting data generated from all materials analyses, whether they are lithic, faunal, floral, or ^{14}C dating analyses. In this paper, we review the ^{14}C dating results of three caves/rockshelters located in the eastern Great Basin (Danger Cave, Smith Creek Cave, Bonneville Estates Rockshelter), and show how a taphonomic, site formation approach can assist in interpreting the dates to produce occupational histories of the shelters during the TPEH. We attempt to show that human occupations of the TPEH at these shelters were relatively short but repeated, and that the same pulses of occupation can be found in multiple shelters, suggesting that they may relate to increased use of the region during periods of climatic amelioration (i.e., relatively mesic conditions).

THE SITES

Danger Cave, Smith Creek Cave, and Bonneville Estates Rockshelter are located along the border of Nevada and Utah, in the interior of western North America (**Figure 20.1**). All three are along the western margin of the Bonneville basin, which during the terminal Pleistocene was filled by a huge and deep Lake Bonneville. Today the western Bonneville basin is dominated by a seasonally dry playa called the Bonneville Salt Flats (1304 m in elevation), which is surrounded by hills and mountains reaching nearly 4000 m in elevation. Great Salt Lake, located to the east of the Bonneville Salt Flats, is a modern remnant of Pleistocene Lake Bonneville.

Danger Cave

Danger Cave is located on the outskirts of Wendover, Utah, and is situated adjacent to the Bonneville Salt Flats (1315 m in elevation), just above the Gilbert Shoreline, which likely formed during the Younger Dryas period, 10,500-10,000 ^{14}C BP (Oviatt et al. 2005). The cave is an

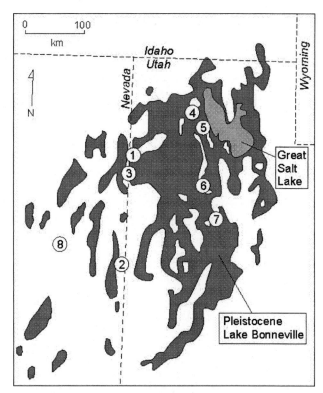

20.1 Map showing location of archaeological sites mentioned in text (1, Danger Cave; 2, Smith Creek Cave; 3, Bonneville Estates Rockshelter; 4, Hogup Cave; 5, Homestead Cave; 6, Camels Back Cave; 7, Old River Bed; 8, Sunshine Well), greatest extent of Lake Bonneville during the late Pleistocene, and other late Pleistocene lakes.

oval chamber about 20 m wide at its mouth and 40 m deep from front to back. Based on a variety of proxy records, Lake Bonneville receded from Danger Cave around 12,500-11,500 ^{14}C BP (Godsey et al. 2005; Oviatt 1997; Rhode et al. 2006).

Archaeological excavation at Danger Cave began in the early 1940s with Elmer Smith's initial tests and continued in 1949-1953 with Jesse Jennings' large-scale excavations (Jennings 1957). Jennings and his team unearthed a series of well-stratified and well-preserved cultural layers spanning the Holocene, from more than 10,000 ^{14}C BP to historic times. Excavation methods used during these early projects were too coarse to permit the tying of specific features, artifacts, and ecofacts to precise strata, and ^{14}C dating methods were in their infancy so that most of Jennings' first series of dates are now considered ballpark figures at best. Since the late 1960s, excavations in Danger Cave were renewed several times to obtain finer-scaled paleoecological and archaeological data. In the late 1960s Fry (1976; Harper and Alder 1972) excavated a trench in the back of the cave, and in 1986 Madsen and Rhode (1990; Rhode and Madsen 1998) excavated a preserved block of sediment near the mouth of the cave. These excavations led to much new information regarding the Archaic occupations of the cave post-dating about 8500 ^{14}C BP, but failed to uncover extensive cultural strata predating 9000 ^{14}C BP. In 2002 and 2004, Madsen and Rhode led a team of researchers again into Danger Cave, this time to re-expose preserved profiles where deposits dating to the TPEH were thought to exist. Their efforts focused on relocating the rear profiles of Jennings' 1949-1953 excavation (called the 143 face) and Fry's 1968 back trench (Rhode et al. 2006).

The TPEH sediments within Danger Cave can be separated into two occupation zones that Jennings (1957) referred to as DI and DII, with DI being the earliest. In Jennings' 143 face, DI is capped by wind-blown sand that is archaeologically sterile (Sand 2), and this is in turn overlain by DII, an occupation zone consisting of three distinct strata. From the bottom-up, these include (1) a layer of charred vegetation (F115), (2) a hardened layer of calcrete, and (3) a layer of well-preserved pickleweed chaff (F117 and F119) (Rhode et al. 2006:222). In the re-exposed back trench profile, Rhode et al. (2006) did not find the DI occupation zone; however, they did identify Sand 2 and DII, represented by a series of stratified features of well-preserved pickleweed chaff and ash.

In contrast to the back trench, Rhode et al. (2006:221) exposed two DI hearth features (called F111 and F112) in the preserved 143 face. These ash lenses are unexcavated portions of hearths that appear in Jennings' stratigraphic profile (Jennings 1957:Figure 54; Rhode et al. 2006:Figure2). For DII, Rhode et al. (2006) exposed a heavily charred layer of bones, vegetal debris, and ungulate and rodent dung (F31), and a pickleweed-rich feature (F30) that most likely represents human processing of seeds.

The DI and DII features appear to have resulted from different kinds of activities, some human and some non-human. For DI, Jennings (1957) described six hearth features, as follows:

These little hearths were in no sense modeled or fabricated fire-places. They were simply areas of blackened or faintly reddened sand where fires of sticks or twigs had been built upon the flat surface. A thin lens of ash or ash and charcoal covered the discolored zone. The diameter of each was approximately 2 feet. In cross-section the sand beneath the fire might be discolored in a lenticular or crescentic pattern from 1 to 3 inches deep at the center, where the heat was greatest, thinning to finally pinch out at the surface at the periphery (1957:54).

Apparently, no such hearth features were encountered in DII (Jennings 1957:64), but Jennings did describe a small pit feature and an extensive deposit of charred remains that appears to have burned naturally.

TPEH human occupations at Danger Cave span from approximately 10,300 to 8300 ^{14}C BP (Rhode et al. 2007). No diagnostic artifacts were recovered during Rhode et al.'s (2006) most recent field studies, so assemblage characterizations must rely on Jennings' (1957) original descriptions. Early excavations of DI produced a small assemblage of cultural remains including one lanceolate-shaped point (that may be a stemmed point) and several scrapers, possible ground-stone artifact fragments, and knotted pieces of twine. DII yielded a much more extensive assemblage primarily consisting of large side-notched and corner-notched points, a variety of flake tools, ground-stone artifacts, basket fragments, twine fragments, and chewed quids. DII also produced human coprolites filled with pickleweed seeds (Fry 1976; Rhode et al. 2006).

Smith Creek Cave

Smith Creek Cave is located about 30 km north of the town of Baker, Nevada. It is situated at an elevation of about 2040 m above sea level, 460 m above the high shoreline of Pleistocene Lake Bonneville. Smith Creek Cave has a large, southeast-facing mouth that is roughly 18 m high and 50 m wide. Its depth from front to back is about 30 m. Although high above the Bonneville shoreline, the cave would have been visible to humans traveling along the lakeshore during the TPEH (Bryan 1979). Bryan directed controlled archaeological excavations at Smith Creek Cave in 1968, 1971, and 1974. Detailed descriptions of the findings of these excavations can be found in two reports (Bryan 1979, 1988).

Artifact-bearing deposits of TPEH age occur in two discernible stratigraphic layers, a grey ash and silt stratum overlain by a deposit of dung, rubble, and silt (Bryan 1979:183-184). The grey ash and silt stratum is more specifically described as a "compact layer, no more than about 12 cm thick, of grey ash, twigs, other plant remains often wadded together with balls of hair, charcoal, silt, and dung in a fine rubble matrix" (Bryan 1979:184). Across the cave the grey ash grades into a deposit of loose rubble and pink silt that also contains artifacts but appears to be significantly reworked. The grey ash and presumably the silt are thought to be aeolian in origin (Bryan 1979:183). The overlying dung, rubble, and silt stratum is "finely stratified," compact, and up to 16 cm thick (Bryan 1979:184-185). Its contents include sheep dung, charcoal, juniper twigs and berries, bone fragments, and artifacts; Bryan (1979) interprets these as representing alternating sheep and human occupations.

Bryan (1979) recovered stone artifacts from both the grey ash and silt stratum and dung, rubble, and silt stratum, and he called them the "Mt. Moriah occupation zones." These include stemmed bifacial points, bifaces, flake tools, fragments of twine, chewed quids, and some worked pieces of bone. Associated faunal remains that appear to be the result of human activity include mountain sheep and pronghorn, but jackrabbits, hares, and rodents also occur, as do isolated bird, fish, and reptile bones (Bryan 1979:185-186).

At least eight charred features interpreted as hearths are described by Bryan (1979:187-190) (**Figure 20.2**). Most of these originated stratigraphically from the dung,

20.2 A view of the Smith Creek Cave excavation, along the west side of the cave at its mouth, where the main Mt. Moriah occupation zones occurred. The grey ash and silt stratum is exposed around the excavator's shadow (photo courtesy of Alan Bryan).

rubble, and silt zone and penetrated downward into the grey ash and silt zone. These features were typically unlined concentrations of ash, charcoal, charred bone, and sometimes charred dung. Some of them, like Hearth 7, also contained uncharred plant remains or uncharred perishable artifacts.

Bonneville Estates Rockshelter

Bonneville Estates Rockshelter is located about 50 km south of the town of West Wendover, Nevada, and is situated upon the high shoreline complex of Pleistocene Lake Bonneville (1580 m in elevation). Bonneville Estates is an open shelter that has a southeast-facing mouth reaching 10 m high and 25 m wide. From front to back, it is about 15 m deep at its deepest point. The lake receded from its high shoreline around 14,500 ^{14}C BP, so at that time the rockshelter became "high and dry" and open for human and animal use.

Initial excavations at Bonneville Estates were carried out in 1988 by P-III Associates, an archaeological consulting firm from Utah (Schroedl and Coulam 1989). Our team began full-scale excavations in 2000 and continued to excavate the rockshelter's sediments through 2006 (Goebel et al. 2003; Graf 2007; Rhode et al. 2005). TPEH cultural deposits have been uncovered in two excavation areas that are called the west block and east block. Details on the site's stratigraphy and dating, lithic artifact assemblages, faunal assemblages, and paleobotanical assemblages can be found in Graf (2007), Goebel (2007), Hockett (2007), and Rhode and Louderback (2007).

Briefly, the sediments within Bonneville Estates Rockshelter consist of a series of silt-and-rubble deposits interdigitated with well-preserved organic layers. The organic layers appear to be the result of a variety of depositional agents: some are clearly the result of human activities, some are the result of animal activities, and some are the result of a combination of the two. Cultural deposits throughout the profile contain not just stone artifacts, but also well-preserved faunal remains (many with cutmarks and other recognizable signs of having been butchered by humans), hearths with charred plant macrofossils, and perishable artifacts like basket fragments, cordage fragments, bone awls and needles, beads and pendants. Through our excavations, we have discerned six cultural components that range in age from the terminal Pleistocene to historic times (i.e., the 20th Century AD).

TPEH occupations at Bonneville Estates Rockshelter span from approximately 11,000 to 9,500 ^{14}C BP (although a single hearth feature with no associated diagnostic artifacts has been ^{14}C dated to 8800 ^{14}C BP) (Graf 2007). In the west block we have identified three stratigraphically distinct cultural layers, and in the east block we have identified a corresponding series of

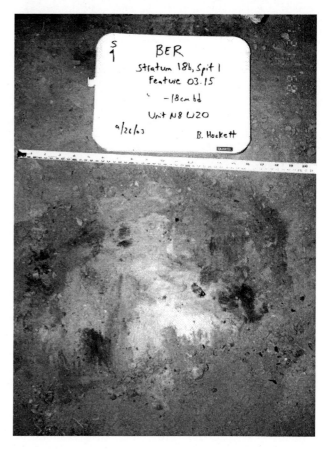

20.3 Hearth feature 03.15 (stratum 18b), one of the oldest ^{14}C-dated hearths at Bonneville Estates Rockshelter (photo by Bryan Hockett).

hearths in a set of loose silt-and-rubble deposits. Assemblages for the west block and east block contain diagnostic Great Basin stemmed points and point fragments, side and end scrapers, gravers, bone awl/needle fragments, and fragments of cordage (Goebel 2007); however, no diagnostic points have been found in association with the oldest hearth, ^{14}C dated to about 11,010 ^{14}C BP. Associated faunal remains that can be attributed to human activity include artiodactyls (pronghorn, deer, and mountain sheep), sage grouse, hare, black bear, and grasshopper (Hockett 2007). Through 2006, we had excavated more than 14 hearth features that predate 8800 ^{14}C BP; typically these are unlined oval-shaped hearths (50-80 cm in diameter) that have shallow basins and are filled with wood charcoal and ash underlain by reddened silt and rubble (or unintentionally charred organics of an earlier event) (**Figure 20.3**). Their sizes, shapes, and contents fit Jennings' (1957) descriptions of TPEH hearths at Danger Cave. Charred plant macrofossils found within the Bonneville Estates hearths are the result of using plants for fuel; however, Rhode and Louderback (2007) suggest that some charred seeds found in the hearths could be evidence of human consumption (although no ground-stone artifacts have been found in the TPEH deposits at Bonneville Estates).

THE RADIOCARBON DATES

Tables 1-3 present reported ^{14}C dates for Danger Cave, Smith Creek Cave, and Bonneville Estates Rockshelter. Although 98 ^{14}C dates are listed in these tables, only about half can be reliably used to estimate ages of the TPEH cultural occupations in these sites. For this reason, detailed taphonomic reviews of the ^{14}C chronologies for the three sites are required here.

Danger Cave

Forty-six ^{14}C dates have been reported for the lower strata at Danger Cave. In **Table 20.1** and the discussion below, these are presented by cultural component. We argue that the DI cultural occupation dates to some brief interval between 10,350-10,000 ^{14}C BP, lower DII dates to about 10,100-10,000 ^{14}C BP, and middle/upper DII dates to less than 8600 ^{14}C BP.

Twenty ^{14}C dates come from samples ascribed stratigraphically to DI, however, some are unreliable and some are of natural, not cultural, origin. The two oldest dates (11,453 ± 600 [C-609] and 11,151 ± 570 ^{14}C BP [C-610]) are unreliable solid carbon dates (not gas dates) and rejected here, following Rhode and Madsen (1990). A third date (10,270 ± 650 ^{14}C BP [M-204]) is on naturally deposited sheep dung collected from below the DI occupation surface (situated at the top of Sand 1) (Jennings 1957). Additionally, five samples ranging from 11,000 ± 700 to 9050 ± 180 ^{14}C BP (M-118, M-119, Beta-19611, Tx-89, and Tx-88) are from Sand 2, above the DI occupation surface (Jennings 1957:60), and provide age estimates for naturally occurring materials (either sheep dung or uncharred twigs and sticks) but not the cultural occupation. This leaves 12 samples that either come from the DI occupation surface or are clearly the product of human activities (i.e., they are human coprolites). Even among these samples, though, there is a major discrepancy between dates obtained from hearth charcoal and dates obtained from other materials. The hearth charcoal dates (10,270 ± 650 [M-202], 10,310 ± 40 [Beta-168656], and 10,270 ± 50 ^{14}C BP [Beta-158549]) and dispersed charcoal date (10,150 ± 170 ^{14}C BP [Tx-87]) dates all overlap at one standard deviation and suggest an age of occupation bracketed between 10,350-10,000 ^{14}C BP for the DI occupation surface. A single AMS ^{14}C date of 10,600 ± 250 (AA-20859) on jackrabbit bone collagen seems too old; this sample may be the result of some non-human agent. Associated uncharred twigs are either significantly older (10,600 ± 200 ^{14}C BP [Tx-85]) or younger (8970 ± 150 ^{14}C BP [Tx-86]), and the five human coprolites attributed to DI yielded ^{14}C ages range from 8680 ± 50 to 3020 ± 59 ^{14}C BP (Beta-187446, Beta-97898, Beta-187444, Beta-187083, Beta-187445). As Rhode et al. (2006) suggest, these coprolites are aberrantly young and likely redeposited from later-aged sediments. Lastly, more information is needed for the sample that produced the ^{14}C date of 9780 ± 210 (Beta-19336) to evaluate whether it relates to the DI cultural occupation.

Only five ^{14}C samples can be attributed to lower DII (**Table 20.1**). One of these is a solid carbon date (8960 ± 34 ^{14}C BP [C-640]) and cannot be trusted, and two are on naturally occurring twigs, ash, and dung that produced disparate ages of 10,130 ± 250 (GaK-1899) and 6960 ± 210 ^{14}C BP (GaK-1895). Given that these two were bulk samples of natural materials, they do not provide specific age estimates and should be disregarded (Rhode et al. 2006). This leaves just two ^{14}C dates to use to establish the age of the lower DII cultural occupation: 10,080 ± 130 (Beta-19333) and 10,050 ± 50 ^{14}C BP (Beta-169848). The 10,050 age estimate is on charcoal from a heavily charred deposit, and the 10,080 age estimate overlaps it at one standard deviation (although its sample material has not been reported [Madsen and Rhode 1990]).

The middle and upper parts of DII are more securely dated. Disregarding a solid carbon date (9789 ± 630 ^{14}C BP [C-611]) and two pickleweed chaff dates (9900 ± 400 [GaK-1900] and 9590 ± 160 ^{14}C BP [GaK-1896]), which according to Rhode et al. (2006) are bulk samples that cannot be used to specifically date this stratum, all of the remaining ^{14}C samples post-date 8600 ^{14}C BP. Of the ten human coprolites sampled, seven fall between about 8400 and 8100 ^{14}C BP, while three are significantly younger (6020 ± 50 [Beta-187452], 5060 ± 40 [Beta-189085], and 5030 ± 40 ^{14}C BP [Beta-187451]). These younger coprolites are probably redeposited from above-lying DIII or DIV (which have been shown to date to roughly 6000-3000 ^{14}C BP [Madsen and Rhode 1990]). The rest of the ^{14}C dates conform with the bulk of the coprolite dates, indicating that middle and upper DII probably

Material Dated	^{14}C Age	Sample Number	Reference	Stratigraphic context
DI				
Sheep dung	11,453 ± 600	C-609[1]	Jennings 1957:60	Sand 2 [split sample with M-118]
Uncharred plant stem	11,151 ± 570	C-610[1]	Jennings 1957:54	Top of Sand 1; DI occupation surface
Sheep dung	11,000 ± 700	M-118[2]	Jennings 1957	Sand 2
Lepus bone collagen	10,600 ± 250	AA-20859	Mullen 1997	DI
Twigs, F108[3]	10,600 ± 200	Tx-85	Tamers et al. 1964	Top of Sand 1; DI occupation surface
Charcoal, F108	10,270 ± 650	M-202	Jennings 1957	Top of Sand 1; DI occupation surface
Uncharred twigs	10,400 ± 700	M-119	Jennings 1957	Sand 2
Charcoal, F111	10,310 ± 40	Beta-168656	Rhode et al. 2006	Top of Sand 1; DI occupation surface
Charcoal, F112	10,270 ± 50	Beta-158549	Rhode et al. 2006	Top of Sand 1; DI occupation surface
Charred sheep dung	10,270 ± 650	M-204	Jennings 1957	Sand 1, below cultural occupation
Charcoal, vegetal debris	10,150 ± 170	Tx-87	Tamers et al. 1964	Top of Sand 1; DI occupation surface
Sheep dung	9920 ± 185	Beta-19611	Madsen and Rhode 1990	Stratum 3 (1986 excavation); equivalent to Sand 2
Not reported[4]	9780 ± 210	Beta-19336	Madsen and Rhode 1990	Stratum 2 (1986 excavation); equivalent to surface of Sand 1
Twigs and sticks	9740 ± 210	Tx-89	Tamers et al. 1964	Lower Sand 2
Sheep pellets	9050 ± 180	Tx-88	Tamers et al. 1964	Lower Sand 2
Charred dung, twigs	8970 ± 150	Tx-86	Tamers et al. 1964	Contact between Sand 1 and Sand 2—D1 occupation surface
Human coprolite	8680 ± 50	Beta-187446	Rhode et al. 2006	DI
Human coprolite	3310 ± 60	Beta-97898	Rhode et al. 2006	DI
Human coprolite	3270 ± 40	Beta-187444	Rhode et al. 2006	DI
Human coprolite	3030 ± 40	Beta-187083	Rhode et al. 2006	DI
Human coprolite	3020 ± 50	Beta-187445	Rhode et al. 2006	DI
Lower DII				
Twigs, ash, and dung	10,130 ± 250	GaK-1899	Harper and Alder 1972	Lower DII
Not reported[4]	10,080 ± 130	Beta-19333	Madsen and Rhode 1990	Stratum 5 (1986 excavation); equivalent to lower DII
Charcoal, F115[5]	10,050 ± 50	Beta-169848	Rhode et al. 2005	Lower DII
Charred rat dung	8960 ± 340	C-640[1]	Jennings 1957	Top of Sand 2
Twigs and ash	6960 ± 210	GaK-1895	Harper and Alder 1972	Lower DII
Middle/Upper DII				
Pickleweed chaff	9900 ± 400	GaK-1900	Harper and Alder 1972	Fry's layer F12; Upper DII
Pit feature, charcoal	9789 ± 630	C-611[1]	Jennings 1957:64	Middle of DII
Pickleweed chaff	9590 ± 160	GaK-1896	Harper and Alder 1972	Layer F12 (1968 excavation); Upper DII
Pickleweed chaff	8570 ± 40	Beta-193123	Rhode et al. 2006	Stratum 04-11 (2004 excavation); equivalent to Fry's Layer F12; Middle DII

Table 20.1 (continued)

Material	Date	Lab No.	Reference	Stratum
Not reported[4]	8440 ± 50	Beta-190887	Rhode et al. 2006	Stratum 6 (1986 excavation); equivalent to middle DII
Not reported[4]	8410 ± 50	NSRL-11436	Rhode et al. 2006	Stratum 8 (1986 excavation); equivalent to upper DII
Pickleweed chaff	8380 ± 60	Beta-193124	Rhode et al. 2006	Stratum 04-10 (2004 excavation); equivalent to Fry's Layer F12; Middle DII
Human coprolite	8380 ± 40	Beta-187449	Rhode et al. 2006	DII
Human coprolite	8300 ± 40	Beta-187450	Rhode et al. 2006	DII
Charcoal, F119[6]	8270 ± 40	Beta-168857	Rhode et al. 2006	Upper DII
Not reported	8200 ± 50	Beta-190866	Rhode et al. 2006	Stratum 7 (1986 excavation); equivalent to middle/upper DII
Human coprolite	8190 ± 50	Beta-187448	Rhode et al. 2006	DII
Human coprolite	8160 ± 40	Beta-189084	Rhode et al. 2006	DI
Human coprolite	8130 ± 50	Beta-187447	Rhode et al. 2006	DII
Human coprolite	8100 ± 40	Beta-187453	Rhode et al. 2006	DII
Human coprolite	8100 ± 40	Beta-187454	Rhode et al. 2006	DII
Not reported	7920 ± 80	Beta-23653	Rhode et al. 2006	Stratum 9 (1986 excavation); equivalent to upper DII
Pine nut hull	7410 ± 120	AA-3623	Madsen and Rhode 1990	Stratum 10 (1986 excavation); equivalent to upper DII
Human coprolite	6020 ± 50	Beta-187452	Rhode et al. 2006	DII
Human coprolite	5060 ± 40	Beta-189085	Rhode et al. 2006	DII
Human coprolite	5030 ± 40	Beta-187451	Rhode et al. 2006	DII

[1] Solid carbon date that is unreliable (Madsen and Rhode 1990:96).
[2] Cited as M-116 in Jennings (1957:60).
[3] Sample split with M-202 (Rhode et al. 2006).
[4] Probably charcoal (D. Madsen, pers. commun., 2007).
[5] F31 in Jennings (1957).
[6] F30 in Jennings (1957).

20.1 Radiocarbon Dates for the Terminal Pleistocene and Early Holocene Deposits at Danger Cave.

reflect human use of Danger Cave from about 8600 to 7400 ^{14}C BP.

Smith Creek Cave

Sixteen ^{14}C dates have been reported for the Pre-Archaic, "Mt. Moriah" component in Smith Creek Cave (**Table 20.2**). These range from 14,220 ± 650 to 9280 ± 160 ^{14}C BP, but most cluster between about 11,000 and 10,000 ^{14}C BP.

Bryan (1979) dismissed the oldest and youngest dates as aberrant. The date of 14,220 ± 650 ^{14}C BP (RIDDL-796) was on artiodactyl hair and is inconsistent with dates from the cultural component as well as an underlying layer of bristlecone pine needles ^{14}C dated to about 13,000-12,500 ^{14}C BP (Bryan 1988:67). The young dates of 9280 ± 160 (GaK-5446) and 9800 ± 190 ^{14}C BP (GaK-5444) were not properly pretreated at the Gakushuin University laboratory (GaK) in Japan; re-dating of the latter charcoal sample at the Birmingham laboratory (Birm) produced a date of 10,740 ± 130 ^{14}C BP (Birm-702) (Bryan 1979:186). By deleting these obviously discordant dates, Bryan concluded that the Mt. Moriah component at Smith Creek Cave began by 12,000 ^{14}C BP, "before the earliest evidence for Clovis on the High Plains" (1988:70).

This interpretation, however, does not hold up against closer scrutiny of the remaining ^{14}C age estimates, especially when considering the context of the finds and the variety of ^{14}C-dated materials. In his reports, Bryan notes several times that the grey ash zone and its features were mixed. Most telling is the following statement: "The range of dates, combined with the fact that so many perishables were preserved in the completely burned wood ash (burning must have preceded deposition of the perishables), as well as the evidence for intrusive hearths and sheep beds, all show that considerable bioturbation had occurred after the ash layer started to accumulate" (Bryan 1988:68). This certainly explains the aberrantly old date of 14,220 ± 650 ^{14}C BP, and it helps explain how a piece of uncharred cordage dated to 10,420 ± 100 ^{14}C BP (TO-1173) could occur in the grey ash and silt layer, which underlies the dung, rubble, and silt layer thought to date to as early as 11,140 ± 200 ^{14}C BP (Tx-1637). To us, these inconsistencies in the dates as well as the evidence for extensive bioturbation in the grey ash and silt calls into question the primary association of the features, artifacts, and ^{14}C-dated ecofacts from this layer. The only ^{14}C dates that should be used to define the age of the cultural component are dates on wood charcoal from hearths or direct dates on perishable artifacts (i.e., the S-twist cordage).

Material Dated	^{14}C Age	Sample Number	Reference	Stratigraphic context
Artiodactyl hair	14,220 ± 650[1]	RIDDL-796	Bryan 1988	Grey ash and silt
Wood twigs	12,150 ± 120[2]	Birm-752	Bryan 1988	Grey ash and silt
Camelid hair	12,060 ± 450	RIDDL-797	Bryan 1988	Grey ash and silt
Scattered charcoal	11,680 ± 160	Tx-1421	Bryan 1979	Grey ash and silt
Charcoal, TP-8 1st hearth	11,140 ± 200	Tx-1637	Bryan 1988:71	Dung, rubble, and silt; dug into grey ash
Bovid hair	10,840 ± 250	RIDDL-795	Bryan 1988	Grey ash and silt
Charcoal, TP-8 2nd hearth	10,660 ± 220	GaK-5442	Bryan 1979	Dung, rubble, and silt; dug into grey ash
Charcoal, TP-8 3rd hearth	10,630 ± 190	GaK-5443	Bryan 1979	Dung, rubble, and silt; dug into grey ash
Charcoal, TP-8 3rd hearth	10,570 ± 160	GaK-5445	Bryan 1979	Dung, rubble, and silt; dug into grey ash
Average of two	*10,590 ± 122*			
Charcoal, TP-8 3rd hearth	9280 ± 160[3]	GaK-5446	Bryan 1979	Dung, rubble, and silt; dug into grey ash
Charcoal, TP-8 4th hearth	10,740 ± 130	Birm-702	Bryan 1979	Dung, rubble, and silt; dug into grey ash
Charcoal, TP-8 4th hearth	10,460 ± 260	GaK-5444b[4]	Bryan 1979	Dung, rubble, and silt; dug into grey ash
Average of two	*10,647 ± 116*			
Wood (cellulose)	10,700 ± 180	Birm-917	Bryan 1979	Grey ash and silt
S-twist cordage	10,420 ± 100	TO-1173	Bryan 1988	Grey ash and silt
Charcoal, TP-6 hearth 12[5]	10,330 ± 190	Tx-1638	Bryan 1979	Dung, rubble, and silt; dug into grey ash
Charcoal, TP-6 hearth 9	9940 ± 160	Tx-1420	Bryan 1979	Dung, rubble, and silt; dug into grey ash

[1]Bryan (1988) argues that this date must be aberrant since it is older than dates of about 13,000 BP recovered from lower-lying bristlecone pine needle layer.
[2]This age estimate was incorrectly reported as 12,200 ± 300 in Figure 5 of Bryan (1979:187).
[3]Bryan (1979) interprets this as aberrantly too old (sample was not pretreated correctly).
[4]This sample was re-run from earlier sample dated to only 9800 ± 160 (GaK-5444); Bryan (1979) attributed discrepancy to lack of alkaline pretreatment in earlier run.
[5]This hearth was called hearth 7 in Figure 5a of Bryan (1979:188).

Table 20.2 Radiocarbon Dates for the Terminal Pleistocene and Early Holocene Deposits at Smith Creek Cave.

Material Dated	^{14}C Age	Sample Number	Reference	Stratigraphic context
West Block				
Bone	15,240 ± 50	UCIAMS-22180	Graf 2007	Stratum 19
Bone	11,960 ± 60	Beta-209265	Graf 2007	Stratum 19
Charcoal, F04.13b[1]	12,390 ± 40	Beta-195045	Graf 2007	Stratum 19
Charcoal, F04.13b	12,330 ± 40	Beta-195046	Graf 2007	Stratum 19
Bone, F04.13b	11,530 ± 40	Beta-209265	Graf 2007	Stratum 19
Charcoal, F0413b	10,970 ± 60	Beta-200874	Graf 2007	Stratum 19
Charcoal, F03.16/04.13	12,270 ± 60	AA-58595	Graf 2007	Stratum 19/18b
Charcoal, F03.16/04.13	10,690 ± 70	AA-58590	Graf 2007	Stratum 19/18b
Charcoal, F03.17	12,180 ± 60	AA-58587	Graf 2007	Stratum 19
Bone, F03.17	10,900 ± 50	UCIAMS-22176	Graf 2007	Stratum 19
Bone, F03.17	10,830 ± 40	Beta-210524	Graf 2007	Stratum 19
Charcoal, F03.17	10,640 ± 60	Beta-200875	Graf 2007	Stratum 19
Average of three	*10,821 ± 31*			
Charcoal, F05.06	11,010 ± 40	Beta-207009	Graf 2007	Stratum 18b
Charcoal, F03.15a	10,800 ± 60	AA-58594	Graf 2007	Stratum 18b
Charcoal, F03.15a	10,760 ± 70	AA-58592	Graf 2007	Stratum 18b
Average of two	*10,773 ± 46*			
Charcoal, F04.14	10,540 ± 40	Beta-195047	Graf 2007	Stratum 18b
Charcoal, F03.14	10,405 ± 50	AA-58593	Graf 2007	Stratum 18b
Charcoal, F01.01	10,130 ± 60	Beta-170444	Graf 2007	Stratum 18a
Charcoal, F01.01	10,080 ± 50	Beta-164229	Graf 2007	Stratum 18a
Charcoal, F01.01	10,040 ± 70	Beta-170443	Graf 2007	Stratum 18a
Average of three	*10,090 ± 34*			
Charcoal, F05.02	9580 ± 40	Beta-207010	Graf 2007	Stratum 17b'
Charcoal, F03.13	9440 ± 50	AA-58589	Graf 2007	Stratum 17b'
Charcoal, F03.13	9430 ± 50	AA-58588	Graf 2007	Stratum 17b'
Average of two	*9435 ± 35*			
East Block				
Charcoal, E4-12-C5/D3-10-C10[2]	10,380 ± 40	Beta-195013	Graf 2007	Stratum 12
Charcoal, E4-12-C5/D3-10-C10	10,340 ± 60	Beta-203504	Graf 2007	Stratum 12
Average of two	*10,367 ± 33*			
Charcoal, E3-13-C4	10,250 ± 50	Beta-206278	Graf 2007	Stratum 12
Charcoal, D5-10-C8a/C8d	10,380 ± 60	AA-58600	Graf 2007	Stratum 12
Charcoal, D5-10-C8a/C8d	10,050 ± 50	Beta-182935	Graf 2007	Stratum 12
Charcoal, D4-12-C9b	10,030 ± 50	Beta-182934	Graf 2007	Stratum 12
Charcoal, D4-10-C8/C7b	10,560 ± 50	Beta-182931	Graf 2007	Stratum 12
Charcoal, D4-10-C8/C7b	9990 ± 50	AA-58598	Graf 2007	Stratum 12
Charcoal, E4-10-C3/E5-10-C7	9580 ± 40	Beta-195042	Graf 2007	Stratum 10
Charcoal, E4-10-C3/E5-9-C5b	9570 ± 40	Beta-195044	Graf 2007	Stratum 10
Average of two	*9595 ± 48*			
Charcoal, E6-10-C10	9520 ± 60	Beta-161891	Graf 2007	Stratum 10
Charcoal, E6-10-C10	9440 ± 80	AA-58599	Graf 2007	Stratum 10
Average of two	*9493 ± 48*			
Charcoal, A4-9-C1	8830 ± 60	Beta-203507	Graf 2007	Stratum 9

[1]Hearth feature designation for west block.
[2]Hearth feature designation for east block.

Table 20.3 Radiocarbon Dates for the Terminal Pleistocene and Early Holocene Deposits at Bonneville Estates Rockshelter.

When considered in this way, there are nine ^{14}C dates from six features and one artifact that provide an accurate representation of the age of the Mt. Moriah component at Smith Creek Cave. These range from 11,140 ± 200 to 9940 ± 160 ^{14}C BP, and probably reflect repeated short-term visits to the cave by Paleoindians. The first of these events may have occurred around 11,140 ^{14}C BP, however, this date does not conform to five other dates ranging from 10,660 ± 220 (GaK-5442) to 10,570 ± 160 ^{14}C BP (GaK-5445) from the same hearth complex (excavated in TP8). Confirmation of this early ^{14}C date should be sought either by dating additional charcoal from this particular concentration or by dating the yucca quid that Bryan (1979:188) reportedly found within it. Until this happens, we conclude that the Mt. Moriah component at Smith Creek Cave unequivocally dates to between 10,700-9900 ^{14}C BP, but could date to as early as 11,200 ^{14}C BP.

Bonneville Estates Rockshelter

Thirty-six ^{14}C dates have been reported for the TPEH strata at Bonneville Estates Rockshelter (**Table 20.3**). These are presented in **Table 20.3**, first by excavation (west block and east block) and second by stratigraphic layer. Together the ^{14}C dates suggest that humans repeatedly utilized Bonneville Estates Rockshelter between about 11,000 and 9400 ^{14}C BP, and possibly again briefly around 8800 ^{14}C BP.

West block TPEH deposits are divided into four strata, 19, 18b, 18a, and 17b'. Stratum 19 is the deepest stratum to have produced stone artifacts and burn features interpreted as hearths; however, these occur in an area where the stratum is unsealed from upper-lying stratum 18b. Age estimates for stratum 19 range from 15,240 ± 50 (UCIAMS-22180) to 10,640 ± 60 ^{14}C BP (Beta-200875). The dates of 15,240 ± 50 and 11,960 ± 60 ^{14}C BP (Beta-209265) were obtained from bones not associated with any lithic artifacts or features; they are probably paleontological (Graf 2007). The remaining ten dates are from wood charcoal or bone samples from two hearth features. Six of these dates range from 12,390 ± 40 (Beta-195045) to 10,970 ± 60 ^{14}C BP (Beta-200874), while the other five range from 10,900 ± 50 (UCIAMS-22176) to 10,640 ± 60 ^{14}C BP (Beta-200875). We agree with Graf (2007) that the earlier set of dates likely represents the time that stratum 19 was deposited by natural agents like woodrats (Neotoma sp.), raptors, and carnivores, and that the later set of dates likely represents the time that humans occupied the surface of stratum 19, lighting one or two fires (features 03.16/04.13 and 03.17) and discarding a few pieces of stone debitage. The wood charcoal samples that produced the older dates are probably natural fragments of dried vegetation that became incorporated into the humans' fires. Feature 04.13b, for example, is stratigraphically within stratum 19 and likely represents burning of natural vegetation when the upper hearth (03.16/04.13) was dug into it and burned. Given this interpretation, we conclude that all of the feature 04.13b dates are of natural, not cultural, origin. We tentatively conclude that stratum 19 is essentially a natural deposit of organics, and that human activities occurred only on or near its surface. The timing of these activities is best represented by the young ^{14}C dates from features 03.16/04.13 and 03.17; together they indicate an age of 10,900-10,600 ^{14}C BP.

Strata 18b, 18a, and 17b' are more securely ^{14}C dated (**Table 20.3**) (Graf 2007). For stratum 18b we have four hearth features that range in age from 11,010 ± 40 (Beta-207009) to 10,405 ± 50 ^{14}C BP (AA-58593). The hearth dated to 11,010 ± 40 ^{14}C BP came from an area of the west block where stratum 19 was not deposited, thereby explaining the apparent reversal in dates. For stratum 18a we have one hearth feature with three dates ranging from 10,130 ± 60 (Beta-170444) to 10,040 ± 70 ^{14}C BP (Beta-170443). For stratum 17b' we have two hearth features that range from 9580 ± 40 (Beta-207010) to 9430 ± 50 ^{14}C BP (AA-58588). The single ages on the hearths from strata 18b and 17b' (especially the oldest one at 11,010 ± 40 ^{14}C BP) need to be confirmed with additional analyses.

East block strata dating to the TPEH (strata 12, 10, and 9) were loose and more massively bedded than west block strata, making it difficult to distinguish individual cultural occupations. Seven hearth features, however, were identified during the excavation, and ^{14}C dates on these features generally conform to their stratigraphic positions. The lowest two hearths, found in the silts of stratum 12, yielded ^{14}C dates that range from 10,380 ± 40 (Beta-195013) to 10,250 ± 50 ^{14}C BP (Beta-206278). Immediately above these were three hearths likely dating to between 10,050 ± 50 (Beta-182935) and 9990 ± 50 ^{14}C BP (AA-58598); dates of 10,560 ± 50 (Beta-182931) and 10,380 ± 60 ^{14}C BP (AA-58600) on two of these hearths are probably too old given their stratigraphic position above the lowest pair of hearths (Graf 2007). Although undoubtedly TPEH in age, it is possible that the use of old pine wood in these latter two hearths accounts for the dates that appear "too old" in comparison with the other dates that are consistently several centuries younger. The two hearths from stratum 10 date to between 9580 ± 40 (Beta-195042) and 9440 ± 80 ^{14}C BP (AA-58599), and the hearth feature recovered from the lower part of stratum 9 dates to 8830 ± 60 ^{14}C BP (Beta-203507).

CALIBRATION AND ANALYSIS OF THE ^{14}C CHRONOLOGIES OF DANGER CAVE, SMITH CREEK CAVE, AND BONNEVILLE ESTATES ROCKSHELTER

Of the 98 ^{14}C dates reviewed above, only about half are appropriate for interpreting the ages of the TPEH cultural occupations preserved in Danger Cave, Smith Creek Cave, and Bonneville Estates Rockshelter (**Table 20.4**). These are the dates from cultural features, artifacts, and ecofacts that are consistent with other ^{14}C dates and stratigraphic position. Interpreting these "good" dates nonetheless is a complex procedure. As Bryan (1988) has pointed out, we cannot simply average all of the dates from a single cultural stratum to determine the age of that stratum, nor can we count every date from a cultural stratum as representing an individual occupation event, since some features, ecofacts, or artifacts may have multiple dates. Here we use "occupation event" as a unit of observation, calculating a single ^{14}C-BP age for each feature, perishable artifact, or human coprolite dated. When a feature or artifact had multiple ^{14}C dates, we averaged those that are considered accurate, deleting from the analysis any dates recognized in the discussion above as being discordant or incorrect (**Tables 20.1-20.3**). We then calibrated the resulting dates (at one sigma) using the Calpal Online Radiocarbon Calibration program, interpreting an age for each event in calendar years before present (cal BP) (**Figure 20.4**).

Danger Cave		Smith Creek Cave		Bonneville Estates Rockshelter	
^{14}C BP	Cal BP	^{14}C BP	Cal BP	^{14}C Age	Cal BP
10,310 ± 40	12,360-12,014	11,140 ± 200	13,260-12,856	11,010 ± 40	13,018-12,818
10,270 ± 650	12,679-10,966	10,660 ± 220	12,788-12,177	10,821 ± 31*	12,852-12,722
10,270 ± 50	12,268-11,872	10,647 ± 116*	12,721-12,366	10,773 ± 46*	12,803-12,700
10,150 ± 170	12,185-11,469	10,590 ± 122*	12,679-12,260	10,690 ± 70	12,739-12,607
10,080 ± 130	11,950-11,413	10,420 ± 100	12,539-12,121	10,540 ± 40	12,648-12,363
10,050 ± 50	11,750-11,419	10,330 ± 190	12,483-11,741	10,405 ± 50	12,511-12,161
8570 ± 40	9552-9523	9940 ± 160	11,772-11,262	10,367 ± 33*	12,474-12,124
8440 ± 50	9509-9435			10,250 ± 50	12,172-11,842
8410 ± 50	9485-9361			10,090 ± 34*	11,832-11,489
8380 ± 60	9465-9324			10,050 ± 50	11,750-11,419
8380 ± 40	9460-9340			10,030 ± 50	11,710-11,392
8300 ± 40	9396-9264			9990 ± 50	11,593-11,346
8270 ± 40	9368-9172			9595 ± 48*	11,079-10,822
8200 ± 50	9249-9077			9580 ± 40	11,056-10,814
8190 ± 50	9240-9066			9493 ± 48*	11,008-10,698
8160 ± 40	9191-9045			9435 ± 35*	10,710-10,613
8130 ± 50	9151-9028			8830 ± 60	10,091-9774
8100 ± 40	9091-9011				
8100 ± 40	9091-9011				
7920 ± 80	8932-8655				
7410 ± 120	8340-8090				

* Average date of single hearth or bone with multiple ^{14}C dates.

Table 20.4 Reliable ^{14}C ages dating cultural occupation events at Danger Cave, Smith Creek Cave, and Bonneville Estates Rockshelter.

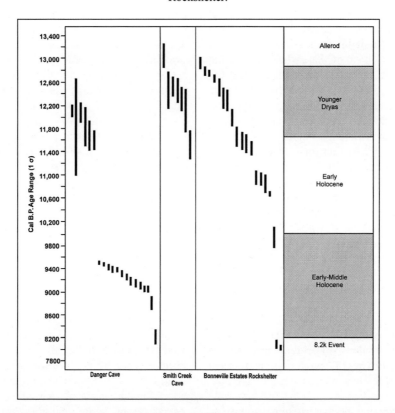

20.4 Calibrated ^{14}C ages that accurately measure the ages of the early cultural occupations at Danger Cave, Smith Creek Cave, and Bonneville Estates Rockshelter (from Table 4). According to multiple proxy records, during the Younger Dryas period the western Bonneville basin filled with water, forming the Gilbert shoreline. Gradual drying occurred during the early Holocene, followed by extremely hot and arid conditions during the early-middle Holocene. During the "8.2k" event, climate was more mesic than during the previous 1500 years.

Danger Cave appears to have had two major pulses of human occupation. The first (the brief DI and lower DII occupations, separated by a few decades to centuries) probably began no earlier than 12,200 cal BP and ended 11,400 cal BP. After a hiatus of about 1500 years, a second period of occupation (middle-upper DII) occurred from about 9450 cal BP to 9000 cal BP. Human use of Danger Cave continued through the middle Holocene (Madsen and Rhode 1990; Rhode and Madsen 1998).

Humans used Smith Creek Cave most intensively from 12,800 to 12,100 cal BP. One hearth, however, may date to earlier than this (13,200-12,900 cal BP), and another hearth dates to later (11,800-11,300 cal BP). The early hearth needs to be re-dated to confirm its presumed age. Human occupation ceased entirely by 11,300 cal BP, until Fremont foragers began to use the cave during the late Holocene (Bryan 1979).

Human occupation of Bonneville Estates Rockshelter may have begun by 13,050 cal BP, but as in the case of Smith Creek Cave, the single ^{14}C date on the earliest hearth needs confirmation. After this, humans appear to have repeatedly visited Bonneville Estates from 12,850 to 11,350 cal BP (Graf 2007), lighting small fires, butchering carcasses of artiodactyls, sage grouse, and other animals, repairing nets, and resharpening stone tools (Goebel 2007; Hockett 2007). Humans may have used the rockshelter less frequently from 12,150 to 11,850 cal BP, and not at all from 11,350 to 11,100 cal BP. A second pulse of activity occurred from 11,100 to 10,650 cal BP. Cultural remains of this later occupation are virtually identical to those of the earlier occupation, suggesting that human activities in and around the rockshelter remained unchanged. One last hearth feature chronicles a possible human occupation of Bonneville Estates about 10,150-9800 cal BP. No artifacts or faunal remains were found in or around this hearth, so if humans did use the rockshelter during this time, their stay must have been brief and ephemeral. Besides this, we can find no evidence to show that humans used Bonneville Estates Rockshelter between 10,650 and 8380 cal BP (Graf 2007), a lengthy hiatus of at least 2200 years. When they re-entered the rockshelter after 8380 cal BP, they were making large side-notched points to hunt artiodactyls and grinding up lots of small seeds with ground-stone artifacts.

DISCUSSION AND CONCLUSIONS

Based on the results of this analysis, the following conclusions can be made concerning the human populations of the eastern Great Basin during the terminal Pleistocene and early Holocene.

1. Initial human occupation of the western Bonneville basin occurred as early as 12,850 cal BP and possibly earlier. Humans certainly used Bonneville Estates Rockshelter by 12,850 cal BP, Smith Creek Cave by 12,800 cal BP, and Danger Cave by 12,250 cal BP. At Smith Creek Cave and Bonneville Estates, older hearths have been ^{14}C dated to 13,260-12,856 cal BP and 13,018-12,818 cal BP, respectively. These dates, however, need to be confirmed with additional analyses. Confirmation of these two dates would suggest that humans first used these two caves at a time coeval with some of the earliest Clovis sites of North America.

2. In the western Bonneville basin from 12,850 to 10,650 cal BP, humans exclusively made and used Great Basin stemmed points. Stemmed points have been found associated with the early hearths at Smith Creek Cave and Bonneville Estates Rockshelter, and a possible stemmed point came from the DI cultural occupation at Danger Cave (Jennings 1957:109; Jennings 1978:39, Figure 20). Clovis fluted points have not been found in a primary context in any of these or any other caves or rockshelters in the region. Whether stemmed points were produced earlier than 12,850 cal BP, however, can not be confirmed with the present evidence. As noted above, at Smith Creek Cave the earliest hearth (13,260-12,856 cal BP) needs confirming dates, and at Bonneville Estates the earliest hearth (13,018-12,818 cal BP) is not associated with any bifacial points (stemmed or fluted). From this we formulate the following working hypothesis. In the eastern Great Basin, stemmed points were only produced after 12,850 cal BP, while fluted points were only produced before 12,850 cal BP.

3. The strong pulse of early human use of Danger Cave, Smith Creek Cave, and Bonneville Estates Rockshelter, 12,850-10,560 cal BP, coincided with a period of cool, possibly wet climate. Multiple proxy records suggest that the western Bonneville basin was well-watered during this time. Lake Bonneville transgressed to an elevation of about 1288 m (4250 ft.) and constructed the Gilbert shoreline 12,350-11,950 cal BP (Oviatt et al. 2005). Although lake levels fell shortly after this time (Broughton et al. 2000), water continued to flow into the western Bonneville basin for another 2000 years, primarily from the south along the Old River Bed (Oviatt et al. 2003). A similar hydrologic record exists for the Sunshine Well site (located in Long Valley west of the Bonneville basin, central Nevada), which was well-watered during the TPEH until about 9650-9550 cal BP. Paleobotanical data suggest that the continued flow of water in the region was as much a consequence of cooler temperatures as it was increased summer precipitation (Madsen et al. 2001; Rhode 2000). The presence of sage grouse, marmot, and black bear in the TPEH deposits at Bonneville Estates conforms to these other proxy data as well (Hockett 2007). This period of relatively cool, wet climate has been attributed to the Younger Dryas (Madsen 2002; Oviatt et al. 2003), but clearly these conditions as well as the use of the region's caves persisted for some centuries after the close of this climatic period (11,500 cal BP).

4. All three sites record a significant hiatus in human occupation during the early Holocene. Early human use

of the caves ceased by 11,350 cal BP, except at Bonneville Estates where it continued until 10,650 cal BP. Danger Cave does not appear to have been occupied again until 9450 cal BP, Bonneville Estates Rockshelter does not appear to have been occupied again until 8380 cal BP (with the exception of a possible ephemeral event around 10,000 cal BP), and Smith Creek Cave was not re-occupied until the late Holocene. Similarly, other caves in the region (e.g., Hogup Cave and Camels Back Cave) were not occupied at all until after 9450 cal BP (Aikens 1970; Schmitt and Madsen 2005). Perhaps the lack of early Holocene human occupations in the region's caves before 9450 cal BP is merely a sampling problem, but we suspect that it may be related to regional climate change. Multiple lines of evidence suggest that the eastern Great Basin became very hot and dry during this time, especially around 10,000 cal BP. The Old River Bed dried up completely by 10,000 cal BP (Oviatt et al. 2003; Young et al. 2005), marshes dried up in the small basins to the west of the Bonneville basin by 9500 cal BP (Huckleberry et al. 2001; Thompson 1992), and xeric-adapted small mammals replaced montane-and-mesic-adapted small mammals between 10,000 and 9200 cal BP at places like Bonneville Estates, Homestead Cave, and Camels Back Cave (Grayson 1998; Hockett 2007; Schmitt et al. 2002). Oviatt et al. (2003) have argued that human use of the Old River Bed continued well into the early Holocene, and that there was no gap between the Pre-Archaic and Early Archaic occupations there. This interpretation, though, is based on relative dating of surface artifact finds and needs to be confirmed with ^{14}C dates on buried and sealed cultural deposits. For now we conclude that there is no unequivocal evidence for continuous human occupation of the western Bonneville basin through the early-middle Holocene, and that between about 10,650 and 9450 cal BP humans responded to aridification by ceasing to use all but the best-watered areas of the eastern Great Basin landscape, or by abandoning the region entirely.

5. Human occupations of Danger Cave and Bonneville Estates Rockshelter resumed during the early-middle Holocene; at Danger Cave this occurred by 9450 cal BP and at Bonneville Estates by 8250 cal BP. Danger Cave was probably re-occupied at such an early time because of its location adjacent to a productive marsh, where pickleweed seeds and other marsh-adapted plants and animals could be intensively collected, even during the driest times. When humans re-appeared there, they certainly were taking full advantage of these resources (Rhode et al. 2006). The lack of a nearby spring or marsh likely precluded use of Bonneville Estates in this fashion. Instead, re-occupation of Bonneville Estates may correlate to a wet period that occurred 8200-8000 cal BP (Hockett 2007). When humans resumed their use of Bonneville Estates at this time, they were focusing much of their subsistence on the hunting of artiodactyls like pronghorn and bison (Hockett 2007). The presence of these animals in the rockshelter's faunal assemblage implies the presence of abundant grasses and sagebrush on the hillslopes flanking the western Bonneville basin; this vegetation community likely resulted from an increase in precipitation, specifically summer rainfall. This hypothesized mesic episode may relate to the so-called "8.2-k" event, a 160-year-long period of cooling in the North Atlantic that may have led to significant climate change in many regions of the world (Alley and Ágústsdóttir 2005; Schmidt and LeGrande 2005; Thomas et al. 2007). As Graf (2007) and Hockett (2007) have pointed out, from this time onward through the middle Holocene, intensity of human use of Bonneville Estates Rockshelter seems to have kept pulse with fluctuating climate. For example, another middle Holocene occupation at Bonneville Estates dating to about 7200-6800 cal BP (Graf 2007) may correspond to another "anomalously" wet period indicated by regional paleobotanical records (Rhode 2000; see also Madsen et al. 2001).

Without doubt, the caves and rockshelters of the Great Basin have provided and will continue to provide an important resource to archaeologists studying the region's terminal Pleistocene and early Holocene archaeological record. Cultural chronologies are based largely on evidence from these enclosed sites, and virtually everything we know about early subsistence behavior has been gleaned from the well-preserved faunal and floral remains that they contain. Given the relative importance of the cave record in this part of the world, it is especially important that we always interpret their contents through a taphonomic perspective focusing on natural as well as cultural formation processes. A broad array of taphonomic agents is known to have contributed to the deposition of the materials found in these natural shelters, making it especially difficult to interpret resulting ^{14}C ages. Chronologies of cultural events must be developed through the dating of materials clearly relevant to those cultural events. Every dated sample needs to be carefully evaluated in terms of its material, origin, potential for redeposition, and potential for contamination. By focusing just on ^{14}C ages from samples clearly of human origin (i.e., charcoal from hearths, perishable artifacts, human coprolites), we can build accurate chronologies of human occupations in these sites, and can more precisely relate these to climatic events and better understand the evolution of human adaptations. As Dave Madsen (personal communication, 2007) likes to remind us, "if people would just remember the mantra, 'when in doubt, throw it out,' we would all be way ahead."

Acknowledgements. Many students have contributed to the excavations at Bonneville Estates Rockshelter and cataloging of its material remains. Special thanks are due to them as well as to David Madsen, who has offered much food for thought since we began to work at Bonneville Estates in 2000. Thanks also to Y. A. Gomez Coutouly for translating our abstract from English into French.

REFERENCES

AIKENS, C. M. (1970) *Hogup Cave*. University of Utah Anthropological Papers No. 93. Salt Lake City.

ALLEY, R. B.; ÁGÚSTSDÓTTIR, A. M. (2005) The 8k Event: Cause and Consequences of a Major Holocene Abrupt Climate Change. *Quaternary Science Reviews* 24:1123-1149.

BECK, C.; JONES, G. T. (1997) Pleistocene/Early Holocene Archaeology of the Great Basin. *Journal of World Prehistory* 11:161-236.

BROUGHTON, J. M.; MADSEN, D. B.; QUADE, J. (2000) Fish Remains from Homestead Cave and Levels of the Past 13,000 Years in the Bonneville Basin. *Quaternary Research* 53:392-401.

BRYAN, A. L. (1979) Smith Creek Cave. In TUOHY, D. R.; RENDALL, D. L., eds. *The Archaeology of Smith Creek Canyon*, pp. 164-251. Nevada State Museum Anthropological Papers No. 17. Carson City.

BRYAN, A. L. (1988) The Relationship of the Stemmed Point and Fluted Point Traditions in the Great Basin. In WILLIG, J. A.; AIKENS, C. M.; FAGAN, J. L., eds. *Early Human Occupation in Far Western North America: The Clovis-Archaic Interface*, pp. 53-74. Nevada State Museum Anthropological Papers No. 21. Carson City.

ELSTON, R. G.; ZEANAH, D. W. (2002) Thinking Outside the Box: a New Perspective on Diet Breadth and Sexual Division of Labor in the Prearchaic. *World Archaeology* 34:103-130.

FRY, G. (1976) Analysis of Prehistoric Coprolites from Utah. *University of Utah Anthropological Papers* No. 97. Salt Lake City.

GODSEY, H. S.; CURREY, D. R.; CHAN, M. A. (2005) New Evidence for a Late Pluvial Event and Links to Global Climate Trends from Lake Bonneville, Utah. *Quaternary Research* 63:212-223.

GOEBEL, T. (2007) Pre-Archaic and Early Archaic Technological Activities at Bonneville Estates Rockshelter: A First Look at the Lithic Artifact Record. In GRAF, K.; SCHMITT, D., eds. *Paleoindian or Paleoarchaic? Great Basin Human Ecology at the Pleistocene-Holocene Transition*, in press. University of Utah Press, Salt Lake City.

GOEBEL, T.; GRAF, K. E.; HOCKETT, B. S.; RHODE, D. (2004) Late Pleistocene Humans at Bonneville Estates Rockshelter, Eastern Nevada. *Current Research in the Pleistocene* 20:20-23.

GRAF, K. E. (2007) Stratigraphy and Chronology of the Pleistocene to Holocene Transition at Bonneville Estates Rockshelter, Eastern Nevada. In GRAF, K. E.; SCHMITT, D., eds. *Paleoindian or Paleoarchaic? Great Basin Human Ecology at the Pleistocene-Holocene Transition*, in press. University of Utah Press, Salt Lake City.

GRAYSON, D. K. (1983) The Paleontology of Gatecliff Shelter: Small Mammals. In THOMAS, D. H., ed. *The Archaeology of Monitor Valley: 2. Gatecliff Shelter, pp.* 98-126. Anthropological Papers Vol. 59, pt. 1. American Museum of Natural History, New York.

GRAYSON, D. K. (1998) Moisture History and Small Mammal Community Richness during the Latest Pleistocene and Holocene, Northern Bonneville Basin, Utah. *Quaternary Research* 49:330-334.

HARPER, K. T.; ALDER, G. M. (1972) Paleoclimatic Inferences Concerning the Last 10,000 Years from a Resampling of Danger Cave, Utah. In FOWLER, D. D., ed. *Great Basin Cultural Ecology: A Symposium*, pp. 13-23. Desert Research Institute Publications in the Social Sciences No. 8. Reno.

HOCKETT, B. (1991) Toward Distinguishing Human and Raptor Patterning on Leporid Bones. *American Antiquity* 56:667-679.

HOCKETT, B. (1994) A Descriptive Reanalysis of the Leporid Bones from Hogup Cave, Utah. *Journal of California and Great Basin Anthropology* 16:106-117.

HOCKETT, B. (2007) Nutritional Ecology of Late Pleistocene-to-Middle Holocene Subsistence in the Great Basin: Zooarchaeological Evidence from Bonneville Estates Rockshelter. In GRAF, K. E.; SCHMITT, D., eds. *Paleoindian or Paleoarchaic? Great Basin Human Ecology at the Pleistocene-Holocene Transition*, in press. University of Utah Press, Salt Lake City.

HUCKLEBERRY, G.; BECK, C.; JONES, G. T.; HOLMES, A.; CANNON, M.; LIVINGSTON, S.; BROUGHTON, J. M. (2001) Terminal Pleistocene/Early Holocene Environmental Change at the Sunshine Locality, North-Central Nevada, U.S.A. *Quaternary Research* 55:303-312.

JENKINS, D.; CONNOLLY, T.; AIKENS, C. M. (2004) *Early and Middle Holocene Archaeology of the Northern Great Basin*. Anthropological Papers of the University of Oregon No. 62. Eugene.

JENNINGS, J. D. (1957) *Danger Cave*. Anthropological Papers of the University of Utah No. 27. Salt Lake City.

JENNINGS, J. D. (1978) *Prehistory of Utah and the Eastern Great Basin*. Anthropological Papers of the University of Utah No. 98. Salt Lake City.

JONES, G. T.; BECK, C.; JONES, E. E.; HUGHES R. E. (2003) Lithic Source Use and Paleoarchaic Foraging Territories in the Great Basin. *American Antiquity* 68(1):5-38.

MADSEN, D. B. (2002) Great Basin Peoples and Late Quaternary Aquatic History. In HERSHELER, R.; CURREY, D. R.; MADSEN, D. B., eds. *Great Basin Aquatic Systems History*, pp. 387-405. Smithsonian Institution Press, Washington DC.

MADSEN, D. B.; RHODE, D. (1990) Early Holocene Pinyon (Pinus monophylla) in the Northeastern Great Basin. *Quaternary Research* 33:94-101.

MADSEN, D. B., RHODE, D.; GRAYSON, D. K.; BROUGHTON, J. M.; LIVINGSTON, S. D.; HUNT, J.; QUADE, J.; SCHMITT, D. N.; SHAVER, M. W. III (2001) Late Quaternary Environmental Change in the Bonneville Basin, Western USA.

Palaeogeography, Palaeoclimatology, Palaeoecology 167:243-271.

MULLEN, C. O. (1997) *Mammalian Response to Pleistocene/Holocene Environmental Change in the Great Basin: the Jackrabbit's Tale.* Ph.D. Dissertation, University of Nevada, Reno.

NIALS, F. (1999) *Geomorphic Systems and Stratigraphy in Internally-Drained Watersheds of the Northern Great Basin: Implications for Archaeological Studies.* Technical Papers of the University of Nevada Sundance Archaeological Research Fund No. 5. Reno.

OVIATT, C. G. (1997) Lake Bonneville Fluctuations and Global Climate Change. *Geology* 25:155-158.

OVIATT, C. G.; MADSEN, D. B.; SCHMITT, D. N. (2003) Late Pleistocene and Early Holocene Rivers and Wetlands in the Bonneville Basin of Western North America. *Quaternary Research* 60:200-210.

OVIATT, C. G.; MILLER, D. M.; MCGEEHIN, J. P.; ZACHARY, C.; MAHAN, S. (2005) The Younger Dryas Phase of Great Salt Lake, Utah, USA. *Palaeogeography, Palaeoclimatology, and Palaeoecology* 219:263-284.

RHODE, D. (2000) Holocene Vegetation History in the Bonneville Basin. In MADSEN, D. B., ed. Late Quaternary Paleoecology in the Bonneville Basin, pp. 149-163. *Utah Geological Survey Bulletin* 130, Salt Lake City.

RHODE, D.; LOUDERBACK, L. (2007) What is the Evidence for Dietary Plant Use in the Bonneville Basin during the Pleistocene/Holocene Transition? In GRAF, K. E.; SCHMITT, D., ed. *Paleoindian or Paleoarchaic? Great Basin Human Ecology at the Pleistocene-Holocene Transition,* in press. University of Utah Press, Salt Lake City.

RHODE, D.; MADSEN, D. B. (1998) Pine Nut Use in the Early Holocene and Beyond: The Danger Cave Archaeobotanical Record. *Journal of Archaeological Science* 25:1199-1210.

RHODE, D.; GOEBEL, T.; GRAF, K.; HOCKETT, B.; JONES, K. T.; MADSEN, D. B.; OVIATT, C. G.; SCHMITT, D. N. (2005) Latest Pleistocene-Early Holocene Human Occupation and Paleoenvironmental Change in the Bonneville Basin, Utah-Nevada. In PEDERSON, J.; DEHLER, C., ed. *Interior Western United States*, pp. 211-230. Geological Society of America Field Guide 6. Boulder, Colorado.

RHODE, D., MADSEN, D. B.; JONES, K. T. (2006) Antiquity of Early Holocene Small-Seed Consumption and Processing at Danger Cave. *Antiquity* 80(308):328-339.

SCHMIDT, G. A.; LEGRANDE, A. N. (2005) The Goldilocks Abrupt Climate Change Event. *Quaternary Science Reviews* 24:1109-1110.

SCHMITT, D. N.; MADSEN, D. B. (2005) *Camels Back Cave.* University of Utah Anthropological Papers No. 125. Salt Lake City.

SCHMITT, D. N.; MADSEN, D. B.; LUPO, K. D. (2002) Small-Mammal Data on Early and Middle Holocene Climates and Biotic Communities in the Bonneville Basin, USA. *Quaternary Research* 58:255-260.

THOMAS, E. R.; WOLFF, E. W.; MULVANEY, R.P.; STEFFENSEN, J. P.; JOHNSEN, S. J.; ARROWSMITH, C.; WHITE, J. W. C.; VAUGHN, B.; POPP, T. (2007) The 8.2ka Event from Greenland Ice Cores. *Quaternary Science Reviews* 26:70-81.

THOMPSON, R. S. (1992) Late Quaternary Environments in Ruby Valley, Nevada. *Quaternary Research* 37:1-15.

WILLIG, J. A.; AIKENS, C. M. (1988) The Clovis-Archaic Interface in Far Western North America. In WILLIG, J. A.; AIKENS, C. M.; FAGAN, J. L., eds. *Early Human Occupation in Far Western North America: The Clovis-Archaic Interface*, pp. 1-40. Nevada State Museum Anthropological Papers No. 21. Carson City.

YOUNG, D. C.; DUKE, D. G.; WRISTON, T. (2005) *Cultural Resources Inventory of High Probability Lands near Wildcat Mountain: the 2005 Field Season on the US Air Force Utah Test and Training Range-South, Tooele County, Utah. Hill Air Force Base, Ogden, Utah, Project No. U-05-FF-0512m.* Far Western Anthropological Research Group, Davis, CA.

A GIS PERSPECTIVE ON ROCK SHELTER LANDSCAPES IN WYOMING

Mary Lou LARSON

George C. Frison Institute of Archaeology and Anthropology, University of Wyoming, Department 3431, 1000 East University Avenue, Laramie WY 82071, U.S.A.; mlarson@uwyo.edu

Abstract. Long term intensive research on rockshelters and caves throughout the world has yielded some of the best preserved and most informative data about human existence. Many of these investigations include multiple locations within a region that have produced significant comparisons between shelters. The use of geographic information systems (GIS) in rockshelter research allows new and exciting perspectives about the landscape within which rockshelters are found. This paper discusses GIS projects for the rockshelters in north central Wyoming. The two GIS projects include data on 138 previously recorded shelters in the Bighorn Mountains and Basin many of which have been excavated intensively and 179 shelters recorded in a survey of Paint Rock canyon. Comparison of some of the physical features of these shelters provides new ways to look at ancient human occupation of rockshelters.

Keywords: GIS, rockshelters, Bighorn Mountains, Wyoming, archaeology.

Résumé. La recherche intensive à long terme sur des abris et des cavernes dans le monde entier a rapporté certaines des données mieux préservées et la plupart des instructives au sujet de l'existence humaine. Plusieurs de ces investigations incluent les endroits multiples dans une région qui ont produit des comparaisons significatives entre les abris. L'utilisation des géographiques information systèmes (GIS) dans la recherche des abris permet de nouvelles et passionnantes perspectives au sujet du paysage dans lequel des abris sont trouvés. Cet article discute des projets de GIS pour les abris au Wyoming central du nord. Les deux projets de GIS incluent des données sur 138 abris précédemment enregistrés dans Bighorn Mountains et Basin d'Amérique du nord beaucoup dont ont été excavés intensivement et 179 abris enregistrés dans Paint Rock Canyon. La comparaison de certains des dispositifs physiques de ces abris fournit de nouvelles manières de regarder le métier humain antique des abris.

Mots-clés: GIS, abris, Bighorn Mountains d'Amérique du nord, Wyoming, archéologie.

Long term intensive research in rockshelters and caves throughout the world has yielded some of the best preserved and most informative data about human existence (Galanidou 2000; Walthall 1998). Many of these investigations include multiple locations within a region that have produced significant comparisons between time periods and cultural groups (e.g., Laville et al. 1980; Regional Synthesis section, this volume) Despite their regional focus, few studies have gathered data through systematic, reconnaissance survey (e.g., Borrero, this volume; Komšo and Blečić, this volume) and even fewer include "non-sites" or rockshelters and caves without archaeological deposits as part of their sample (Montagnari Kokelj, this volume). The creation of geospatial data sets with comparable information on all shelters in a region presents a new approach to the study of rockshelters and caves (Montagnari Kokelj, this volume; Montagnari Kokelj et al. 2003). Using data collected during major rockshelter excavations conducted over the last 40 years, revisits to previously recorded shelters and intensive survey for unrecorded rockshelters, this article discusses some of the ways that geographic information systems are starting to be used to provide a geo-spatio-temporal perspective on ancient human occupation of the rockshelters of the Bighorn region of northern Wyoming.

My study area includes the Bighorn Basin, Bighorn Mountains and the western portion of the Powder River Basin (**Figure 21.1**). The region is characterized by great temperature extremes (http://www.wrds.wyo.edu/wsc/wsc.html). Temperatures range from as low as minus 48 degrees Celsius in winter to as high as 43 degrees in summer at Worland in the southern Bighorn Basin and a more equitable minus 44 to 31 degrees Celsius at Burgess Junction located in the northern Bighorn Mountains (**Figure 21.1**). Precipitation is generally low, but is mediated by elevation of the reporting station. Worland receives only 17 cm of precipitation a year while Burgess Junction reports 55 cm of precipitation on average.

Northern Wyoming is home to hundreds of rockshelters, many of which contain stratified archaeological deposits. Numerous shelters have been excavated by archaeologists from the University of Wyoming (Frison 1962, 1965, 1968, 1976, 1991; Huter 2001; Miller 1988; Shaw 1982). Wilfred Husted (1969) reported on the excavations of several well-known rockshelters along Bighorn reservoir on the Wyoming-Montana border. The Mummy Cave site (Hughes 2003; Husted and Edgar 2002) located in northwestern Wyoming while not included in our sample provides an excellent comparison of shelters excavated outside of our study region.

Most of the Bighorn shelters were excavated in the 1960s and 1970s, but have been revisited on several occasions since to obtain radiocarbon for dating to more clearly

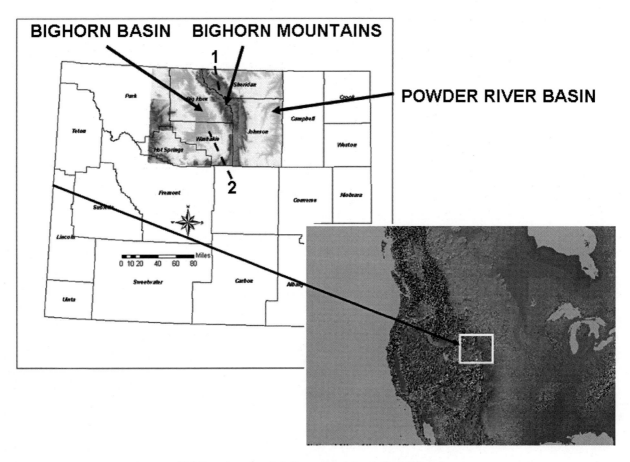

21.1 Location of rockshelter research in northern Wyoming.
Map of North American continent from the National Atlas of the United States - http://www.nationalmap.gov. Climate reporting stations: 1) Burgess Junction and 2) Worland.

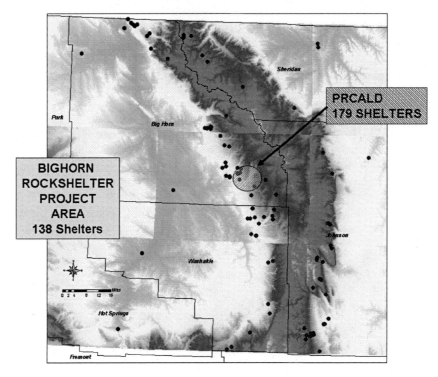

21.2 The Bighorn shelter and Paint Rock Archaeological Landscape District (PRCALD) project areas.
Plotted sites are from Bighorn shelter sample.

understand sites stratigraphy and to better record their contents and location (e.g., Finley, this volume; Finley et al. 2000; Kornfeld, this volume; Nelson 2004; Prasciunas et al. 2004). Many of these rockshelters have yielded stratified deposits and the more spectacular excavated shelters typify how most archaeologists and the public think about Bighorn rockshelters today. However, it is when the smaller shelters are combined with the large spectacular shelters that we can begin to understand prehistoric rockshelter use in the Bighorns.

Time periods represented within many of the excavated cave sites range from Early Paleoindian period to historic occupations that began just over 100 years ago (Frison 1991; Kornfeld, this volume). All caves carry a sequence of chronologically diagnostic projectile points, and/or radiocarbon dates (Kornfeld, this volume). Many of the caves also have rock art. Rock art in the region is primarily Late Prehistoric (1500-250 BP) in age based on radiocarbon dates (Francis and Loendorf 2002:183). Much of rock art of the Bighorn study area is attributed to Great Plains cultural groups and extends as far back as Archaic times or earlier (Francis and Loendorf 2002:185, 195).

Starting in the 1990s and continuing to today, investigations at several rockshelters and areas in the Bighorn region have involved returning to many of the older shelters to answer questions and beginning survey to locate new unrecorded shelters. The two primary projects concerned with rockshelters of northern Wyoming and southern Montana **(Figure 21.2)** are:

1) The Bighorn rockshelter project, which focuses on the shelters that were excavated and recorded during the last 35 years including a few on the eastern slope of the Bighorn mountains and in the Powder River Basin, and the Bighorn Canyon and Pryor Mountain rockshelters of northern Wyoming and southern Montana.
2) The Paint Rock Canyon Archaeological Landscape District or PRCALD, which combines intensive investigations of previously recorded shelters with archaeological survey to locate new shelters. PRCALD is an archaeological landscape district listed on the United States National Register of Historic Places.
Over the past five years, my students and I have been involved in the creation of geographic information systems (GIS) for each project.

OVERVIEW OF THE ARCHAEOLOGY AND CHRONOLOGY OF THE BIGHORN ROCKSHELTER PROJECT

This section discusses the archaeology from the 138 shelters recorded for the Bighorn rockshelter project. Our evidence shows occupation of the shelters throughout the time that humans have existed in the Rocky Mountain region (11,500[1] BP – present). A few of the larger shelters have a continuous record of occupation throughout prehistory, while many others contain stratified levels from several different time periods. My discussion considers the Bighorn shelters that contain components dating to Paleoindian, Archaic (Early, Middle, and Late) and Late Prehistoric/Protohistoric times.

Paleoindian Archaeology (11,500 – 7700 BP)

Fourteen sites have at least one component recorded for the time between 11,000 and 7500 years ago. These 14 sites provide evidence of occupation of the Bighorn region at the time when the first people were moving into the New World. Besides radiocarbon dates, excavators have relied on the presence of diagnostic projectile points to establish the age of different levels in the shelters.

Excavations at Ditch Creek in the southwest of the study area have yielded a Clovis point. Two Moon shelter (Kornfeld, this volume), contains a Folsom (~10,200 – 11,000 BP) level and an Agate Basin level (~10,000 BP) in a stratigraphically separate layer immediately above the Folsom level. The most common Paleoindian projectile points found in the mountain rockshelters are those from the Late Paleoindian Foothill/Mountain Tradition **(Figure 21.3A)**, dated between about 9500 and 7700 BP (Frison 1991:27). Additionally, Late Paleoindian storage pits were excavated in 8300 year old deposits at the Medicine Lodge Creek site (Frison 1991:342-343), another stratified rockshelter within the Bighorn Shelter study area.

Archaic Archaeology (7700-1500 BP)

The archaeology of the Bighorn Mountains and foothills is perhaps most famous for its Early Archaic archaeology. We have seventeen shelters within the Bighorn rockshelter sample with material dating to this time (7700 – 4500 BP) (Frison 1962, 1976, 1991; Frison and Huseas 1968; Miller 1988; Shaw 1982). Early Archaic (7700-4500 BP) shelters contain numerous features, both fire hearths and storage pits as well as diagnostic side-notched projectile points **(Figure 21.3A)**.

Middle Archaic components are found in the largest number of shelters within our sample of sites with components located in 19 shelters. The Middle Archaic (4500 – 3000 BP) occupations are marked by a change in projectile point styles to the McKean complex that includes lanceolate McKean, stemmed Duncan, and the shouldered with expanding stem Hanna variants (Wheeler 1954). Other projectile point types, particularly those with Great Basin affiliations, are known as well. Middle Archaic components in Bighorn shelters often contain slab-lined features (commonly fire hearths/ovens), ground stone, and a diversity of fauna (bison, mountain sheep, and deer) (Frison 1991:89; Hughes 2003).

---

[1] Dates are given in uncorrected radiocarbon years BP

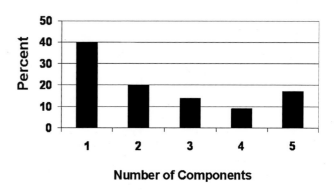

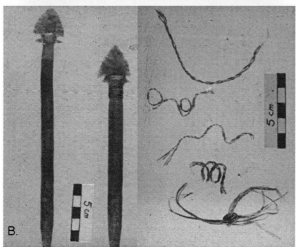

21.4 Percentage of archaeological components per shelter for Bighorn rock shelter data set.

21.3 Artifacts from Southsider Shelter and Spring Creek cave. A. Projectile points from Southside Shelter. Top row – Early Archaic projectile points, bottom row, Late Paleoindian projectile points. B. Hafted Late Archaic corner notched projectile points with foreshafts (left) and cordage (right) from Spring Creek Cave.

Late Archaic (3000 – 1500 BP) occupations of shelters (n=15) contain some of the most intriguing archaeology in the region if only because of fluke of preservation within dry cave sites in the region. Late Archaic artifacts found in shelters include projectile points in the haft, bone foreshafts, basketry, cordage, and many tools such as drills, scrapers, and gravers that were made to manufacture the perishable material found in the caves **(Figure 21.3B)** (Frison 1962 1965, 1968, 1991). Frison et al. (1986) argue for similarities in construction techniques of basketry found in the Bighorn dry cave sites and baskets from the Great Basin.

Late Prehistoric and Protohistoric Archaeology (1500-175 BP)

Finally, the Late Prehistoric Period occurs at fifteen of the rock shelters within the project area. This period is marked by a shift to much smaller projectile points commonly associated with the adoption of the bow and arrow. Pottery attributed to Crow or Mandan groups and carved steatite vessel fragments have been found in the Bighorn Mountains (Adams 1992; Frison 1991:117).

Many of the Bighorn shelters contain numerous components, some of which extend throughout the time of human occupation. Of the sample of shelters with information on dated components obtained from the Wyoming SHPO cultural records office (n= 100), 40% of the shelters contain only one component, but 17% of them contain components from all time periods and 26% have either four or five components **(Figure 21.4)**. As will be discussed below, one of the questions is why so many shelters were occupied and if we are merely seeing the product of over 40 years of intensive archaeological investigations rather than a representative sample of all shelters with or without archaeological deposits.

We are still gathering data for the Bighorn shelter study. In particular, we need to obtain spatial data useful for our GIS and intend to create the same type of data that we have collected for the Paint Rock archaeological district (see below). However, this involves investigating each individual shelter and gathering spatial data from a myriad of sources. The GIS may allow us to obtain information that otherwise would be accessible only by visiting each shelter. For example, it will be possible to use a combination of site photographs and digital elevation models to obtain information on the aspect of the shelter without visiting the site.

PAINT ROCK CANYON ARCHAEOLOGICAL LANDSCAPE DISTRICT ARCHAEOLOGY

The Paint Rock Canyon Archaeological Landscape District project provides a very different spatial resolution on the shelters of the Bighorns, but one that informs on the well known Bighorn Shelter sample as well. Ranging from high elevations above the headwaters of the canyon to where Paint Rock creek empties out into the Bighorn Basin near the town of Hyattville, Paint Rock Canyon represents the broad range of diverse environments that typify the mountain and foothill regions of north central Wyoming **(Figure 21.5)**.

21.5 Paint Rock Canyon looking west towards the Bighorn Basin.

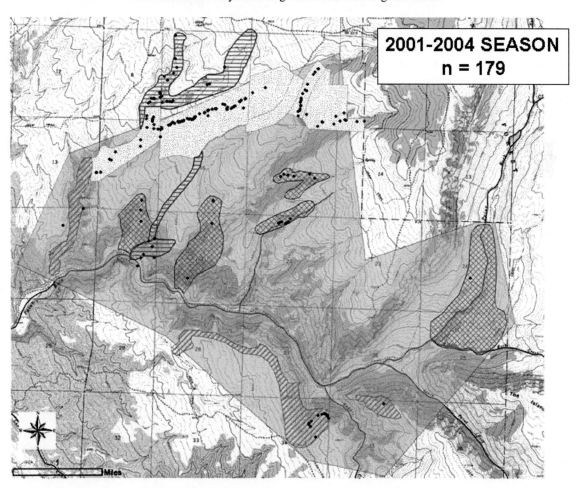

21.6 Paint Rock Archaeological Landscape district overlaid on Allen Draw 7.5' quadrangle showing rock shelters and survey areas. Black triangles show rockshelter location.

The PRCALD survey

The PRCALD project involves archaeological survey for rock shelters recorded in the 1970s and identification of previously unrecorded shelters. Locations of the shelters recorded in the 1970s were plotted on 7.5' USGS quadrangles, but subsequent work in the canyon indicates that the many of the shelters were plotted incorrectly. The density of rock shelters in Paint Rock canyon **(Figure 21.6)** made it virtually impossible to identify which shelter had been recorded on the old site form. The original archaeological site form for each rock shelter provides virtually no information on a shelter's archaeological contents, its setting, or other data that are routinely collected today. Additionally, and critical to the GIS, the original location for each shelter was recorded using the Public Land Survey System (PLSS - consisting of township, range, and section) most often to the nearest quarter section (¼ by ¼ mile or 16,000 square meters). The PLSS was the standard location system used by archaeologists in the 1970s, but it is not a coordinate system that a GIS can use to locate a point on the earth's surface. For all of these reasons, the University of Wyoming began working with funding from the Bureau of Land Management, Worland Field Office to resurvey the canyon and properly record the shelters on Cultural Property forms from the Wyoming SHPO cultural records office (http://www.wyshpo.wy.us/).

The survey identified all shelters within the study area including those with no visible archaeological material on the surface. Survey teams plotted the shelters on USGS 7.5' quadrangles, took a GPS reading in UTMs for the shelter, and photographed each shelter. In addition to the Wyoming Cultural Properties form, surveyors also recorded data specific to rock shelters. These data include information on shelter size, bedrock type, aspect (looking out from the shelter), whether the shelter is hidden by vegetation, the nature of deposition on the floor and in front of the shelter, archaeological materials observed, and evidence of disturbance.

PRCALD GIS

The current PRCALD GIS contains all of the information recorded in the field by surveyors working from 2001 to 2004 as well as other data available from other sources. The GIS allows display of the USGS 7.5 Allen Draw quadrangle, the streams within the Paint Rock vicinity, the individual survey blocks, and the archaeological sites (virtually all rock shelters) recorded during the surveys **(Figure 21.6)**.

One of the strengths of a GIS is its ability to link maps and data. Our GIS is in ArcGIS 9.1, which allows direct linkage to a data table within the software. Each line in the data table holds information for a single shelter recorded in the Paint Rock landscape, For example, shelter 24 (RS01-24) contains prehistoric archaeology, whereas the shelters surrounding it do not **(Figure**

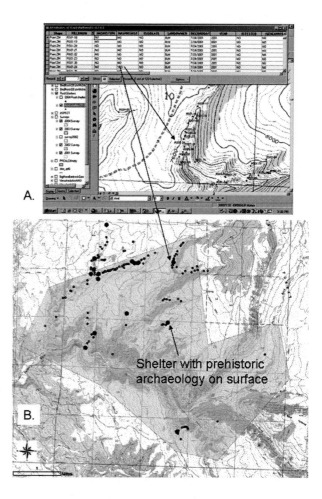

21.7 Paint Rock Canyon Archaeological Landscape district showing ArcGIS 9.1 map and data table (top) and rock shelters with those containing archaeological material on the surface of the shelter highlighted (bottom). Arrows on both figures point to the same shelter highlighted on the data table.

21.7A). The GIS highlights both the line in the data table and the item on the map simultaneously **(Figure 21.7A)**. Another example of the use of GIS to understand, display, and visualize patterns can be seen in the display of all of the recorded shelters with those shelters where survey crews observed prehistoric archaeology on the surface of the shelter highlighted **(Figure 21.7B)**.

Analysis of PRCALD Rock Shelters

Analysis of the data collected in the 2001 through 2004 surveys is still in its preliminary stages. We have begun to look at the data entered into the GIS and have some initial results to present. The analysis relies in large part on the collected wisdom of years of site location and predictive modeling studies (e.g., Judge and Sebastian 1988; Kvamme 1985, 1990; Westcott and Brandon 2000; Mehrer and Westcott 2006). In these studies, locations with what are thought to be optimal environmental conditions for human settlement (e.g., low to no slope, southern (or western) aspect, near to water and other resources) are the ones predicted to contain

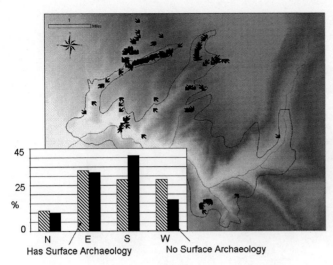

21.8 Aspect of All Paint Rock Canyon Rock Shelters. Bar chart shows percentage of shelters with archaeology visible on the surface (hatched) and the percentage of shelters with no archaeological material visible on the surface (solid).

archaeological sites. Although I have not done a predicted model of PRCALD sites here, what follows is a discussion of those variables common to many predictive models.

The Paint Rock Canyon shelters vary greatly in form and size; ranging from very small to large, shelters with deposition and depth or very shallow indentations in the bedrock along the canyon wall. The depth of shelters varies from a 10 m deep cave to boulders that have fallen down the slope providing a small protected area on one side of the boulder. However, for the Paint Rock Canyon shelters, no statistically significant difference exists between shelter size (height and width at opening, depth, height at back of shelter) for those sites with archaeology on the surface and those without.

Following the premise that aspect (what one can see from the shelter) is one of the factors that humans use to select a shelter for occupation, I expect there will be differences in the aspect of shelters that contain archaeology and those that do not. Investigation of PRCALD shelter aspects yields unexpected patterning. A map of the aspect of PRCALD shelters **(Figure 21.8)** shows no obvious patterning in aspect direction, although a graph of shelters with archaeology on the surface seems to indicate a preference for eastern, southern, and western aspects. However, because the Paint Rock Canyon survey provides information on not just those shelters with archaeology, but shelters with no visible archaeology, it is possible to discover if the aspect played a role in shelter selection **(Figure 21.8)**. When one compares the shelter with archaeology to those without, there is no statistically significant difference between the two types of shelters. Test excavations at shelters with deposition, but no archaeology on the surface may alter the aspect percentages. However, at this point in time, one can only conclude that the majority of all shelters in Paint Rock Canyon face south, east, or west regardless of whether they contain archaeology.

A very different distribution exists for the sample of Bighorn shelters that have aspect recorded for them (n=96). For those Bighorn shelters with known aspect 56% face south, 25% face east, eight percent face west, and 11 percent face north. Compared to the PRCALD sample, southern exposure seems to be a choice for habitation in the Bighorn shelter sample. Without data on the aspect of shelters with no archaeology, it is impossible to say if the selection of south facing shelters is based on a predominance of this shelter type or if prehistoric inhabitants preferred this shelter aspect over others.

Interestingly, in a comparison between Bighorn shelters (the majority of which have archaeology) and PRCALD shelters with visible archaeology (a minority), the aspect of the two samples differs significantly from the other (χ^2 = 26.113. df = 3, p < .001, Cramer's V = .307). Several hypotheses can be proposed to explain this patterning.

First, given that the Bighorn region is one of the hotter places in this Central Rocky Mountains and that summer temperatures easily exceed 40° C both in Paint Rock Canyon and in the Bighorn Basin, the choice of rock shelter aspect may vary seasonally. It may be that the difference in aspect between the Bighorn and Paint Rock sample may indicate shelter use at different seasons of the year, in that shelters in the Basin are occupied during cooler months of the year and shelters in the foothills (Paint Rock Canyon) are occupied during the summer and fall. Second, the Bighorn shelter sample is one of large, often deeply stratified archaeological deposits whereas the PRCALD sample is dominated by very shallow shelters with at this point, unknown deposition, it may be that the two samples represent two or more poses in the structure of Bighorn region settlement. Third, season, temporal, and spatial differences need to be considered to infer a truly reliable model of rockshelter use in the Bighorns and surrounding areas. Finally, a consideration of cultural factors as represented in the rock art on the walls of some of the larger shelters should be included in any consideration of rockshelter use in the basin, foothills, and mountains of northern Wyoming.

The slope of the ground at an archaeological site is another variable that is often considered important to people's selection of one location over another. The terrain at Paint Rock Canyon is very steep and rugged **(Figures 21.5, 21.9)**. As is the case with aspect, there is no statistical difference between the slope of PRCALD shelters with archaeology on the surface and those without. All recorded PRCALD shelters have a mean slope of 26 degrees, while the archaeology bearing shelters have a mean of 25 degrees. Even though crews have surveyed some of the flatter areas above the canyon rim **(Figures 21.5, 21.6)** they have found little archaeology (and few shelters) on the flat lying areas.

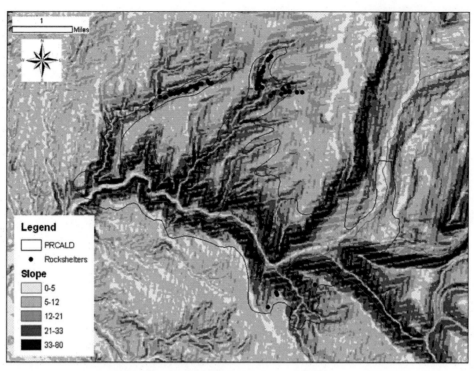

21.9 Slope in degrees of Paint Rock Canyon rock shelters. Areas with the darkest shading have the steepest slope, the lighter shading indicates flatter slopes. Small black dots are rockshelters.

Most of the shelters lie just below the canyon rim, notched into the almost vertical bedrock exposures. Some boulders that serve as shelters have made it into the flatter areas in the bottom of the canyon.

In comparison, Bighorn shelters have an average slope of 14 degrees, much less than the Paint Rock average of 25 degrees for shelters with archaeology. This most likely represents differences in the terrain in which the shelters occur, but again this needs investigation. I imagine if data were collected in the Basin for non-archaeology bearing shelters they too would probably be associated with gentler slopes.

A possible explanation for the differences identified between the two samples becomes apparent when one looks at the number of shelters within each project that contain archaeology. Whereas the vast majority of the sites in the Bighorn shelter project contain prehistoric archaeology (n=125 91%), only 13% (N=24) of the shelters recorded at Paint Rock Canyon have prehistoric archaeology visible on the surface of the shelter. This pattern is an artifact of the differences between how the data have been recorded. Where the Bighorn shelters are known primarily because they have been excavated and contain a rich archaeological record, the Paint Rock Canyon shelters are the product of several years of archaeological reconnaissance survey in the canyon, where the methodology involved the recording of all possible shelters, not just those containing archaeology.

It will be interesting to compare the archaeology within the samples in the two study areas. The reason for occupation of shelters along Paint Rock Canyon may differ significantly from the shelters in the Bighorn Basin. No matter what the reason is, the Bighorn Shelters provide us with detail on the richness of the archaeological use of rock shelters and caves in the study area; the Paint Rock shelter data provides the background upon which to array rockshelter use in northern Wyoming.

CONCLUSIONS

The landscape of rockshelters in northern Wyoming is starting to provide insight into hunter-gatherer use of these features. Whereas the Bighorn shelter sample gives us information about the occupation and re-occupation of many shelters throughout the time humans lived in the Rocky Mountains, the Paint Rock sample provides data about shelter use in a single canyon. Kvamme (2006:18) argues that rockshelters should be left out of predictive models, noting that rockshelter placement is not determined by human preference. Because the Bighorn and PRCALD samples contain just rockshelters, the issue of human choice in which shelters to occupy, how and why to use a shelter can be addressed through predictive models as well as other spatial analyses. It appears that the common factors used to predict site selection (e.g., aspect, slope) do not work as well as one would predict within the Paint Rock Canyon sample. Whereas the Bighorn shelter sample coincides more closely with the perceived wisdom of site selection. Other factors, some of which are obtainable through GIS analysis, can be used to provide the answer to why the two project samples differ. The Bighorn and Paint Rock shelter studies promise to provide a new perspective on the scale and resolution of rock shelter use by prehistoric hunter gatherers in the area.

Acknowledgements. Numerous students through the years have contributed to the data collection, data entry, and data analysis of the Bighorn and PRCALD shelters – I wish to thank them all collectively for their hard work in putting these data together. Mike Bies of the Worland BLM field office has been and continues to be the driving force behind our investigations in the Bighorns.

REFERENCES

ADAMS, R. A. (1992) *Pipes and bowls: steatite artifacts from Wyoming and the region.* Unpublished Master's thesis, Department of Anthropology, University of Wyoming, Laramie.

FINLEY, J.B.; D. A. BYERS; C. FINLEY; M. KORNFELD; and G.C. FRISON (2003).*The Black Mountain Archaeological District: 2002 Field Studies.* Technical Report No. 24, George C. Frison Institute of Archaeology and Anthropology, University of Wyoming, Laramie.

FINLEY, Judson; M. KORNFELD; and G. C. FRISON (2000). *Black Mountain Archaeological District: Preliminary Report for the 1999 Field Season.* Technical Report No. 16b Department of Anthropology, University of Wyoming, Laramie.

FRANCIS, Julie E. and Lawrence LOENDORF (2002). *Ancient visions.* University of Utah Press, Salt Lake City.

FRISON, G. C. (1962). Wedding of the Waters cave: a stratified site in the Bighorn Basin of northern Wyoming. *Plains Anthropologist* 7:246-265.

FRISON, G. C. (1965). Spring Creek Cave, Wyoming. *American Antiquity* 31:81-94.

FRISON, G. C. (1968). Daugherty Cave, Wyoming. *Plains Anthropologist* 13:253-295.

FRISON, G. C. (1976). The chronology of Paleo-Indian and Altithermal Period groups in the Bighorn Basin, Wyoming. In *Culture change and continuity: essays in honor of James Bennett Griffin,* edited by Charles E. Cleland, pp. 147-173. Academic Press, New York.

FRISON, G. C. (1991) *Prehistoric Hunters of the High Plains.* Academic Press, San Diego.

FRISON, G. C.; J. M. ADOVASIO; and R. C. CARLISLE (1986). Coiled Basketry From Northern Wyoming. *Plains Anthropologist* 31(112):163-167.

FRISON, G. C. and M. HUSEAS (1968). Leigh Cave, Wyoming, Site 48WA304. *Wyoming Archaeologist* 11(3):20-33.

GALANIDOU, N. (2000) Patterns in Caves: Foragers, Horticulturalists, and Use of Space. *Journal of Anthropological Archaeology* 19:243-275.

HUGHES, S. (2003). *Beyond The Altithermal: The Role of Climate Change in the Prehistoric Adaptations of Northwestern Wyoming.* Unpublished Ph.D. dissertation, Department of Anthropology, University of Washington, Seattle.

HUSTED, W.M. (1969) *Bighorn Canyon Archaeology.* Smithsonian Institution River Basin Surveys, Publications in Salvage Archaeology, No. 12. Washington, DC.

HUSTED, W. M.; and R. EDGAR (2002). *The Archaeology of Mummy Cave, Wyoming: An Introduction to Shoshonean Prehistory.* National Park Service, Midwest Archaeological Center Special Report No. 4 and Southeast Archaeological Center Technical Reports Series No. 9. Lincoln, Nebraska.

HUTER, Pamela (2001) *Assessment of Changing Diet Breadth at Southsider Shelter.* Unpublished Master's thesis, Department of Anthropology, University of Wyoming, Laramie.

KORNFELD, M.; J. B. FINLEY; J. KENNEDY; J. DURR; and G. C. FRISON (1997) *Black Mountain archaeological district: Preliminary report for the 1996 field season.* Technical Report No. 16, Department of Anthropology, University of Wyoming, Laramie.

JUDGE, W. J. and L. SEBASTIAN EDS. (1988) *Quantifying the Present and Predicting the Past: Theory, method, and application of archaeological predictive modeling.* U.S. Department of the Interior, Bureau of Land Management, Washington D.C.

KVAMME, K. L. (1985) Determining empirical relationships between the natural environment and prehistoric site locations: a hunter-gatherer example. In *For Concordance in archaeology: Bridging data structure, quantitative techniques, and theory,* edited by C. Carr, pp.208-238. Westport Publishers, Kansas City.

KVAMME, K. L. (1990) The fundamental principles and practice of predictive archaeological modeling. In *Mathematics and Information Science in Archaeology; A flexible framework,* edited by A. Voorrips pp. 257-295. Holos Verlag, Bonn.

KVAMME, K. L. (2006) There and back again: Revisiting archaeological locational modeling. In *GIS and Archeological Site Location Modeling,* edited by M. W. Mehrer and K. L. Westcott, pp. 3-40. CRC Taylor and Francis, Boca Raton.

LAVILLE, H.; J. P. RIGAUD; and J. SACKETT (1980) *Rockshelters of the Perigord.* Academic Press, New York.

MILLER, Karen G. (988) A comparative analysis of cultural material recovered from two Bighorn Mountain archaeological sites: A record of 10,000 years of occupation. Unpublished Master's thesis, Department of Anthropology, University of Wyoming, Laramie.

MONTAGNARI KOKELJ. E. (2003) GIS and caves: An example from the Trieste Karst (northeastern Italy). In *General Sessions and Posters: Session 1 Method and Theory,* edited by Le secrétariat du Congrés. BAR International Series 1145

NELSON, K.; J. FINLEY; M. KORNFELD; and G. C. FRISON (2004) *Preliminary report of the 2003 Field Season at the Black Mountain archaeological District.* Technical Report No. 32, George C. Frison Institute of Archaeology and Anthropology, University of Wyoming, Laramie.

PRASCIUNAS, M.; M. L. LARSON; M. KORNFELD; R. KELLY; and G. C. FRISON 2004 *Results of 2003 Investigations in Paint Rock Canyon archaeological district.* Technical Report No. 30, George C. Frison Institute of Archaeology and Anthropology, University of Wyoming, Laramie.

SHAW, L. C. (1982). *Early Plains Archaic Procurement Systems during the Altithermal: the Wyoming Evidence.* Unpublished Master's thesis, Department of Anthropology, University of Wyoming, Laramie.

WALKER, D.N. (1975). *A Cultural and Ecological Analysis of the Vertebrate Fauna from the Medicine Lodge Creek Site (48BH499).* Unpublished Masters's thesis, Department of Anthropology, University of Wyoming, Laramie.

WALTHALL, J.A. (1998) Rockshelters and hunter-gatherer adaptation to the Pleistocene/Holocene transition. *American Antiquity* 63:223-238.

WESTCOTT, K. L. and J. BRANDON (2000) *Practical applications of GIS for Archaeologists: A predictive modeling toolkit.* Taylor and Francis, London.

WHEELER, R.P. (1954) Two new projectile point types: Duncan and Hanna points. *Plains Anthropologist* 1:7-14.

THE GEOLOGIC AND GEOMORPHIC CONTEXT OF ROCKSHELTERS IN THE BIGHORN MOUNTAINS, WYOMING

Judson Byrd FINLEY

Department of Anthropology, Washington State University, Pullman, WA 99164-4910; Mail to: 2416 D Rice Ave NW, Albuquerque, NM 87104, U.S.A.; Email: hpdrifter00@yahoo.com

Abstract. Defining the geologic and geomorphic context of rockshelters in the Bighorn Mountains is fundamental to understanding their formation and interpreting their stratigraphy. Bighorn Mountain rockshelters form in limestone and sandstone where bedrock lithology and morphology, hillslope geometry, and shelter aspect influence weathering and sedimentation. Clear stratigraphic and sedimentological differences contrast sediment sources in limestone rockshelters, whereas subtle textural differences mark contrasting sediment sources in sandstone rockshelters requiring grain-size data from deposits and bedrock control samples. Three geoarchaeological examples from the Bighorn Mountains are presented that illustrate geomorphic and sedimentologically based models of rockshelter formation.

Keywords: rockshelter, stratigraphy, formation processes, Bighorn Mountains

Résumé. Définir le contexte géologique et géo-morphique des abris sous-roche dans les montagnes de "Bighorn" est essentiel pour comprendre leur processus de formation et pour interpréter leur stratigraphie. Les abris sous-roche de la montagne de « Bighorn » sont composés de calcaire et de grés, où la lithologie et la morphologie du soubassement, la géométrie des grés, et la physionomie des abris influencent l'effritement et la sédentarisation. De nettes différences stratigraphiques et sédimentologiques contrastent avec les sources de dépôt dans les abris sous-roche en calcaire alors que de subtiles différences de texture montre des sources contrastées de dépôt dans les abris sous roche en grés qui nécessitent des données très pointues provenant d'échantillons de contrôle issues des soubassements. Trois exemples archéologiques illustrent des modèles géo-morphiques et sédimentologiques de la formation des abris sous-roche.

Mots-clés: abris sou-roche, stratigraphie, processus de formation, montagnes des Bighorn.

Geoarchaeological investigations of rockshelters in the Bighorn Mountains of north-central Wyoming have been limited but have the potential to make significant contributions to understandings of site formation processes and environmental change. Analysis of rockshelters in the northern Plains and Middle Rocky Mountains has played a key role in regional cultural history (Frison 1976, 1991; Husted and Edgar 2001; Mulloy 1958). The extensive sample is currently being used to address unanswered archaeological questions of early Paleoindian settlement, Archaic adaptive strategies, social interactions between the Great Basin and Bighorn Basin, and environmental change (see Kornfeld, this volume). To date, little paleoenvironmental research has been conducted in the Bighorn Mountains, and no detailed framework exists for interpreting human forager adaptations to changing environmental conditions. This paper presents a framework for using rockshelter deposits as one tool for environmental reconstruction in the Bighorn Mountains.

More specifically, this paper focuses on geologic and geomorphic variables affecting rockshelter formation with particular emphasis on the importance of characterizing and contrasting autogenic (i.e., roof fall or éboulis) and allogenic (i.e., aeolian, colluvial, and alluvial) sedimentation processes. Understanding variability in physical characteristics and geomorphic processes responsible for deposition of autogenic and allogenic sediments is essential for building climatic linkages. Initially, I discuss Paleozoic bedrock geology as a fundamental control influencing rockshelter development as a location for sediment accumulation and human occupation. Next, I describe the relevant geologic deposits and formations of the Bighorn Basin that serve as potential sources for aeolian sediments in rockshelter stratigraphy. I place considerable emphasis on geomorphic variability within Bighorn Mountain rockshelters highlighting differences between limestone and sandstone settings. Here I explore archaeologically relevant sedimentary facies models of autogenic and allogenic sedimentation. Facies models aid in understanding depositional environments and emphasize identification of key mechanisms of sediment transportation, deposition, and diagenesis (Boggs 2001:262-266), also primary goals of geoarchaeological analysis (Stein 2001). Facies models highlight climatic-geomorphic mechanisms involved in rockshelter formation processes that have implications for analyzing environmental change. Finally, I provide three rockshelter examples where geologic and geomorphic variability result in contrasting formation processes and stratigraphy. These formation processes provide a foundation for reconstructing human adaptive strategies from the archaeological record in rockshelters.

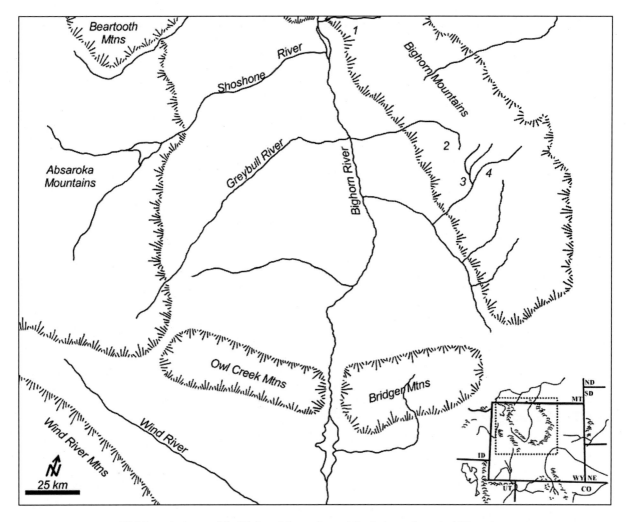

22.1 Location map of the Bighorn Mountains and Basin in north-central Wyoming.
Sites mentioned in text: 1) Eagle Shelter, 2) BA Cave, 3) Medicine Lodge Creek, 4) Southsider Shelter

GEOLOGY OF THE BIGHORN MOUNTAINS AND BASIN

An understanding of bedrock geology is essential to defining the context of regional rockshelter formation. The Bighorn Mountains are the easternmost range of the Middle Rocky Mountain physiographic province (**Figure 22.1**) (Fenneman 1931), a series of northwest-southeast trending folded and faulted Laramide structures. This is the youngest of the Middle Rocky Mountain ranges uplifted largely during the Eocene Epoch approximately 50 million years ago (Ma) (Brown 1993). Superpositioning of streams during uplift incised deep canyons along the western slope of the range exposing thick (>500 m) sequences of Paleozoic sedimentary rocks. Important bedrock units are Madison (Mississippian) limestone, Amsden (Mississippian-Pennsylvanian) limestone and sandstone, and Tensleep (Pennsylvanian-Permian) sandstone (Boyd 1993), all exposed in an extensive belt along the foothills of the Bighorn Mountains. The majority of rockshelters occur in Madison limestone and Tensleep sandstone. There is no regional evidence to indicate rockshelters on the landscape and accumulating deposits prior to the late Pleistocene about 70,000 year ago (ka). Thus, Bighorn Mountain rockshelters are a late Quaternary phenomenon.

Massive sedimentary units deposited during the Mesozoic Era are exposed along the basin-foothills interface. Most notable of these units are the Triassic redbeds of the Chugwater Formation and badlands of the dinosaur-bearing Morrison Formation (Picard 1993). Sandstone, siltstone, and shale associated with the Cretaceous Western Interior Seaway (Steidtmann 1993) also crop out along the basin margin. Eocene claystone and shale dominate the basin interior. The Quaternary history of the region is one of cyclical mountain glaciation with basin deposition and erosion leading to widespread terrace development (Madole et al. 1987; Reheis et al. 1991). Fine-grained Quaternary alluvium is stored within most stream systems draining the western slope of the mountains.

The sedimentary geology of the Bighorn Basin links rockshelter formation to regional paleoenvironmental change as the locally extensive sedimentary units are a significant source for aeolian sediments. A cool, semi-arid, high-elevation desert climate dominates the basin and foothills environment producing sparse vegetation cover. Persistent annual westerly winds result in considerable aeolian sediment erosion and transport. During the late Quaternary, periodically decreased effective precipitation reduced vegetation cover amplifying aeolian sediment transport. Rockshelters are effective sediment traps evidencing late Quaternary environmental change as discrete aeolian units.

BEDROCK CONTROLS AND GEOMORPHIC SETTINGS

Three geomorphic settings are common for Bighorn Mountain rockshelters. A paleokarst formed in the Madison limestone during Mississippian-Pennsylvanian times (Sando 1974, 1988). Gradient-controlled active karsts formed via joint-controlled phreatic flow during the early Pleistocene when the Bighorn River was at a higher base level (Huntoon 1985; Sutherland 1976). Sandstone rockshelters primarily occupy footslopes relatively low in alluvial valleys, and fluvial erosion is the dominant process creating rockshelter cavities (Donahue and Adovasio 1990; Goldberg and Macphail 2006:169-174). Where Quaternary downcutting has occurred, age of many sandstone rockshelters and associated sedimentary deposits is a function of distance above modern stream channels.

The Madison Paleokarst

The Madison limestone formed during the Mississippian Period approximately 350-330 Ma (Boyd 1993). As the Mississippian Sea retreated to the west and the limestone platform was exposed, meteoric flow dissolved evaporites low in the Madison creating a paleohydraulic conduit for west-flowing water. This channelized, subsurface flow formed an extensive karst system widespread across the then extant western margin of the North American continent (Sando 1974, 1988). Infilling began as limestone ceilings within the solution features collapsed and fine-grained matrix composed of terrigenous sediments from the overlying Pennsylvanian-Amsden formation filled voids. As Pennsylvanian sea levels rose, beach sands were redeposited disconformably on Madison limestone and filled open paleokarst chambers. Shale and siltstone completely filled karst chambers by the late Pennsylvanian where it remained buried until post-Laramide uplift and stream incision exposed it throughout the interior of western North America. The paleokarst is now commonly exposed in cliffs and high-angle hillslopes of the Bighorn Mountains. The poorly cemented, highly jointed breccia of paleokarst fill weathers easily to form pockets, cavities, and incipient shelters. Once established, microclimates promote weathering in rockshelters via ice wedging and dissolution (Goldberg and Macphail 2006:169-174).

Gradient-Controlled Active Karsts

Although archaeological associations are rare, important Pleistocene paleontological sites occur in active karst systems. Natural Trap Cave, a 25-m-deep sinkhole, began collecting Rancholabrean fauna nearly 120 ka evidencing change in late Quaternary biotic communities and extinction of megafauna at the Pleistocene-Holocene boundary (Gilbert and Martin 1984). Known archaeological remains in active karst are limited to open cave-mouths (Sutherland 1976).

Significant relationships that act as important geomorphic controls exist between the structural geology of the Bighorn Mountains and active karst passages (Sutherland 1976). For example, bedrock joints on Little Mountain are perpendicular to the strike of the dominant local anticline. Most well-developed cave passages follow this trend, and a steep gradient to the Bighorn River produces a significant hydraulic head that pushes water through the passages. The elevation of some cave entrances approximates the highest stream terrace in the Bighorn Basin indicating that active karsts likely began forming during the early Pleistocene (Sutherland 1976). The mid-Pleistocene Pearlette ash, found in the interior of Horsethief Cave, provides a limiting age for active karst formation of about 600 ka (Sutherland 1976). Caves in the Trapper-Medicine Lodge Creek area are likely contemporaneous with the Horsethief-Bighorn system (Huntoon 1985).

Sandstone Rockshelters

Few examples of formation processes in sandstone shelters exist in the geoarchaeological literature. Those that have been published (Abbott 1997; Donahue and Adovasio 1990; Kibler 1998) discuss the importance of Quaternary fluvial erosion to initial rockshelter development. Granular disintegration of sandstone bedrock is the primary mode of autogenic sedimentation. With the exception of the Ditch Creek site, which formed in Flathead (Cambrian) sandstone, most sandstone rockshelters in the Bighorn Mountains formed in the Tensleep formation. This mineralogically mature sandstone was deposited in terrestrial and shallow marine environments from the mid-Pennsylvanian through early Permian periods (Boyd 1993). Grain-size is dominantly fine to very fine quartz sand cemented with anhydrite or dolomite, making it more easily weathered than silica-cemented sandstones. As I will show in a later section of this paper, grain size and cement composition are important variables to rockshelter sedimentation.

Most sandstone rockshelters we have observed occur in lower topographic stream valley settings, as opposed to limestone shelters that occur high in canyon walls. With exception of the Medicine Lodge Creek site (Finley 2007), the maximum age of stratigraphic and archaeological deposits is related to distance above the modern stream channel. Rockshelters observed proximal to modern stream channels contain little or no recent sediments or massive deposits of fine-grained alluvium. In either case, archaeological deposits are rare or absent indicating a recent age for rockshelter genesis or sediment recycling in active floodplain environments. Sandstone rockshelters occupying higher footslope positions offer potentially long temporal occupations if in relatively shallow geologic contexts (see discussion of Southsider Shelter below).

FACIES MODELS AND SEDIMENTATION PROCESSES

In both limestone and sandstone settings, knowing the important geomorphic variables and having the ability to associate geomorphic processes with rockshelter formation provides the framework for interpreting stratigraphy. I utilize a conceptual framework developed by Farrand (1985, 2001) and Woodward and Goldberg (2001) that partitions sediments into autogenic and allogenic categories. In this scheme autogenic sediments are viewed as self-derived, or are those produced through weathering of the rockshelter bedrock. Autogenic sedimentation is strictly a function of weathering processes and bedrock controls, particularly bedding, jointing, or fracturing. Autogenic sediments are the typical rocky deposits archaeologists often associate with rockshelters. Autogenic sedimentation in sandstone rockshelters is also dependent on bedrock characteristics but is more likely to occur as single-grain, sand-sized particles rather than cobbles or boulders (Donahue and Adovasio 1990).

Traditionally rockshelter sedimentologists have linked autogenic sedimentation to paleoenvironments through physical weathering: colder environments enhanced rates of physical weathering and autogenic sedimentation (Laville et al. 1980). Thus, a relative frequency curve of autogenic sediments is a proxy for environmental change (Fryxell et al. 1968; Huckleberry and Fadem 2007). Recent work in the Great Basin, however, has demonstrated that over shorter time-scales, a simple one-to-one correlation between climate and rock-weathering rates may not be warranted since greater diurnal and annual temperature extremes produce weathering rates equivalent to sustained, long-term, climatic cooling (Madsen 2000).

Allogenic sediments are introduced into rockshelters from external sources, transported via alluvial, colluvial, or aeolian processes. Understanding allogenic sedimentation is crucial to rockshelter geoarchaeology since these deposits link rockshelters to landscape-scale geomorphic processes and paleoenvironmental change (Woodward and Goldberg 2001). True alluvial stratigraphy is rare in Bighorn Mountain rockshelters and occurs only in those adjacent to perennial streams such as the Medicine Lodge Creek site (Finley 2007). Colluvial sedimentation in both limestone and sandstone sites is common and dependent on hillslope length and angle above the rockshelter, as well as bedrock morphology, which must allow sediments to reach the interior. Aeolian deposition is not restricted by morphological constraints, although the size of the opening and topographic constraints such as aspect can be an important control.

I apply the concepts proposed by Farrand (1985, 2001) and Woodward and Goldberg (2001) to identify key differences between depositional regimes in limestone and sandstone rockshelters (Finley 2001). The logic of facies models (Boggs 2001:262-266) builds on these concepts to define the primary processes resulting in rockshelter sedimentation. The differences between autogenic and allogenic sediments identified in the facies models can then be extended to climatic-geomorphic processes that evidence conditions influencing rockshelter formation, paleoenvironmental change, and, ultimately, human adaptive strategies.

Limestone Settings

The Madison paleokarst is the predominant geomorphic setting for Bighorn Mountain rockshelters, although they are also created by fluvial erosion, differential weathering and collapse of hardrock slabs along major joints or bedding planes, and under massive fall blocks. In addition to identifying the bedrock context, when considering the potential contribution of different sediment sources it is necessary to account for geomorphic variables that include 1) hillslope position, 2) hillslope shape and angle above the shelter, 3) number and placement of sediment pipes or other sources of hillslope deposition, and 4) height above stream bottom.

During autogenic sedimentation in paleokarst settings, angular clasts are contributed regularly to sedimentary deposits via cement dissolution of the surrounding breccia matrix. Concentrated roof fall represents numerous conditions including increased physical weathering rates, decreased input of fine-grained allogenic sediments, or stochastic collapse of rockshelter ceilings. The climatic controls of these processes are not well understood, thus paleoenvironmental interpretations of autogenic deposits must be made cautiously and include supporting evidence. The kind and degree of jointing in limestone bedrock influences autogenic sedimentation in non-paleokarst settings, and rates of roof fall production are generally lower. Autogenic sedimentation may be more directly related to climate under these conditions.

Allogenic sediments in limestone rockshelters are limited to aeolian and colluvial deposition. Alluvial stratigraphy has not been observed in the rockshelter sample, which can be attributed to patterned cliff and free-face hillslope positions well above modern drainages. Important controls of aeolian sedimentation are aspect and size of shelter opening. GIS-based analysis of regional rockshelter distribution patterns (Larson, this volume) shows that most sites have east or south aspects. Strong, westerly, diurnal and seasonal winds carry aeolian sediments from the basin and foothills margin. Aeolian transport models indicate that winds cresting ridge tops separate and lose velocity before dropping into valley bottoms (Wiggs et al. 2002). This process results in sediment deposition and likely explains aeolian rockshelter sedimentation.

The primary variables accounting for colluvial deposition are slope shape and angle above the rockshelter, as well as number and placement of openings that channel sediments into rockshelter interiors. Rainsplash and overland flow carry sediments across relatively steep, shallow, and poorly vegetated limestone hillslopes (Carson and Kirkby 1972). Long, high-angle, convex hillslopes above rockshelters obviously have greater potential to contribute colluvial sediments than do short, low-angle, concave hillslopes. Bedrock morphology must include channels or pipes that funnel colluvium towards rockshelter interiors. Conversely, a well-developed apron in front of the rockshelter may channel colluvium over the brow towards the interior.

The geomorphic processes accounting for autogenic and allogenic sedimentation in limestone rockshelters can be extended to a general facies model that describes the sedimentological and stratigraphic characteristics of the distinct depositional regimes. Autogenic sediments have higher percentages of gravel and overall larger mean grain-size (these variables are autocorrelated), generally higher proportions of sand in relation to silt and clay, and significantly larger calcium carbonate values associated with limestone debris of the host rock. Depending on the proportion of fine-grained sediments, autogenic deposits are generally loosely consolidated with either grain- or clast-supported matrices. Allogenic aeolian sediments have lower gravel percentages, smaller mean grain-size, more silt and clay than sand, and lower calcium carbonate values. Aeolian deposits often have laterally discontinuous pinchout geometry. Heavy mineral assemblages may identify a basin source for aeolian deposits (Farrand 1985, 2001). Like autogenic deposits, allogenic colluvium may have high gravel percentages making it difficult to distinguish the two in some cases. Clast shape, roundness, and porosity are key variables differentiating roof fall and colluvial gravels (Farrand 1985, 2001). Allogenic colluvium has larger percentages of silt and clay, as well as lower carbonate values. Colluvium is more consolidated due to a larger fine-grained component derived from reworked aeolian sediments and pedogenesis and may have laterally discontinuous pinchout geometry.

Sandstone Settings

Because relatively little is known of formation processes in sandstone rockshelters, and they occur in large numbers in the Bighorn Mountains, understanding the variability in their formation is needed. Preliminary observations of stratigraphy in sandstone rockshelters shows that, unlike in limestone settings, roof fall is generally not abundant. The dominant autogenic process results from dissolution of anhydrite and dolomite cement that binds together individual quartz grains. Thus, autogenic sediments are contributed to deposits as single, sand-sized quartz grains rather than the coarse gravels and cobbles typical of limestone roof fall. The grain-size of autogenic sediments varies little from the coarser-grained fraction of allogenic sediments, particularly the fine and very fine sand component of aeolian and colluvial sediments. This subtle difference between autogenic and allogenic sediments complicates traditional analysis of sandstone rockshelters.

In a simple facies model of sandstone rockshelter formation processes autogenic sedimentation is characterized by grain-size values similar to that of the host rock. Deviations towards finer-grained sediments (more silt and clay) indicate either allogenic sedimentation or in situ weathering of autogenic sediments (i.e., pedogenesis). Deviations towards coarse grain-size distributions may accompany colluvium with a large gravel component. The same processes acting on aeolian and colluvial sedimentation in limestone rockshelters operate on sandstone rockshelters. Wind dynamics and shelter aspect effect aeolian sedimentation. Hillslope characteristics and bedrock morphology determine the ability of colluvium to enter rockshelter interiors. Depositional geometry and architecture of allogenic sediments may also indicate sources external to rockshelter openings. I use Southsider Shelter as an example to illustrate how bedrock controls influence grain-size differences between autogenic and allogenic sediments.

CASE STUDIES OF VARIABILITY IN ARCHAEOLOGICAL CONTEXTS

In the following section I apply the concepts discussed above to briefly illustrate the formation histories of three rockshelters (**Figure 22.1**). The geomorphic contexts and processes discussed in the facies models combine to create rockshelter deposits that are unique in practically every case. BA Cave and Eagle Shelter illustrate variability within paleokarst settings that is based on geomorphic variables such as slope position, slope shape and angle above the rockshelter, and bedrock morphology, while Southsider Shelter illustrates the subtle differences between autogenic and allogenic sedimentation in sandstone rockshelters.

BA Cave (Finley 2001; Finley et al. 2005) contains excavated stratigraphic deposits approximately 1 m deep that date to the last 5 ka. Primary formation processes are roof fall and aeolian sedimentation resulting in a highly stratified deposit (**Figure 22.2a**). Bedrock pipes or conduits do not channel slopewash into the rockshelter, thus colluvium is not present. Furthermore, while BA Cave is situated at the base of a large, high-angle free face, the overlying hillslope is a low-angle, convex slope

22.2 Examples of rockshelter stratigraphy from BA Cave (a), Eagle Shelter (b), and Southsider Shelter (c).

that is mostly exposed bedrock and does not provide a regular sediment supply to the rockshelter. While aeolian stratigraphy is discrete, roof fall is dispersed throughout the profile. Concentrated roof fall may indicate decreased allogenic sediment inputs, increased autogenic sedimentation, or stochastic roof fall events. The observed contrast between autogenic roof fall and allogenic aeolian sediments in BA Cave may provide an important link to late Holocene environmental change in the Bighorn Basin (Finley et al. 2005).

Eagle Shelter (Finley and Kelly 2006) exemplifies rockshelter formation in a paleokarst setting dominated by allogenic sedimentation. This site has a deep sedimentary deposit (>3 m) that spans the last 14 ka (**Figure 22.2b**). Like BA Cave, Eagle Shelter sits at the base of a cliff face. The overlying hillslope is a long, high-angle, convex slope that delivers overland flow to the rockshelter through bedrock channels on either side of the opening and over the rockshelter brow. In this case, colluvium contains large, angular gravel clasts equivalent in size to those contributed through autogenic sedimentation making it difficult to distinguish autogenic sediments from allogenic colluvium. The key stratigraphic variation between allogenic sediments occurs as discrete aeolian units deposited regularly throughout the profile. Because of the dominant colluvium, well-developed aeolian strata indicate three possible conditions: decreased hillslope contributions with stable aeolian deposition, stable hillslope contributions with increased aeolian sedimentation, or decreased hillslope contributions with increased aeolian contributions. Like BA Cave, this stratigraphic record indicates potentially significant late Pleistocene and Holocene environmental changes (Finley and Kelly 2006).

Southsider Shelter (Frison 1991) contains a stratified deposit approximately 1.5 m deep that spans the last 10 ka (**Figure 22.2c**). Sediments with textures finer than the sandstone host rock are common indicating that much of the deposit is either allogenic aeolian sediment and/or colluvium or is the product of in situ weathering (i.e., pedogenesis). Colluvium is a likely explanation for most allogenic sediments as bedrock conduits channel sediment into the rockshelter at both sides of the opening, and, likewise, significant sediment is delivered to the interior from the long, high-angle, convex hillslope overlying the brow. Eroded hillslope sediments with aeolian contributions and modified by pedogenesis may have finer textures contributing to grain-size distributions observed in the rockshelter deposits. Preliminary stratigraphic interpretations from this site indicate a long period of mid-Holocene surface stability that is consistent with the record of soil development and hillslope processes at the nearby Laddie Creek site (Reider and

Karlstrom 1987). While we have a great deal to learn about formation processes in sandstone rockshelters, this example provides important baseline data.

CONCLUSION

Understanding the geologic and geomorphic setting of Bighorn Mountain rockshelters is an essential point of departure for examining formation processes and interpreting stratigraphy. Bedrock geology provides a fundamental geomorphic control for rockshelter development that shapes the character of subsequent stratigraphic deposits. Knowledge of Bighorn Basin geology also informs of potential aeolian sediment sources and linkages to regional environmental change. Limestone paleokarst and sandstone geomorphic settings act as an important control influencing the character of stratigraphic deposits. Geomorphic variables such as aspect, hillslope position, and hillslope shape and angle above the rockshelter likewise contribute to the unique character of rockshelter deposits. Concepts of autogenic and allogenic sedimentation provide the essential framework for characterizing variability in rockshelter deposits. Variation in stratigraphic deposits can then be linked to facies models concerning autogenic and allogenic sedimentation in limestone paleokarst and sandstone settings.

It is essential to place archaeological remains within their full spatiotemporal and biophysical context. Geoarchaeological analyses of rockshelter deposits that employ facies models not only help provide this context but may also provide insight into regional paleoenvironmental change. Facies models link rockshelter formation to local and regional environmental change through geomorphic process-response models based in climatic geomorphology (Bull 1991). A comprehensive view of past environmental change is one goal of geoarchaeological research at rockshelters that works to place human adaptive strategies within appropriate environmental frameworks.

Acknowledgements. This paper benefited from the serious considerations of Oskar Burger, Gary Huckleberry, and Janet Spector. Charlotte Nogier graciously provided a French translation of the abstract.

REFERENCES

ABBOTT, J. T. (1997). Stratigraphy and geoarchaeology of the Red Canyon Rockshelter, Crook County, Wyoming. *Geoarchaeology* 12:315-335.

BOYD, D. W. (1993). Paleozoic history of Wyoming. In *Geology of Wyoming,* edited by A.W. SNOKE, J.R. STEIDTMANN, and S.M. ROBERTS. Laramie, Wyoming, Wyoming State Geological Survey, p 164-187.

BOGGS, S. Jr. (2001). *Principles of Sedimentology and Stratigraphy,* 3rd Ed. Prentice Hall. Upper Saddle River, New Jersey.

BROWN, W. G. (1993). Structural style of Laramide basement-cored uplifts and associated folds. In *Geology of Wyoming,* edited by A.W. SNOKE, J.R. STEIDTMANN, and S.M. ROBERTS. Laramie, Wyoming, Wyoming State Geological Survey, p. 312-373.

BULL, W. B. (1991) *Geomorphic Responses to Climatic Change.* Oxford University Press.

CARSON, M. A., and M. J. KIRKBY (1972). *Hillslope Form and Process.* Cambridge, University Press.

DONAHUE, J. and J. M. ADOVASIO (1990). Evolution of sandstone rockshelters in eastern North America: A geoarchaeological perspective. In *Archaeological Geology of North America*, edited by J. DONAHUE and N.P. LASCA. Boulder, Colorado, Geological Society of America, vol. 4, p 231-252.

FARRAND, W. R. (1985). Rockshelters and cave sediments. In *Archaeological Sediments in Context*, edited by J.K. STEIN and W.R. FARRAND. Orono, Maine, Center for the Study of Early Man, University of Maine, p. 21-39.

FARRAND, W. R. (2001). Archaeological sediments in rockshelters and caves. In *Sediments in Archaeological Contex*t, edited by J.K. STEIN and WILLIAM R. FARRAND. Salt Lake City, University of Utah Press, p 29-66.

FENNEMAN, N. M. (1931). *Physiography of the Western United States*, 1st Ed. New York, McGraw-Hill Book Company.

FINLEY, J. B. (2001*). Late Holocene Environments and Rockshelter Formation Processes in the Bighorn Mountains, Wyoming.* Unpublished Master's thesis, Department of Anthropology. Pullman, Washington State University, p. 99.

FINLEY, J. B. (2007). Stratigraphy, sedimentology, and geomorphology. In *The Medicine Lodge Creek Site,* edited by G.C. Frison, and D.N. Walker. Laramie, Wyoming, University of Wyoming, p 131-152. (In Press).

FINLEY, J.B.; M. KORNFELD, B.N. ANDREWS, G.C. FRISON, C.C. FINLEY, and M.T. BIES (2005). Rockshelter archaeology and geoarchaeology in the Bighorn Mountains, Wyoming. *Plains Anthropologist* 50:227-248.

FINLEY, J. B. and R. L. KELLY (2006). *Late Quaternary geoarchaeology and environmental history of Eagle Shelter (48BH657), Bighorn Mountains, Wyoming, USA.* Manuscript in Preparation.

FRISON, G.C. (1976). The chronology of Paleo-Indian and Altithermal cultures in the Bighorn Basin, Wyoming. In *Culture change and continuity: essays in honor of James Bennet Griffin,* edited by C.E. CLELAND. New York, Academic Press, p. 147-173.

FRISON, G. C. (1991). *Prehistoric Hunters of the High Plains,* 2nd Ed. San Diego, Academic Press.

FRYXELL, R. T. B.; R.D. DAUGHERTY; C.E. GUSTAFSON, H.T. IRWIN; B.C. KEEL and G.S. KRANTZ (1968). *Human skeletal materials and artifacts from sediments of Pinedale (Wisconsin) glacial age in southeastern Washington, United States. VIIIth International Congress of Anthropological and Ethnological Sciences*, Tokyo.

GILBERT, B. M. and L. D. MARTIN (1984). Late Pleistocene fossils of Natural Trap Cave, Wyoming, and the climatic model of extinction. In *Quaternary Extinctions; A Prehistoric Revolution*, edited by P.S. MARTIN and. R.G. KlEIN. Tucson, University of Arizona Press, p 138-147.

GOLDBERG, P. and R. I. MACPHAIL (2006). *Practical and Theoretical Geoarchaeology*. Oxford, Blackwell.

HUNTOON, P. W. (1985). Gradient controlled caves, Trapper-Medicine Lodge area, Bighorn Basin, Wyoming. *Ground Water* 23:443-448.

HUCKLEBERRY, G.A. and C. FADEM (2007). Environmental Change Recorded in Sediments from the Marmes Rockshelter Archaeological Site, Southeastern Washington State, U.S.A. *Quaternary Research* (in press)

HUSTED, W. M. and R. EDGAR (2002). *The archaeology of Mummy Cave, Wyoming: an introduction to Shoshonean prehistory*. Midwest Archaeological Center and Southeast Archaeological Center Special Report No. 4, Lincoln, Nebraska.

KIBLER, K. W. (1998). Late Holocene environmental effects on sandstone rockshelter formation and sedimentation on the Southern Plains. *Plains Anthropologist* 43:173-186.

LAVILLE, H.; J. P. RIGAUD, and J. SACKETT (1980). *Rockshelters of the Perigord: Geological Stratigraphy and Archaeological Succession*. New York, Academic Press.

MADOLE, R.F.; W.C. BRADLEY; D.S. LOEWENHERZ; D.F. RITTER, N.W. RUTTER and C.E. THORN (1987). Rocky Mountains. In Geomorphic Systems of North America, edited by W.L. GRAF. Boulder, *Centennial Special Volume No. 2, Geological Society of America*, p. 211-257.

MADSON, D. B. (2000). Late Quaternary Paleoecology in the Bonneville Basin. *Utah Geological Survey Bulletin* 130. Salt Lake City.

MULLOY, W. T. (1958). A preliminary historical outline for the Northwestern Plains. *University of Wyoming Publications*, 22, 1-235.

PICARD, D. M. (1993). The early Mesozoic history of Wyoming. In *Geology of Wyoming*, edited by A.W. SNOKE, J.R. STEIDTMANN and S.M. ROBERTS. Laramie, Wyoming, Wyoming State Geological Survey, p 210-239.

REHEIS, M. C.; R C. PALMQUIST; S. AGARD; C. JAWOROWSKI; B. MEARS, Jr.; R.F. Madole; A.R. NELSON and G. D. OSBORN (1991). The Quaternary history of some Southern and Central Rocky Mountain basins: Bighorn Basin, Green Mountains-Sweetwater River area, Laramie Basin, Yampa River Basin, northwestern Uinta Basin. In *Quaternary Nonglacial Geology of the Conterminous United States*, edited by R. B. MORRISON. Boulder, Geological Society of America, p 407-440.

REIDER, R.G. and E.T. KARLSTROM (1987). Soils and stratigraphy of the Laddie Creek Site (48BH345), an Altithermal-age occupation in the Big Horn Mountains, Wyoming. *Geoarchaeology* 2:29-47.

SANDO, W. J. (1974). Ancient solution phenomena in the Madison Limestone (Mississippian) of north-central Wyoming. *Journal of Research of the U.S. Geological Survey* 2:133-141.

SANDO, W. J. (1988). Madison Limestone (Mississippian) paleokarst: a geological synthesis. In *Paleokarst*, edited by K.C. LOHMAN; P.W. CHOQUETTE and N.P. JAMES. New York, Springer-Verlag, p 256-277.

STEIDTMANN, J. R. (1993). The Cretaceous foreland basin and its sedimentary record. In *Geology of Wyoming*, edited by A.W. SNOKE, J.R. STEIDTMANN, and S.M. ROBERTS. Laramie, Wyoming State Geological Survey, p. 250-271.

STEIN, J. K. (2001). A review of site formation processes and their relevance to geoarchaeology. In *Earth Sciences and Archaeology,* edited by P. GOLDBERG; V.T. HOLLIDAY and C.R. FERRING, New York, Kluwer Academic/Plenum, p. 37-54.

SUTHERLAND, P. W. (1976). *The geomorphic history of Horsethief Cave, Bighorn Mountains, Wyoming*. Unpublished Master's thesis, Department of Geology. Laramie, Wyoming, University of Wyoming, p. 73.

WIGGS, G.F. S.; J.E. BULLARD, B.G. GARVEY and I.P. CASTRO (2002). Interactions between airflow and valley topography with implications for aeolian sediment transport. *Physical Geography* 23:366-380.

WOODWARD, J. C. and P. GOLDBERG (2001). The sedimentary records in Mediterranean rockshelters and caves: Archives of environmental change. *Geoarchaeology* 16:327-354.

CLOSED SITE INVESTIGATION IN THE AMERICAN NORTHEAST: THE VIEW FROM MEADOWCROFT

J. M. ADOVASIO

Mercyhurst Archaeological Institute, Mercyhurst College, Erie, Pennsylvania 16546, USA; adovasio@mercyhurst.edu

Abstract. Though not as well known as other portions of North America for cave and rockshelter research, the American Northeast contains one of the most intensively investigated closed site loci ever excavated. Meadowcroft Rockshelter, in southwestern Pennsylvania, has been under nearly continuous study since 1973 and the research protocols employed at that site are widely considered to represent the state-of-the-art. The constantly evolving research methodology used at Meadowcroft is summarized and its implication for rockshelter/cave research in the Northeast and, more broadly, other areas are addressed.

Keywords: Meadowcroft rockshelter, History of research, Northeastern America, Excavation Methods and Protocols.

Résumé. Quoique ne bien connu autant que l'autres sections de l'Amérique du Nord pour la recherche conduise dans les cavernes ou les abris, le nord-est Américain contenit un de le plus intensivement étudié abris dans l'histoire d'excavation archéologique. L'Abri Meadowcroft, situé vers le sud-ouest de Pennsylvanie, a était étudié presque sans interruption depuis 1973. Les protocoles de recherche utilisées a cette localité étaient considérées presque universellement de représenter la nec plus ultra. Les implications de la méthodologie continument évoluées a Meadowcroft étaient sommeriez pour le nord-est de l'Amérique, spécifiquement, et le monde, généralement.

Mots-clés: L'Abri Meadowcroft, Histoire de la recherche, Le Nord-est de l'Amérique, Méthodologie et protocoles d'excavation.

This paper is not meant to summarize the history of multidisciplinary closed site research in the Northeast, nor is it designed to synthesize broadly or even selectively the results of such research. It does not represent an attempt to critique such studies as they are conducted by others. The explicit purpose of this paper is to examine briefly a single archaeological project not from the standpoint of the substantive data generated by it but rather from the perspective of its overall research philosophy, organizational structure and methodological goals and execution. The paper conveys the author's notions and those of his colleagues in this project on how one kind (i.e., brand or variety) of closed-site research can be conducted in the Northeast.

THE MEADOWCROFT/CROSS CREEK ARCHAEOLOGICAL PROJECT

History of Research

When the author joined the University of Pittsburgh's Anthropology Department in 1972, it was decided to revive the temporarily defunct Archaeological Research Program (later to become the Cultural Resource Management Program) in western Pennsylvania. To that end, it was decided that a geographical area of manageable dimensions (i.e., not more than 20,000 ha; [49,419 acres]) should be located within an appropriate working distance from Pittsburgh. This area would serve as the focal point of a comprehensive, multi-year research enterprise. In addition to relative ease of accessibility and overall size, other desirable attributes for the prospective study area included the potential for relatively long term prehistoric occupation and the absence of any previous extensive research activities in that area. As originally conceived, "long-term prehistoric occupation" meant that the study area should offer the potential for the recovery of Early Archaic or Middle Archaic through terminal Woodland materials in stratified contexts. In selecting the prospective study area, it was not considered or deemed important that it afford the specific probability for locating, much less excavating, Paleoindian sites. An area was sought within which little previous work had been done precisely because there was no wish to "rework old problems," nor did the participants wish to begin this project with any preconceptions based on previously excavated or published data.

Only one other "attribute" was considered essential for the projected study area—it was to have at least one relatively undisturbed closed site—either a cave or, more likely, a rockshelter. The reasons for this were both personal and practical. Virtually all of the author's previous research experience in the North American West was in closed site contexts, and it was felt that such sites in the East afforded the potential for better preservation of organic materials than is normally encountered in open site loci. Moreover, such sites are often occupied and reoccupied intermittently for thousands of years and can be used therefore as chronological anchor points or "lynch pins" for a given study area.

Due to research commitments in Cyprus in the winter of 1972–1973, there was no time to mount anything approximating a systematic survey or reconnaissance for prospective project areas in southwestern Pennsylvania. Consequently, word was "circulated" that the University of Pittsburgh was interested in locating rockshelter or cave sites in that portion of the state. In April 1973, Dr. P. R. Jack, historian, archaeologist, and then chairman of the

Social Science Department at California State College, California, Pennsylvania, informed this writer of the rockshelter since known as Meadowcroft.

Meadowcroft Rockshelter is located on the property of Meadowcroft Village, a restored, predominantly nineteenth century, rural community operated by the nonprofit Meadowcroft Foundation. The village was developed by Albert and Delvin Miller (Vice president and President, respectively, of the Meadowcroft Foundation) on a portion of their family farm, presently some 1976.8 ha (800 acres) in extent. The farm and the rockshelter have been in the continuous possession of the Miller family since 1795.

The location of the deposits within the confines of Meadowcroft Rockshelter and the specific protection afforded by Albert Miller even prior to the creation of the village resulted in the fact that the rockshelter had escaped serious despoliation.

The archaeological potential of the shelter was long suspected by Albert Miller although he refrained from any excavations until 1955. In that year, his enlargement of an animal burrow yielded lithic debitage, shell, and faunal remains confirming his suspicions of aboriginal occupation at the shelter. Efforts to interest professional archaeologists in the site resulted in its recording as 36WH297 in 1968. For one reason or another no further excavations were begun, and the site remained (with the exception of the Miller test "hole") untouched until 1973.

Once informed of the existence of the site, arrangements were made to visit it in April 1973. The potential of the rockshelter itself, among the largest extant in southwestern Pennsylvania, as well as the surrounding Cross Creek drainage was obvious. Permission to initiate excavations was immediately sought from and very shortly granted by Albert and Delvin Miller. Work commenced at the rockshelter and elsewhere in the Cross Creek drainage in June 1973 as part of the Department of Anthropology's Summer Field Program in Archaeology. Over the first 6 years of the project, the staff spent 466 days (12–14 hours each) and about $1,500,000 (1970s money).

Additional and extensive excavations were conducted in 1983, 1985, and 1987 under the continuing auspices of the Cultural Resource Management Program of the University of Pittsburgh. Further research was conducted in 1993, 1994, and 1997 under the aegis of Mercyhurst Archaeological Institute, Mercyhurst College, Erie, Pennsylvania. All fieldwork at the site was directed by the author.

The Study Area

Meadowcroft Rockshelter is located 48.3 km by air, 78.8 km by road southwest of Pittsburgh and 4 km northwest of Avella in Washington County, Pennsylvania (**Figure 23.1**). The site is situated on the north bank of Cross Creek, a small tributary of the Ohio River that lies 12.2 km to the west.

Geologically, Meadowcroft is located in the unglaciated portion of the Allegheny Plateau, west of the Ridge and Valley province of the Appalachian Mountains and northwest of the Appalachian Basin. The surface rocks of this region are layered sedimentary rocks of Middle to Upper Pennsylvanian age (Casselman Formation). The prevailing lithologies are shale, quartz sandstone, limestone, and coal in decreasing order of abundance. Deformation is very mild with a regional dip of 3–5° to the southeast.

Topographically, the region around Meadowcroft is maturely dissected. More than one-half of the 14,164.3 ha Cross Creek watershed is composed of moderate to steep valley slopes with upland and valley bottom areas in the minority. Maximum elevations in the Cross Creek drainage are generally above 396 m. At the divides on the east, elevations are above 426 m. Elevations at stream level are 310 m at Rea on the South Fork, 276 m at Avella and 193 m normal pool level at the confluence of the Ohio.

Within the Cross Creek watershed, the main stem of Cross Creek flows for 31.3 km. The maximum north-south width of the watershed is approximately 15 km. The prevailing stream pattern is dendritic with numerous small creeks and runs supplying the main stem of Cross Creek. The drainage is northwestward to westward toward the West Virginia border and the Ohio River.

Strategy

From its inception, the Meadowcroft/Cross Creek Project was a multidisciplinary undertaking. Indeed, the overriding philosophy of the entire project necessitated a multidisciplinary perspective. Almost from the commencement of the project in 1973 and certainly by 1974, the central goal or theme of the operation was the systematic acquisition, analysis, and integration of any and all data bearing on the archaeology, history, paleoecology, geology, geomorphology, pedology, hydrology, climatology, and floral and faunal succession of the entire Cross Creek drainage. Moreover, this data gathering, analysis, and interpretation was to be executed with as great a degree of precision and employing the most sophisticated methodologies of which any of the project staff were cognizant. Additionally, and to us critically, this research was carried out virtually without temporal or fiscal constraints. In short, the project was designed to epitomize so-called "state of the art" data gathering and analytical methodologies and procedures.

As the foregoing suggests, the Meadowcroft/Cross Creek Project was *always* essentially empirical in orientation and in some ways "old line" in philosophy. Meadowcroft Rockshelter and the Cross Creek drainage were never

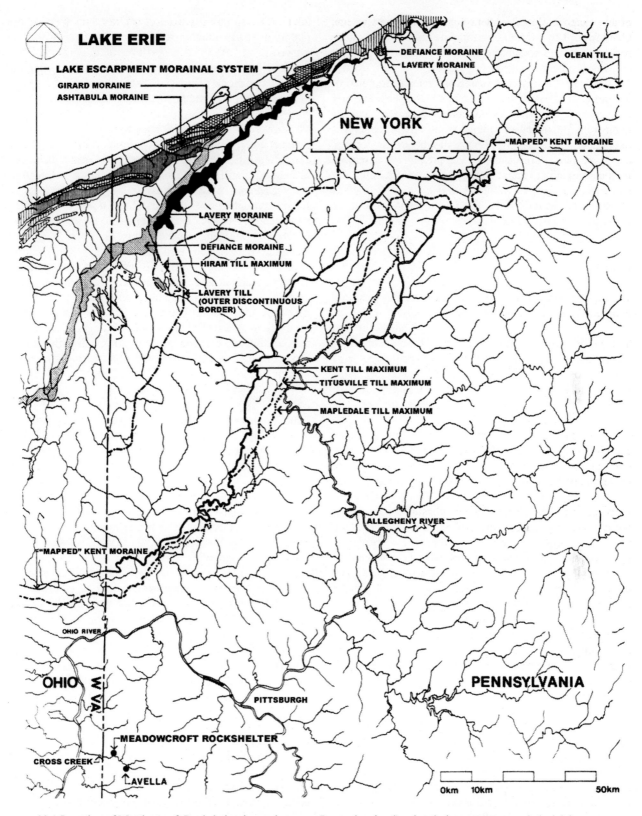

23.1 Location of Meadowcroft Rockshelter in southwestern Pennsylvania, showing drainage systems and glacial features.

approached—whether in 1973 or 1997—as vast laboratories in which to test explicit hypotheses or in which to seek logico-deductive covering laws or nomothetic truths about aboriginal behavior. Those who view archaeology as a vehicle or medium for the pursuit of such ends should not be denigrated for it, but theoretical considerations of this sort were *never* an a priori concern of the project under discussion.

The assembly of a multidisciplinary staff who were not only individually competent but who had both the time and energy to stay with and contribute to the project in all of its phases until its completion was a primary concern. Meadowcroft was not to be an archaeological excavation with a part-time consulting geologist, faunal expert, or palynologist thrown in to achieve quasi-scientific "legitimacy." Rather, the research project was designed to use the drainage as a focal point in which specialists in a wide variety of fields could collect data that was of interest to themselves, germane to their collective research concerns, and contributory to the archaeology of the study area. Fortunately, it was possible to assemble such a group and to maintain its integrity as a functioning entity to the present. Without this sort of long-term cooperation and continuity, the results of the efforts in Cross Creek would have been far different and much less holistic.

Tactics

The implementation of the basic research strategy, simple though it was in theory, was exceedingly complex in practice. The scale of the project planned for the rockshelter and in the drainage required the assembly of the necessary logistics facility and the capacity to staff and to operate it. A number of basic decisions subsequently were made in June 1973 that would (in retrospect) drastically affect the future of the project.

It was decided at the outset to perform a "total coverage" archaeological survey of the potentially inhabitable portions of the study area. Although certain areas of the Cross Creek drainage had been modified by farming, road construction, and strip mining, over 75% of it seemed to be pristine. Initially, a sampling strategy was considered, but this was discarded in favor of the more conceptually naive and primitive but certainly more absolutely representative "total coverage" approach. This decision in itself and the size of the drainage ordained that the survey phase would take four field seasons, each of three months duration, if not more.

A second consideration concerned the overall tactics for the excavation of the rockshelter. A design was conceived to elicit information not only on the aboriginal utilization of the site but also on its evolution through time as a geomorphological entity.

The specific excavation and sampling procedures employed at Meadowcroft are too lengthy to discuss in detail here. They have been summarized elsewhere (e.g., Adovasio et al. 1979–1980a, 1979–1980b; Carlisle and Adovasio 1982), and they will be treated exhaustively in the final project report. However, a summary of the procedures is warranted as they reflect not only the input of the archaeological staff but also the dictates and concerns of the geologists, paleontologists, paleobotanists, and other cooperating individuals.

Prior to excavation, a complete floral inventory was taken of the extant vegetation within and around the rockshelter itself as well as on the contiguous upland slopes and talus. All vegetation within 20 m east and west of the midpoint of the modem rockshelter overhang was then systematically stripped to the level of minute roots and twiglets. This in turn revealed the extant topography as well as the modem surficial distribution of roof spalls, historic and prehistoric features and artifacts.

Meadowcroft Rockshelter was then mapped with an alidade and plane table, and a grid system originally consisting of 2 m square units was established. All horizontal coordinates were reckoned relative to this grid. A permanent elevation datum and where needed subsidiary data were affixed to the north wall of the rockshelter from which all vertical measurements on all observable phenomena were taken. The combined fixed elevation datum and grid provided absolute, readily coded, and computerized Cartesian coordinates on all artifacts, features, and phenomena encountered during the course of excavation. The grid was subdivided where necessary into 1.0 m, 0.50 m and, 0.25 m units and was used solely as a recording device. Throughout the 1990s and to the present, the piece-plotting of artifacts and ecofacts employed an electronic total station surveying instrument.

Excavations commenced by opening a south to north trench that traversed the distance from outside to inside the drip line. This trench was subsequently expanded as the situation dictated to its ultimate configuration (**Figure 23.2**).

The location and expansion of the original trench was predicated on the position of major roof blocks on the eastern and western edges of the main occupation area of the site. For a brief period in 1973, work was occasionally stalled by subsurface roof spalls of varying sizes. It quickly became apparent that heavy-duty rock drilling and reduction equipment would be necessary to cope with this problem. With the cooperation of the Millers, the site was supplied with electrical power that grew from a modest single-outlet 120 volt line to a vast complex that supplied power to the diverse electrical equipment such as rock saws, diamond-tipped drills, ventilation systems, and a very sophisticated daylight and color-corrected lighting system which allowed the excavators to control the hue, intensity, and chroma of the lights in any part of the site. Land-line phone service has been available in the rockshelter from 1974 to the present. Initially, the phone was used to connect a Textronix Graphics Terminal to the mainframe at the University of Pittsburgh via modem. As far as the author is aware, this was the first time that either a site-specific color corrected lighting system or a computer connection had ever been employed in closed-site excavations in the world.

Excavations at Meadowcroft were conducted in natural levels and where possible by microstrata within natural levels. Where micro-strata were absent, arbitrary 1 cm, 5 cm, or 10 cm levels within natural strata of considerable

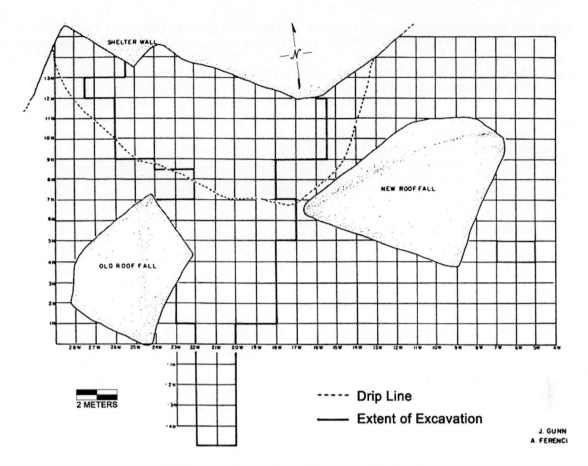

23.2 Plan map of excavations at Meadowcroft Rockshelter.

thickness were employed to facilitate vertical control. In cases where micro-strata were thinner than the edge of a trowel blade, single-edged razor blades were employed in the excavations.

All eleven of Meadowcroft's major strata and its multitude of micro-strata were defined initially by subjective criteria including texture, apparent composition, friability, degree of compaction, and in a more limited basis, color. Objective quantifiable verification of the integrity of these units was provided by geochemical, grain size, and compositional analysis. Coarser than silt-sized material was processed via conventional techniques while silt and clay sized material was analyzed via a Coulter Counter. Once again, as far as this writer is aware, this was the first ever use of a Coulter Counter in the analysis of sediment from a closed site in the Americas.

All fill from all strata except the deepest occupational unit, Stratum IIa, was dry-processed through 0.6 cm (¼ in) mesh screen. Fill from Stratum IIa was wet-processed with water through 0.3 cm (1/8 in) mesh screen. In order to recover materials smaller than 0.3–0.6 cm (about–¼ in), a constant volume sample (CVS) of fill [2900 cm^3] was taken from each natural stratum, microstratum or arbitrary 5 cm or 10 cm unit within each major stratum from every excavated square on the site. If the CVS sample was derived from stratum fill or from a feature unrelated to firing, it was processed using water flotation through graded sieves, the smallest of which was 200μ. If the sample was derived from a fire feature, it was again processed with graded sieves using hydrogen peroxide (H_2O_2) flotation. Although relatively expensive, the use of hydrogen peroxide for flotation proved well worth the investment as it minimized damage to charred items and acted as an excellent deflocculant. The combination of the dry/wet screening and flotation procedures detailed above resulted in the recovery of over two million separate items, a gross breakdown of which is presented elsewhere in this report.

Cultural features were three-dimensionally mapped, and their coordinates were then computerized. Additionally, stratum interface maps and profiles were drawn every 1 m or 0.5 m throughout the excavation. As the presence and distribution of roof spall clearly determined the amount of open floor space available to the aboriginals visiting the shelter, special care was taken to map virtually all of the spalls that were larger than a "silver dollar." Although tedious in the extreme, the recording and computerization of the coordinates of roof spalls and cultural features as well as associated artifactual and nonartifactual materials rendered the production of large scale "floor" maps a relatively simple task. It also provided the means of generating paleotopographical "surface" maps which clearly showed the evolution of the site's topography through time (Donahue and Adovasio 1990)

Seventeen kinds of standardized field forms were used for data recording, and over 30,000 pages of field notes were amassed at the rockshelter between 1973 and the present. This verbal record is buttressed by some 20,000 black and white and color prints/slides that encompass every phase of the operation from its inception through site closure at the end of each field phase of the project.

From the end of the 1973 season to the present, the excavations at Meadowcroft were completely enclosed beneath a substantial, permanent structure that afforded not only protection from the elements during the "off season" but also the ability to continue work in any weather. As noted above, the structure housed a computer terminal and was equipped with a daylight color-corrected lighting system.

During the multi-season excavations at Meadowcroft Rockshelter proper, some 60.5 m^2 of surface area inside the drip line and 46.6 m^2 outside the drip line were excavated resulting in the removal of over 230 m^3 of fill. Because of the inordinate amount of rock fall in the site, nearly all of the excavation was conducted with trowels or smaller instruments.

Specific geological sampling procedures employed at Meadowcroft included the extraction of 12 continuous sediment columns from selected localities across the site. These columns were cut from the surface of the site to sterile Stratum I; in two cases, they extended into sterile Stratum 1. Bulk samples of about 1000 g were extracted at 5 or 10 cm sampling intervals in each column. Where sediment changed composition, that is, at stratum interfaces, samples were taken on both sides of the change. The sample columns were placed to insure complete coverage of all major strata at the site from east to west. Samples were also taken both inside (north) and outside (south) the drip line.

Augmentations to the bulk samples extracted from the continuous sediment columns taken from 1976 onward were made in the form of "splits" derived from all of the constant volume samples from every stratum or microstratum in all excavation units at the site. The extraction of these columns served several purposes. Each bulk sample was divided into fractions, a portion of which was set aside for grain size analysis, analysis of the silt clay fraction, carbonate analysis, palynological assay, geochemical composition, trace element scrutiny, microfaunal study, and at the most minute level, electron microscope analysis of the diagenesis of individual sand grains.

Another geological sampling series was initiated during the 1975 season. An aluminum sampling tray was placed on the sloping wooden roof that protected the excavation. An area of about 25 m^2 was swept daily to collect sand and rock fragments falling from the rockshelter ceiling and wall; the area of collection was divided into 5 m strips extending from the tray up to the cliff face. This sampling was continued year-round until the termination of excavation in 1978. Its purpose was to establish the kind, character, and volume of modem sedimentation at the site by comparing the samples with weather variables such as temperature, humidity, precipitation, and other parameters that affect modern sedimentation. This in turn provided a useful index or gauge by which the composition of the colluvial pile at the site was judged, both in terms of sedimentation mechanisms and rate of accumulation.

To gauge the effects of sheet wash from upland sources at the site, a large holding tank (400 l) and drainage system were emplaced above the eastern margin of the rockshelter in 1976. The holding tank effectively trapped all sediment and water moving at that locus during rainstorms. The drainage area at the sampling locus encompassed some 25 m^2 (268.8 ft^2). The upland area, above the tank was about 0.9 km^2. The sampling procedure facilitated the establishment of an index of the kind and volume of upland materials transported during rainstorms and also offered yet another method of studying the accumulation of the Meadowcroft colluvial pile.

The penultimate geological sampling series undertaken at the rockshelter involved the extraction of a column of rock samples from the base of the Morgantown-Connellsville cliff at 20–50 cm intervals to the top. Thin sections were prepared from these samples and compared on the basis of grain size and composition to samples extracted from roof spalls in the colluvial pile. In this way, the pattern of erosion of the rockshelter face itself could be studied.

The most recent geological sampling procedures implemented at the rockshelter were the collection of a series of micro-morphological samples by Paul Goldberg and his associates (Goldberg and Arpin 1999). These samples were collected and analyzed to explore the presence and/or effects of post-depositional subsurface water percolation and, in fact, no evidence of any such movement was discussed.

Concurrent with the excavations at Meadowcroft were a series of parallel data-gathering operations that included the already briefly discussed archaeological reconnaissance of the Cross Creek drainage. The survey began in 1973 at the confluence of Cross Creek and the Ohio River near Wellsburg, West Virginia; it then proceeded east to southeast toward the headwaters of the drainage. A total of 231 discrete occupational loci were located between 1973 and 1978 ranging in cultural ascription from Paleoindian through Late Woodland. Of these, 17 loci were extensively tested, and two sites were excavated in their entirety. The full-scale excavations included a Late Archaic transitional period village

(36WH293) and an Early Woodland/Middle Woodland burial mound (36WH415) near Avella. The excavation and data-gathering methods employed at all the tested and fully excavated sites included (with only slight modifications) the basic excavation/data acquisition strategies that were employed at Meadowcroft Rockshelter, itself.

Collection of floral and faunal inventories as well as soil series, geological mapping projects, and terrace studies also were undertaken while the archaeological survey was in progress. Backhoe and hand-excavated trenches were cut at critical points along the Cross Creek drainage to examine the alluvial deposition sequence. Deep cores extracted from the floodplain of the South Fork were analyzed for data pertinent to sedimentation and bedrock geology. Complete topographic maps, aerial photos, and motion pictures were employed to trace the terrace system and to assist in the analysis of the geomorphology of the creek system.

Historical research on the Cross Creek drainage focusing on the area in the vicinity of the rockshelter was undertaken in 1974 and was continued intermittently to the present. Initially, this work was begun to provide possible explanations for the deposition of historic period materials in the uppermost strata of the rockshelter. As excavation proceeded, it became evident that at least some of the factors that contributed to the site's utilization historically might also have importance for understanding the prehistoric occupation(s) of the site. On a larger scale, it was obvious that the Cross Creek drainage survey would benefit from an understanding of the historic factors that resulted in the valley's present ecology, both human and natural. Farming, mining, and the general demography of the drainage all had contributed their share to the story; both historical and broader anthropological concepts were therefore incorporated into the overall project design. Over the years, the survey and the Meadowcroft excavation became, each in its own way, historic events that transcended mere "observer" status; the project itself developed as a historical entity, one to be "observed," recorded and understood.

To facilitate all of these data-gathering and field analysis projects, a series of semi-permanent field laboratories and base camps were established in the Cross Creek drainage. The main camp at Meadowcroft Village housed the archaeological field processing laboratories and the personnel of the rockshelter excavation while another complex on the South Fork of Cross Creek supported the field geology/geomorphology labs and the crews of the various open site surveys, tests, and excavations.

Between 1973 and the present, over 500 graduate and undergraduate students from 60 institutions in the United States and abroad participated in one or another phase of the Meadowcroft/Cross Creek field and field laboratory operations. Despite an unavoidable yearly turnover of student field researchers, the basic supervisory and analytical staff remained essentially unchanged for virtually the entire first six years of the field project. This provided a critical element of continuity which immeasurably contributed to the overall efficiency of the entire research enterprise.

Post-field Analyses and Data Management

At the close of each excavation season, the acquired data was transported to the University of Pittsburgh, Mercyhurst College, or to one or another of the cooperating institutions involved in the ongoing analysis phase of the project.

A series of absolute deadlines was established, and the project was committed to a rigid series of interim publications. It was therefore possible in many instances to analyze and to prepare comprehensive preliminary statements on most phases of each yearly field project. Space does not permit delineation of all or even a fraction of the analytical methodologies employed or materials generated during the multi-year Meadowcroft/Cross Creek Project; some commentary is nonetheless warranted on the "philosophy" and execution of the analytical phase of the operation.

In keeping with the original intention to acquire or to extract any data remotely bearing on the project area, no form or method of analysis however time consuming or costly was ruled out of the analysis phase of the project. One of the "results" of this perspective was a nearly constant interchange of ideas among the interdisciplinary analytical staff. This was matched by an almost constant increase in the complexity and rate of the post-field analyses themselves.

The data base generated by the Meadowcroft/Cross Creek Project is truly enormous. Some 20,000 flaked and ground stone, bone, and other perishable artifacts as well as tens of thousands of pieces of flaking debitage constitute the so-called aboriginal artifactual record. Additionally, over 965,000 unmodified animal bones and 1.4 million individual plant remains were recovered. Literally thousands of geological, geochemical, hydrological, pedological, archaeomagnetic, and other samples were collected, a substantial number of which are now analyzed.

The analytical staff included specialists from the so-called "traditional" areas of archaeology and from "ancillary" sciences. There is a conscious and conspicuous emphasis on the analysis of material culture in the old, "outmoded" sense of that term. However, attribute-oriented technological analysis of flaked and ground stone, ceramics, bone, and wooden tools as well as historic artifacts are by design carried to lengths generally considered well beyond what is "traditionally sufficient." As an example, literally every flake equipped with a platform from Meadowcroft and every other site in the Cross Creek drainage has been analyzed and measured for some 16 separate attributes, and all of the resultant data have been computerized for comparison and retrieval. All of the identifiable animal bones from Meadowcroft have been analyzed and coded as have all of the macrofloral remains recovered from the dry and wet screenings. While some sort of sampling strategy could have been employed in these operations, this potential course of action was not chosen. Rather, it was decided in many instances to examine the "recovered data universe" of a particular compositional class of material rather than an artifactual fraction or construct of the total recovered sample. Selective, arbitrary, or random sampling procedures have not been eschewed in some of the analytical operations. In many of the geological and geochemical studies as well as in the microfloral analyses (to name but three examples), no technique other than one which samples the available data is feasible with so large a data base. The main point is that where possible it has been thought preferable to analyze all of a given sample and thereby to be certain of the "representativeness" of that sample within the constraints of probability.

Analysis is of course but one of the two critical aspects of the post-field operations in the Meadowcroft/Cross Creek Project. If the raw data itself is voluminous, the analytical data on each class of recovered material is staggering. The only way of handling this much information is, of course, by extensive computerization, and indeed virtually all of the field and laboratory data generated by the project to date has been computerized. Printouts of discrete data sets—from the 2000 page bone-by-bone faunal inventory to the relatively brief list of charred basketry types and attributes from Meadowcroft—are available and are regularly exchanged by the project staff. Almost instant access to completed data sets has in turn facilitated information exchange among the multidisciplinary team. It has also substantially contributed to the ease of interim manuscript preparation.

The ultimate measure of any project is the ability to integrate the data into a semblance of a unified, cohesive report. This is the phase of the Meadowcroft operation that is currently underway. As data sets such as the invertebrate faunal remains, the modem hydrology of the Cross Creek drainage or edge wear analysis of unifaces from Cross Creek are completed and written, the manuscripts are made available to all others working on the project and on the final report. Oftentimes, this exchange generates more questions and necessitates additional analysis. The result is usually some modification of the particular interpretation in question.

Upon completion of all of the individual segments of analysis and write-up, all manuscripts will be assembled and edited by a core project staff; then and only then will a series of "internal and external correlations" be generated. Put another way, only after all of the individual data sets are analyzed, written, and circulated within the staff will the data be "interpreted" and compared to that available from other areas.

As the final report, although nearing completion, is still in the "assembly of finished segments" stage, the overall biases and perspectives that will influence or determine the "shape" of the overall interpretations is beyond the scope of the present paper. Suffice to note that the goals are culture-historical, neo-processual, and (gasp!) even post-processual, and that more than enough data are available to meet any of these "ends."

RESULTS

The substantive data generated by the Meadowcroft Rockshelter/Cross Creek Project to date have been summarized in numerous publications (Adovasio et al. 1975, 1977a, 1977b, 1979–1980a, 1979–1980b, 1980, 1981, 1982, 1989, 1990, 1992, 1999, 2003; Adovasio, Gunn, Donahue, and Stuckenrath 1978; Adovasio, Gunn, Donahue, Stuckenrath, Guilday, and Lord 1978; Adovasio and Johnson 1980; Adovasio and Pedler 1996, 2000, 2003). As the prefaces or forewords to these documents indicate, they represent but a fraction of the total Meadowcroft data base, and none purport even remotely to be "complete." Nevertheless, these interim statements are indicative of the range, character, and quantity of information that a project like this can produce in the Northeast. The ultimate utility and implications of this project are detailed below.

OVERVIEW

The Meadowcroft Rockshelter/Cross Creek Archaeological Project has been called by some "an exercise in empirical excess," an alliterative observation but one which has certain aspects of truth. If "excess" is taken alone to mean time and money, the costs have indeed been heavy. However, given the goals of the project, this sort of "excess" was unavoidable and, if anything, desired. If, on the other hand, "empirical excess" is taken to mean that this enterprise was purely culture-historical or "ascientific" in the sense that the latter term is employed by some archaeologists, there is strong disagreement. None of the Meadowcroft/Cross Creek staff (despite unguarded protestations to the contrary) is a theoretical or processual nihilist. Indeed, the staff contains more than its normal share of those who seek explanations with a capital "E" and nomothetic generalizations with an equally large "N," "G" or "T" (for Truth). However, most of the Meadowcroft team subscribe to the basic premise that unless data is meticulously and exhaustively gathered and analyzed, *nothing* else is germane. In short, there is complete agreement with the view of David Hurst Thomas (1974:14) when he succinctly states: "This is all that matters in archaeology: objects and contexts. All else is secondary." If nothing else, the Meadowcroft/Cross Creek Project has acquired, recovered, produced, or generated innumerable objects in their contexts and in so doing has produced a data base that is more than amenable to many models, interpretations and explanations.

The Meadowcroft/Cross Creek Project has demonstrated that multidisciplinary closed-site projects like this can and do work in the Northeast just as similar enterprises have succeeded elsewhere in both the Old and New Worlds. While this in itself is scarcely surprising or shocking, it is comforting to a specialist in arid western North American prehistory to see what kinds of information can be extracted from the environs of the colder and wetter Northeast.

It should be stressed on a more serious note that the Meadowcroft project is not now, nor has ever been viewed as *the* model or "guide" to multidisciplinary closed-site investigation projects and how they should be done elsewhere in the Northeast or indeed anywhere else. While great success has been enjoyed in applying the Meadowcroft/Cross Creek methods to projects in Kentucky (Adovasio et al. 1982; Carlisle 1978, 1982; Fitzgibbons et al. 1977; Vento et al. 1980), West Virginia (Applegarth et al. 1978; Johnson et al. 1980; Fitzgibbons et al. 1980), Mississippi (Vento et al. 2000; Yedlowski and Adovasio 1996), and now ongoing projects in Texas there are, of course, many other ways to go about the same thing. As stated at the beginning of this paper and implied in its title, this is one view of closed-site research—no more-and, it is hoped, no less.

Acknowledgements. The excavations at Meadowcroft Rockshelter and the survey and excavations in the Cross Creek drainage were conducted under the auspices of the Archaeological Research Program (subsequently the Cultural Resource Management Program) of the Department of Anthropology, University of Pittsburgh and, after 1990, the Mercyhurst Archaeological Institute, Mercyhurst College. Very generous financial and logistic support for the excavations and attendant analyses was provided by the University of Pittsburgh, the Meadowcroft Foundation, the Aloca Foundation, the National Geographic Society, the Buhl Foundation, the Leon Falk Family Trust, the national Science Foundation, the Soil Conservation Service via contract, Mercyhurst College, and Messrs. John and Edward Boyle of Oil City, Pennsylvania.

REFERENCES

ADOVASIO, J.M.; R.C. CARLISLE, W.C. JOHNSON; P.T. FITZGIBBONS; J. D. APPLEGARTH; J. DONAHUE; R. DRENNAN and J.L. YEDLOWSKI (1982). *The Prehistory of the Paintsville Reservoir, Johnson and Morgan Counties, Kentucky.* Ethnology Monographs No 6. University of Pittsburgh Press, Pittsburgh.

ADOVASIO, J. M.; J. DONAHUE and R. STUCKENRATH (1990). The Meadowcroft Rockshelter Radiocarbon Chronology 1975–1990. *American Antiquity* 55:348–354.

ADOVASIO, J. M.; J. DONAHUE and R. STUCKENRATH (1992). Never Say Never Again: Some Thoughts on Could Haves and Might Have Been. *American Antiquity* 57:327–331.

ADOVASIO, J.M.; J. DONAHUE; R. STUCKENRATH and R.C. CARLISLE (1989). The Meadowcroft Radiocarbon Chronology 1975–1989: Some Ruminations. Paper presented at the First World Summit Conference on the Peopling of the Americas, University of Maine, Orono.

ADOVASIO, J. M.; J. DONAHUE; R. STUCKENRATH and J.D. GUNN (1981). A Response to Dincauze. *Quarterly Review of Archaeology* 2(3):14–15.

ADOVASIO, J. M.; R. FRYMAN; A.G. QUINN and D.R. PEDLER (2003). The Appearance of Cultigens and the Early and Middle Woodland Periods in Southwestern Pennsylvania. In *Foragers and Farmers of the Early and Middle Woodland Periods in Pennsylvania,* edited by P.A. RABER and V.L. COWIN, pp. 67–83. Recent Research in Pennsylvania Archaeology No. 3. Pennsylvania Historical and Museum Commission, Harrisburg.

ADOVASIO, J. M.; J. D. GUNN; J. DONAHUE and R. STUCKENRATH (1975)/ Excavations at Meadowcroft Rockshelter. 1973–1974: A Progress Report. *Pennsylvania Archaeologist* 45(3):1–30.

ADOVASIO, J. M.; J. D. GUNN; J. DONAHUE and R. STUCKENRATH (1977a). Meadowcroft Rockshelter: Retrospect 1976. *Pennsylvania Archaeologist* 47(2–3):1–93.

ADOVASIO, J. M.; J. D. GUNN; J. DONAHUE and R. STUCKENRATH (1977b). Progress Report on the Meadowcroft Rockshelter–A 16.000 Year Chronicle. In *Amerinds and Their Paleoenvironments in Northeastern North America*, edited by W. S. Newman and B. Salwen. pp. 37–159, Annals of the New York Academy of Sciences 228.

ADOVASIO, J. M.; J. D. GUNN; J. DONAHUE and R. STUCKENRATH (1978). Meadowcroft Rockshelter. 1977: An Overview. *American Antiquity* 43(4):632–651.

ADOVASIO, J.M.; J.D. GUNN; J. DONAHUE; R. STUCKENRATH; J. GUILDAY and K. LORD (1978) Meadowcroft Rockshelter. In *Early Man in America from a Circum. Pacific Perspective*, edited by A.L. BRYAN. pp. 140–180. Occasional Papers No. 1. Department of Anthropology, University of Alberta, Edmonton.

ADOVASIO, J.M.; J.D. GUNN; J. DONAHUE; R. STUCKENRATH; J. GUILDAY and K. LORD (1979–1980a). Meadowcroft Rockshelter— Retrospect 1977 (Part 1). *North American Archaeologist* 1(1):3–44.

ADOVASIO, J.M.; J.D. GUNN; J. DONAHUE; R. STUCKENRATH; J. GUILDAY and K. LORD (1979–1980a). Meadowcroft Rockshelter—

Retrospect 1977 (Part 2). *North American Archaeologist* 1(2):99–137.

ADOVASIO, J.M.; J.D. GUNN; J. DONAHUE; R. STUCKENRATH; J.E. GUILDAY and K. VOLMAN (1980). Yes Virginia, It Really Is That Old: A Reply to Haynes and Mead. *American Antiquity* 45(3):588–95.

ADOVASIO, J.M. and W.C. JOHNSON (1981). The Appearance of Cultigens in the Upper Ohio Valley: A View from Meadowcroft Rockshelter. *Pennsylvania Archaeologist* 51 (1–2):63–80.

ADOVASIO, J.M. and D.R. PEDLER (1996). Pioneer Populations in the New World: The View from Meadowcroft Rockshelter. Paper presented at the XIII International Congress of Prehistoric and Protohistoric Sciences, Forlì, Italy.

ADOVASIO, J.M. and D.R. PEDLER (2000). A Long View of Deep Time at Meadowcroft Rockshelter. Paper presented at the 65th Annual Meeting of the Society for American Archaeology, Philadelphia, Pennsylvania.

ADOVASIO, J.M. and D.R. PEDLER (2003). Pre-Clovis Sites and their Implications for Human Occupation Before the Last Glacial Maximum. In *Entering America: Northeast Asia and Beringia before the Last Glacial Maximum*, edited by D. B. MADSEN, pp. 139–158. University of Utah Press, Salt Lake City.

ADOVASIO, J.M.; D.R. PEDLER; J. DONAHUE and R. STUCKENRATH (1999). No Vestige of a Beginning nor Prospect for an End: Two Decades of Debate on Meadowcroft Rockshelter. *In* Ice Age People of North America, R. BONNICHSEN and K. L. TURNMIRE, eds. pp. 416–431. Oregon State University Press, Corvallis.

CARLISLE, R.C. (1978). *An Architectural Study of Some Log Structures in the Area of The Yatesville Lake Dam, Lawrence County. Kentucky.* A Report Submitted to the Department of the Army Huntington District, Corps of Engineers, Huntington, West Virginia under Contract Number DACW-69-78-M-0033.

CARLISLE, R.C. (1982). *An Architectural Study of Some Log Structures in the Area of the Paintsville Lake Dam, Johnson County, Kentucky.* A Report Submitted to the Department of the Army Huntington District Corps of Engineers, Huntington, West Virginia under Contract Number DACW 69-77-C-0132.

CARLISLE, R.C. and J. M. ADOVASIO (editors) (1982). *Meadowcroft: Collected Papers on the Archaeology of Meadowcroft Rockshelter and the Cross Creek Drainage.* Prepared for the Symposium "The Meadowcroft Rockshelter Rolling Thunder Review: Last Act." Forty-Seventh Annual Meeting of the Society for American Archaeology, Minneapolis, Minnesota April 14–17, 1982.

DONAHUE, J. and J.M. ADOVASIO (1990). Evolution of Sandstone Rockshelters in Eastern North America: A Geoarchaeological Perspective. In *Archaeological Geology of North America,* edited by N. P. LASCA and J. DONAHUE, pp. 231–251. Centennial Special Volume 4. Geological Society of America, Boulder, Colorado.

FITZGIBBONS, P.T.; J. M. ADOVASIO and J. DONAHUE (1977). Excavations at Sparks Rockshelter (l5JO19) Johnson County, Kentucky, *Pennsylvania Archaeologist* 47(5) (Special Issue).

FITZGIBBONS, P.T.; J.M. ADOVASIO; W.C. JOHNSON; F.J. VENTO and R.C. CARLISLE (1980). *Surficial Archeology and Aboriginal Lithic Technology of the Burnsville Reservoir, Braxton County, West Virginia.* A Report Prepared by the Cultural Resource Management Program, University of Pittsburgh Under the Supervision of J.M. ADOVASIO, Ph.D. and Submitted to the U.S. Army Corps of Engineers Huntington District Office, West Virginia in Fulfillment of Contract Number DACW-69-79-M-0058.

GOLDBERG, PAUL and TRINA L. ARPIN (1999). Micromorphological Analysis of Sediments from Meadowcroft Rockshelter, Pennsylvania: Implications for Radiocarbon Dating. *Journal of Field Archaeology* 26(3):325–342.

JOHNSON, W.C.; J.M. ADOVASIO and J.P. MARWITT (1980). Fort Ancient on the Frontier: A View from Bluestone Lake West Virginia. Paper Presented at the 45th Annual Meeting of the Society for American Archaeology, Philadelphia.

THOMAS, D.H. (1974). *Archaeology*. Holt. Rinehart and Winston, New York.

VENTO. F. J.; J.M. ADOVASIO and J. DONAHUE (1980). *Excavations at Dameron Rockshelter (15JO23A) Johnson County, Kentucky.* Ethnology Monographs 4.

VENTO, F.J.; J.L. YEDLOWSKI and O. OLANIYAN (2000). *Geoarchaeology*. Prehistory of the Bay Springs Rockshelters, vol. 2. Mercyhurst Archaeological Institute Reports of Investigations Number 1.

YEDLOWSKI, J.L. and J. M. ADOVASIO (1996). *Environment and Excavations*. Prehistory of the Bay Springs Rockshelters, vol. 1. Mercyhurst Archaeological Institute Reports of Investigations Number 1.

THE MADNESS BEHIND THE METHOD: INTERDISCIPLINARY ROCKSHELTER RESEARCH IN THE NORTHEASTERN UNITED STATES

LA DÉMENCE (LA FOLIE) DERRIÈRE LA MÉTHODE : L'INTERDISCIPLINARITÉ DE LA RECHERCHE DES ABRIS ROCHEUX AU NORT-EST DES ÈTATS-UNIS

Jonathan A. BURNS, John S. WAH, and Robert E. KRUCHOSKI

Jonathan A. Burns, M.A., AXIS Research, Inc., Box 393, James Creek, Pennsylvania, U.S.A.; Burns - jburns@axisresearchinc.org; Wah - jswah@axisresearchinc.org; Kruchoski - rkruchoski@axisresearchinc.org

Abstract. The sandstone rockshelters of the northeastern United States are valuable resources for archaeologists seeking contextual and behavioral clues about the recurrent use of sheltered spaces by prehistoric humans over the past 14,000 years. Such clues must be carefully recovered from the complex sediment deposits that have accumulated beneath rock overhangs and on talus slopes. After excavation, the full cultural implications of the evidence are not readily apparent without further detailed analyses. Researchers are faced with the realities of the inherent limitations of archaeological data, decisions about the appropriate methods of spatial analyses, and the challenge of relating their data to past human behavior. A major research concern is linking behavioral theory to the interpretation of the patterned traces of prehistoric hunter-gatherer occupations. Towards this goal, fine-grained excavation data and analyses from recently investigated upland rockshelters in Pennsylvania serve to: 1) illustrate methods to identify natural and cultural formation processes, 2) highlight the idiosyncrasies of rockshelter formation under regional climatic conditions and related excavation requirements, and 3) situate the archaeological interpretation of these sites in the larger context of global hunter-gatherer prehistory.

Keywords: Pennsylvania rockshelters; behavioral patterning; particle-size analysis.

Résumé. Les abris rocheux de grès du Nord-est des États-Unis sont des ressources précieuses pour les archéologues qui sont à la recherche d'indices concernant le contexte et les comportements au sein desquels ces abris rocheux ont été utilisés par les hommes préhistoriques durant les 14000 dernières années. De tels indices doivent être recueillis, avec soin, des dépôts sédimentaires qui se sont accumulés dans les saillies rocheuses et sur les pentes (des talus, des coteaux). Mais une fois les fouilles effectuées, l'ensemble des implications culturelles de ces preuves ne sont pas évidentes/ne se donnent pas à voir sans analyses plus détaillées/ poussées. Les chercheurs sont alors confrontés à la réalité des limites inhérents aux données archéologiques, et au fait de devoir prendre des décisions concernant l'utilisation pertinente des méthodes d'analyses d'un espace donné, comme ils sont confrontés au défi consistant à donner du sens à ces données en restituant les comportements humains passés/ à mettre en relation ces données avec le comportement humain passé. L'un des soucis majeurs de la recherche est de lier la théorie comportementale à l'interprétation des traces typiques/ des traces structurelles laissées par les occupations de chasse et de cueillette des hommes préhistoriques. Afin d'atteindre ce but, les relevés et analyses des fouilles denses/méticuleuses réalisées récemment dans les hauteurs/sur es coteaux de Pennsylvanie servent à : 1) illustrer les méthodes permettant d'identifier les processus de formation naturelle et culturelles, 2) mettre en lumière les particularités de la formation des abris rocheux soumis à certaines conditions climatiques propres à la région et liées aux exigences qu'imposent les fouilles archéologiques, 3) situer l'interprétation de ces sites dans un contexte plus large des activités préhistoriques de chasse et de cueillette.

Mots-clés: abris rocheux au Pennsylvanie; modeler comportemental; l'analyse de taille de particule.

There are few better places for archaeologists to study prehistoric cultural adaptation than deeply stratified rockshelters. As framed stages from which successive groups of prehistoric hunter-gatherers based their subsistence and settlement decisions, rockshelters provide a view of past hunter-gatherer behavior as influenced by very tangible dimensions of space and place. As it is desirable to recover information on the paleoenvironmental context of human occupations—rockshelters are unique in their record of past environments with respect to preservation and process. The challenge of rockshelter archaeology, to relate patterns in spatial data to human behavior from complex palimpsest deposits, can at times be maddening to the researchers who excavate them. Many questions have yet to be answered regarding rockshelters and their past roles in prehistoric human settlement systems.

Rockshelters share some fundamental elements that structure the behaviors of human hunter-gatherers on a global level. These are: 1) the sheltered space, 2) a back wall, 3) a dripline, and 4) a talus slope. The spatial structure of multiple occupations in relation to these physical elements is fundamental to our archaeological interpretations. The archaeological record contained within these sediment traps can be compared on the basis of archaeological spatial patterning around the world; but in order for this to become a reality, excavations must be structured towards this end. This research contributes to the global picture of prehistoric rockshelter habitation by highlighting the archaeology of the sandstone rockshelters of Pennsylvania's Ridge and Valley Province, located in the Middle-Atlantic United States (**Figure 24.1**).

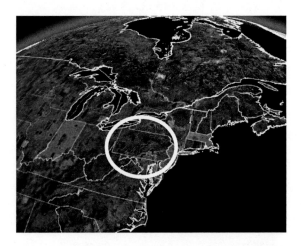

24.1 Geographic location of the State of Pennsylvania in Northeastern North America, circled in white (as modified from Google Earth).

24.2 Locations in the Ridge and Valley Province of the two rockshelter case studies presented in this report, the Mykut and Camelback Rockshelters.

GEOGRAPHIC OVERVIEW

The Ridge and Valley Province is aligned diagonally from the southwest towards the northeast across the center of the state of Pennsylvania (**Figure 24.2**). This physiographic province is a result of differential erosion of folded and faulted rock strata. The region consists of alternating resistant and non-resistant sedimentary rocks which have been laterally compressed during the Paleozoic Era by plate tectonic forces into a series of anticlinal and synclinal folds. It is characterized by long parallel sandstone ridges separated by limestone valleys. Sporadic water and wind gaps are the only breaks in the steep-sided ridges. Located between 39° and 42° latitude, the climate is humid-continental, receiving an average of 97 cm of precipitation per year.

Habitable rockshelters form on upland slopes at contacts between erosion resistant sandstone and softer shale and siltstone. The upland environs are marked by rugged topography and high relief which influenced the mobility and subsistence options available to the prehistoric inhabitants of the region. Additionally, the varied environmental zones associated with differences in elevation and aspect provided a corresponding variety of plant and animal resources within a short distance up or down slope.

ROCKSHELTER ARCHAEOLOGY IN PENNSYLVANIA

The incorporation of rockshelter occupations into regional settlement patterns has been varied across the Middle Atlantic and Northeast regions of North America. Rockshelters comprise a small percentage of excavated sites but have made a disproportionably large contribution to the development of regional cultural frameworks by providing well-preserved stratified occupational debris. Regional manifestations in lithic tool traditions and cultural horizons have been well-dated using associated charcoal and botanical remains from rockshelters.

The earliest documented archaeological investigations of Pennsylvania rockshelters were published in the first half of the twentieth century (e.g., Augustine 1938, 1940; Butler 1947; Clausen 1932; Schrabisch 1926, 1930). During this pioneering period, Max Schrabisch surveyed and trenched rockshelters in the Delaware and Susquehanna drainages searching for chronological cultural trends and distinguishable horizons. By today's standards, his notes were quite general and his field methodology was coarser grained. However, these early investigations fueled the interests of generations of Pennsylvania archaeologists who followed. Throughout the latter half of the twentieth century, although still geared mainly towards issues of culture history, rockshelter excavation in Pennsylvania progressively became more controlled as excavation techniques evolved with a greater appreciation for contextual evidence in interpreting stratigraphy.

Salvage excavations were conducted in Huntingdon County during the 1960's in conjunction with the Raystown reservoir project, granting a glance of several sandstone rockshelters in the impact area (Smith 1966). The most notable of these was Sheep Rock Shelter (Michels and Dutt 1968, Michels and Smith 1967). The archaeology of Sheep Rock Shelter was used to build culture chronologies of central Pennsylvania and beyond. Perishable items like a bark basket, cordage, bone implements, arrow shafts, fire drills, and botanical

24.3 Excavations at Sheep Rock Shelter carried out by Pennsylvania State University and Juniata College Anthropology Departments ca. 1966 (Photo by Jim Filson).

24.4 Excavations at Mykut Rockshelter in Huntingdon County Pennsylvania (1999).

specimens were abundant. The deposits of the rockshelter were over 3.5 meters deep bearing evidence of heavy use by prehistoric cultures throughout the Holocene (**Figure 24.3**).

With chronological control, some generalizations and trends became apparent to researchers. In general, the occupational intensity of rockshelters seems to have increased throughout the Archaic Period and into the Woodland Period; partially due to increased population density—but probably more to do with the organizational changes that accompanied seasonal reoccupation. For hunter-gatherers, rockshelters become part of a running emic mental map of *places* on the landscape associated with elements of shelter and resource procurement that is handed down through successive generations. Rockshelters of the Ridge and Valley were utilized heavily during the Late Archaic. These occupations are traditionally classified by archaeologists as seasonal hunting camps. In 1972, William Turnbaugh (1975) excavated a rockshelter site, 36LY160, along Lycoming Creek in a series of shelters and ledges in the same area coinciding with the geological contact between the Catskill and Pocono Formations. The Late Archaic Lamoka Culture was strongly represented by a tool inventory including projectile points, net sinkers, and adzes. As with many rockshelters of Pennsylvania's Ridge and Valley Province, the site also bore later occupations by ceramic using cultures. Considering the additional organizational changes in settlement and subsistence patterns that came with the increased reliance on agriculture and village life during the Middle and Late Woodland Periods, rockshelters remained important places on the landscape—still heavily utilized during hunting and other wild resource procurement activities.

Despite concerns of culture history and stratigraphy, there was not much behaviorally informative spatial data being recovered using conventional archaeological methods. Still, other rockshelters in Pennsylvania were sporadically "dug" by amateur archaeologists with no research agenda, simply in search of diagnostic artifacts. Until this point in time, 1 m^2 and 5 ft^2 excavation units had been the norm for formal data recovery at most rockshelters in North America. But, just as our understanding of the nature of the archaeological record has evolved, so have our data recovery techniques.

Outside the Ridge and Valley Province, on the western edge of Pennsylvania, detailed multidisciplinary investigations began in the late 1970's at another sandstone rockshelter called Meadowcroft. This rockshelter gained international attention as one of the first of a growing number of archaeological sites in the Americas that push back the timing of the first migrations of humans to the New World to pre-14,000 B. P. The excavations at Meadowcroft stand to this day as the quintessential fully-funded multi-disciplinary rockshelter investigation (see Adovasio, this volume).

MYKUT ROCKSHELTER

During the late 1990's extensive excavations began at Mykut Rockshelter (**Figure 24.4**)—not far from Sheep

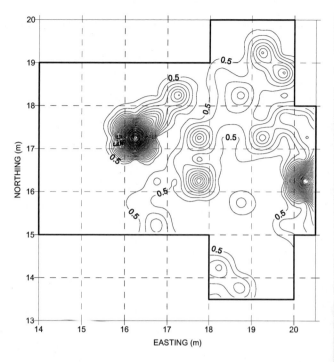

24.5 Sample distributional contour map showing bone distribution by weight for Level 3 of the excavation block at Mykut Rockshelter (contour interval = 0.5g).

Rock. Targeted in conjunction with highway improvements, this sandstone shelter was approached with the intention of recovering detailed spatial data to facilitate future comparative analyses of hunter-gatherer spatial organization (Raber et al. 2005). The Mykut site is situated on southwest-facing slopes at the northern end of Little Valley in Huntingdon County, at the geological contact between the Mississippian-age Mauch Chunk and Pocono formations. More resistant sandstones form the rock overhang, with softer ferruginous siltstones eroded from beneath. The site occupies a key location with respect to routes of local and regional movement, situated at a gap in Terrace Mountain. This gap marks the only break in this northeast-southwest-trending ridge for over 30 km in either direction. Access to the riverine resources of the Raystown Branch of the Juniata River, and to other sites like Sheep Rock Shelter, would have been via this gap.

Approximately 75% of the estimated original site area (about 41 m^2) was excavated in 50 x 50 cm quadrants by 5 cm levels within natural or cultural strata. Point proveniences were recorded for tools and larger artifacts, with the remaining artifacts collected as quadrant lots. Provenience control and recording was maintained during the investigations with a laser total station. Contour density maps (**Figure 24.5**) were plotted for small items like bone fragments and fine debitage for every 5 cm level.

These efforts recovered over 16,000 artifacts, 18,000 faunal specimens, and 23 prehistoric hearth features. Occupations spanning at least 5500 years, from the Late

24.6 Camelback Rockshelter (2004).

Archaic (or earlier) through the Late Woodland period, were evident. The mass of evidence suggests that the basic nature of human activities at the site remained largely unchanged throughout that time span. The processing and consumption of deer was a major focus of activity at the Mykut Rockshelter throughout its occupational history. The methods used at this site served as a methodological catalyst inspiring other detailed excavations of open-air sites in the area with a focus on spatial structure. Now that several sites have been excavated using the same collection methods—archaeologists in this region are beginning to make better use of comparative analytic techniques.

CAMELBACK ROCKSHELTER

Refined excavation methods based on those from Mykut Rockshelter are currently being employed at Camelback Rockshelter (see Burns 2005a), a rare upland location containing a complete Holocene cultural sequence (**Figure 24.6**). Contrasted with the many floodplain sites of the Delaware River, this location offers a rare glimpse of prehistoric use of higher elevation landforms throughout the Holocene. Starting in 2001, detailed excavation methods were employed to capture and discern spatial patterning resulting from behavior as well as natural processes. Over 2000 artifacts have been recovered to date from the 8 m^2 excavation block. Radiocarbon dates on a subset of the charcoal samples are

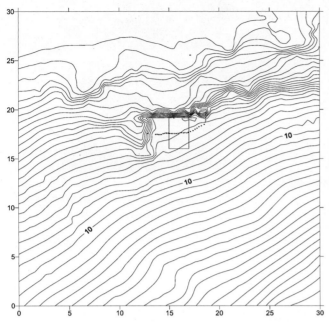

24.7 Surface contour map of Camelback Rockshelter and immediate vicinity, showing the excavation block and the dripline (contour interval = 0.25 m).

forthcoming and will add another independent line of contextual evidence.

The site is situated within the boundary of Big Pocono State Park on the southern slope of Camelback Mountain (part of the Pocono Plateau Escarpment) at an elevation of 445 m A.S.L. (**Figure 24.7**). Camelback Mountain is the most prominent feature along the escarpment with over 300 m of relief. The rockshelter lies 140 m above Pocono Creek 2.2 km to the east, and commands a view of the Delaware Water Gap and Wind Gap. The shelter is formed by an overhang of crossbedded Catskill Formation sandstone above less resistant shale.

Eight contiguous meter units have yielded 31 projectile points representative of the local cultural chronology throughout the Holocene. The depth of Holocene sediments at Camelback Rockshelter is just over 1 m within the dripline—a relatively deep sequence for an upland location in this region. The bulk of the sediments are from roof-fall and trapped local alluvium—sediments moved by overland flow of water.

Excavation at Camelback Rockshelter proceeded by removing collection units measuring 50 cm^2 by 5 cm arbitrary levels within strata, mirroring the slope and contours of distinct sediment horizons. Measurements were controlled using a laser total station (see Dibble 1987; McPherron and Dibble 2002). Artifacts and ecofacts smaller than 1 cm were collected during screening through 3 mm hardware cloth and given a central lot provenience. Artifacts and specimens measuring greater than 1 cm were point-provenienced. An eighth unit is currently being excavated from which all excavated sediments are transported to the laboratory for complete recovery of microdebitage and organic remains.

While protocol for data recovery and spatial control remained largely consistent throughout work at Camelback Rockshelter, some new innovations were adopted. Specifically, data collection techniques were tailored to account for the tabular sandstone slabs encountered as rock fall. These slabs of varying sizes comprise a substantial proportion of the total sediments to be dealt with during excavation at sandstone rockshelters. Before rock fall slabs are removed during excavation, eight measurements are recorded: 1) the Provenience Point; 2) Long Axis Orientation; 3) Long Axis Dip; 4) Long Axis Length; 5) Short Axis Orientation; 6) Short Axis Dip; 7) Short Axis Length; and 8) Thickness. These data can be used to generate proportional plan view and profile maps (see Burns 2005b).

SEDIMENT-SIZE ANALYSIS

Continuous sediment columns were removed in 5 cm levels for sediment size analysis, phytolith studies, and flotation for macrobotanical remains. In the laboratory, we address issues of shelter evolution and the forces and sources of sedimentation operating on the rockshelter deposits. This aids in distinguishing natural from cultural processes and interpreting the degree of behavioral resolution possible. Sediment-size analysis was performed using the pipette method (see Timpson and Foss 1992).

There are two major issues being examined throughout the soils and sediments at Camelback: 1) depositional history of the sediments including sources and relative rates, and 2) post-depositional alteration of sediments, or pedogenesis—an important aspect at this site. This is because soil formation, especially the translocation of clay, tends to mask some of the depositional characteristics of the sediments.

From the sediment-size analysis, slight shifts with depth become evident. The point of sampling in 5 cm increments within horizons is that none of this is visually apparent in the profile while in the field. These shifts in sediment size do not appear to represent long depositional hiati, but may be linked to paleoclimatic fluctuations. Overall, the sediments at Camelback seem to have been deposited constantly and steadily throughout the Holocene. Falling just within the extent of the last glaciation, the shelter formed with the retreat of the ice margin during the Late Pleistocene and continued to fill and evolve.

Figure 24.8 shows the particle size distribution with depth and demonstrates clearly that pedogenesis has been taking place. There is a 6% clay increase in the Bt1-horizon (17-47 cm) and then clay drops off in the less weathered BC-horizon. Evidence of clay illuviation (clay skins or argillans)—necessary to call it an argillic horizon

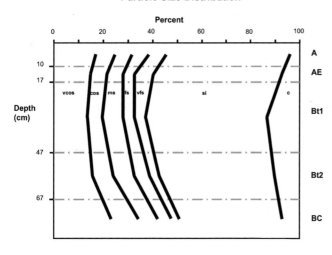

24.8 Schematic particle-size distribution with depth for Camelback Rockshelter.

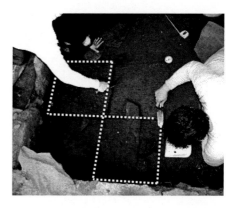

24.9 Excavation units measuring 50cm2 aimed at recovering behaviorally informative spatial data.

is present. Pedogenesis is important because it demonstrates that Camelback is not a totally dry rockshelter but this is typical of the many small rockshelters throughout the region. There is enough water introduced vertically through the profile to move clay particles. This can be advantageous because it also gives us a way to estimate the amount of time that the sediments have been in place and subject to pedogenesis. We estimate the age of the Camelback sediments at a depth of 75 cm to be at least 6,000 years old. Forthcoming radiocarbon dates will corroborate or refute these observations.

SCALE IS EVERYTHING

If we are to separate natural and cultural formation factors, fine-grained collection units are necessary. Crucial to our investigations is an understanding of spatial factors that serve as baselines from which to interpret behavior from the archaeological record. Scale is everything in searching for behaviorally meaningful patterns in archaeological data. It should stand to reason that the overall average size of the human body is the main conditioner of these patterns and sets the minimum analytic unit size capable of detecting behavioral clues.

Of all the metric excavation unit sizes that have been put to use in American Archaeology (i.e., 1 m, 0.75 m, 0.5 m, 0.25 m), ethnoarchaeology tells us we need to be using at most 50 cm analytic units for site structural comparative analyses for all hunter-gatherer sites, especially short-duration occupations (e.g., Jones 1993; Sullivan 1992). Size-sensitive piece-plotting and 50 cm collection lots at 5 cm levels, following the natural contour of the strata (**Figure 24.9**), are arguably the best compromise for upland rockshelters in this region. In addition, 5 cm levels within strata are compatible with our collection techniques for sediment-size analysis. To better understand a rockshelter's depositional history it is important that archaeological levels be directly correlated with the same units of observation used by soil science.

Intrasite patterns may be tracked to search for differences or similarities in the way different groups adapted to each rockshelter through time. An informative analytic pursuit is to graphically display the artifact densities in relation to *in situ* artifacts and cultural features, as well as the features of the sheltered space and associated landscape elements. The larger implications of this approach facilitate comparative analyses between rockshelters and contribute to understanding the global variability and material consequences of prehistoric human adaptation to the constrained spaces of these natural features.

The complex information in the databases that result from these collection techniques is unintelligible to researchers without maps to convey the coded data. The visual inspection of spatial data is a crucial first step in searching for the diagnostic signs of human activities and post-depositional processes. Two types of distributional data provide the building blocks of these procedures: *aggregate lot frequencies* and *piece-plots*. Aggregate lot frequencies can be translated to contour maps with line density customized to suit the nature of specimen class data. Piece-plots are point-provenienced items consisting of an XYZ coordinate linked in a relational database with other information such as orientation, physical condition, and metric measurements. These serve as contextual clues during exploratory spatial analyses. Both data types serve different yet complimentary purposes, and when used together, result in a flexible database for investigating and interpreting independent lines of contextual evidence.

To represent the horizontal distributions of small-sized debris, classed aggregate lot frequencies are displayed for every 5 cm level of the excavation block. Surfer 7, a three dimensional mapping and contouring program by Golden Software, is used to produce graphic displays of the data. Once all the level overlays are built from the aggregate lot frequencies, classed constellations of piece-plotted items can be overlaid or analyzed independently (**Figure 24.10**). Visual inspection allows researchers to focus their attention on specific research questions and

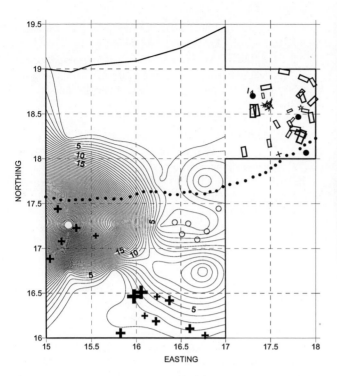

24.10 Combined sample plan view plots from Camelback Rockshelter referred to in the text.

24.11 Cache of lithic bifaces from Camelback rockshelter.

helps choose appropriate analytic techniques for subsequent statistics and tests of association. Upon total area excavation of a rockshelter, the data can then be entered into a GIS universe wherein spatial statistics can be performed to quantify elements of site structure at various scales of resolution.

ROCKSHELTERS AND HUMAN BEHAVIOR

Rockshelters have yet to be well integrated into prehistoric settlement studies in Northeastern North America. Part of this stems from the fact that few site-structural studies have been undertaken in this region. Past behaviors must be confirmed through spatial analysis and the development of pattern recognition techniques.

Rockshelters are places where we can investigate certain prehistoric human behaviors that were carried out in these particular contexts. Caching behavior is a good example. Rockshelters that were used as bivouacs, temporary logistical camps, or seasonal camps have the propensity to become cache spots for items that are used in the extraction of particular resources. Caching is a way to reduce stress, to smooth out incongruence between human groups and resources, to make the environment more predictable, and to save time. Two caches have been recovered at Camelback rockshelter, a cache of seven pecked bola stones made from sandstone cobbles and a cache of 25 lithic bifaces (**Figure 24.11**). The bola stones are specialized gear cached at the dripline by Late Woodland Period horticulturists, presumably for use during hunting forays on the mountain and nearby upland swamps (indicated by the open circles in Figure 10). The bifaces were found in a tight cluster in levels 4 through 6—as if deliberately placed in a pit or depression (N18.6 E17.45 of Figure 10). Just outside the cache, a rejected Lackawaxen style projectile point with a broken tip completes the picture of the intended shape of the standardized performs. The Lackawaxen point style (see Kinsey 1972) is well defined in the Upper Delaware Valley and has been dated from other sites in the area to about 5400 BP. The lithic raw material was transported at least 20 km from sources to the east near the Delaware River. Their presence at Camelback rockshelter is suggestive of a shift from the high residential mobility exhibited by earlier Archaic Period cultures to an increased reliance on logistics and seasonal occupation of fixed points on the landscape.

Another behavior frequently represented at rockshelter sites is the retooling of lithic projectile points (see Keeley 1982). Rather than the context of projectile point use, rockshelters are the settings where broken and worn out projectiles are removed from their shafts and replaced with fresh ones. The condition of the projectile points recovered from both Mykut and Camelback Rockshelters supports this interpretation. Use-life history data recorded for all diagnostic projectile points from these sites show two states of condition: 1) replaced points with heavy re-sharpening and distal impact fractures, and 2) rejected fresh points intended for hafting that exhibit characteristic manufacture errors.

Bone processing for the consumption of marrow or rendering of grease is yet another behavior evident at rockshelter sites. This human habit is represented by highly fragmented bones exhibiting spiral fracturing, anvil cracking, and burning. Considering the bone distribution at Camelback Rockshelter (indicated by the contour lines and the "+" signs in Figure 11), it is evident that bone processing activities were centered near the dripline with larger bone fragments dispersed down the talus slope. This patterning suggests that the prehistoric inhabitants were keeping the waste associated with these activities outside the sheltered space and were tossing the larger debris down slope, resulting in a size-sorting effect that is detectable in the archaeological record (Wandsnider 1996).

CONCLUSIONS

In conclusion, by studying spatial patterns that repeat themselves through time within and among rockshelters, it should be possible to: 1) demonstrate relationships between the natural features of the rockshelter and observed patterns of artifact distributions; 2) detect organizational changes in prehistoric subsistence and mobility behavior; and 3) track changes or stability in human use of space over time. Detailed excavation and comparative spatial analyses coupled with other crucial avenues of inquiry like soil science and geomorphology is the "prescription" for a more universal understanding of the human use of rockshelters. As for the "madness behind the method"—symposium therapy is highly recommended.

Acknowledgements. AXIS (*Archaeological Excavation & Interdisciplinary Science*) Research, Inc. would like to thank Marcel Kornfeld for his encouragement, discussions of methodology, and for inviting us to participate in this forum. Also, thanks to Marcel, Laura Miotti, and Sergey Vasil'ev for organizing this important session and for facilitating an international sharing of perspectives and ideas between rockshelter researchers. Thank you to all the participants who traveled to Lisbon to report their findings. We are grateful to Marie Bihan for translating our abstract into French. And finally, thanks to all those who donated their time and effort to the Camelback Rockshelter project. AXIS Research, Inc. is a registered Non-Profit Public Charity (501 C3) formed to conduct fieldwork and research that addresses past human behavior and interaction with the environment through archaeological investigation, landscape reconstruction, and the examination of paleoclimate and climate change, while fostering the education and practical experience of students of related scientific disciplines. www.axisresearchinc.org

REFERENCES

AUGUSTINE, E. E. (1940) Fort Hill. *Pennsylvania Archaeologist* 10:3, p. 51-58.

AUGUSTINE, E. E. (1938) Important Research on Peck and Martz Rock Shelter Site in Somerset County. *Pennsylvania Archaeologist* 8:4, p. 83-88.

BURNS, J. A. (2005a) What About Behavior?: Methodological Implications for Rockshelter Excavations and Spatial Analysis. *North American Archaeologist* 26:3, p. 267-282.

BURNS, J. A. (2005b) Methods for Upland Rockshelter Excavation: An Analytic Manifesto. In NASH, C. L.; BARBER, M. B., eds. *Uplands Archaeology in the East Symposia VIII and IV*. USA: Archaeological Society of Virginia, p. 313-326. (Special Publication 38-7).

BUTLER, M. (1947) Two Lenape Rockshelters near Philadelphia. *American Antiquity* 12:4, p. 246-245.

CLAUSEN, C. (1932) The Wolves' Den Shelter. *Pennsylvania Archaeologist* 3:2, p. 7-9, 19.

DIBBLE, H. L. (1987) Measurement of Artifact Provenience with an Electronic Theodolite. *Journal of Field Archaeology* 14, p. 249-254.

JONES, K. T. (1993) The Archaeological Structure of a Short-Term Camp. In HUDSON, J., ed. - *From Bones to Behavior: Ethnoarchaeological and Experimental Contributions to the Interpretation of Faunal Remains*. Carbondale: Southern Illinois University, p. 101-114. (Center for Archaeological Investigations, Occasional Papers No. 21).

KEELEY, L. H. (1982) Hafting and Retooling: Effects on the Archaeological Record. *American Antiquity* 47, p. 798-809.

KINSEY, W. F., III (1972) *Archaeology in the Upper Delaware Valley*. Harrisburg: The Pennsylvania Historical and Museum Commission, p. 499. (Anthropological Series No. 2).

MCPHERRON, S. P.; DIBBLE, H. L. (2002) *Using Computers in Archaeology: A Practical Guide*. New York: McGraw-Hill Mayfield. 254 p.

MICHELS, J. W.; DUTT, J., eds. (1968) *Archaeological Investigations of Sheep Rock Shelter, Huntingdon County, Pennsylvania*, Volume 3. University Park: Pennsylvania State University. 505 p. (Department of Anthropology, Occasional Papers).

MICHELS, J. W.; SMITH, I. F. III, eds. (1967) *Archaeological Investigations of Sheep Rock Shelter, Huntingdon County, Pennsylvania*, Volumes 1 and 2. University Park: The Pennsylvania State University. 943 p. (Department of Anthropology, Occasional Papers).

RABER, P. A.; BURNS, J. A.; CARR, B.; MINNICHBACH, N. C.; VENTO, F. J. (2005) *Archaeological Testing and Data Recovery Investigations (Phase II and Phase III Archaeological Studies) Site 36HU143, Mykut Rockshelter, S.R. 3001, Little Valley Road, Todd Township, Huntingdon County, Pennsylvania*. Alexandria: Heberling Associates, Inc. ER No. 94-0607-061. (Report Prepared for The Pennsylvania Department of Transportation, District 9-0 and The Federal Highway Administration).

SCHRABISCH, M. (1926) Aboriginal Rock Shelters of the Wyoming Valley and Vicinity. In DORRANCE, F., ed. *Proceedings and Collections of the Wyoming Historical and Geological Society* Vol. 19, p. 47-218. Wilkes-Barré: The Wyoming Historical and Geological Society.

SCHRABISCH, M. (1930) Archaeology of Delaware River Valley: Between Hancock and Dingman's

Ferry in Wayne and Pike Counties. Harrisburg: Pennsylvania Historical Commission. 181 p.

SMITH, I. F. III (1966) Raystown Reservoir Archaeological Salvage and Survey Program. Report submitted to National Park Service, Northeast Region, NER-895.

SULLIVAN, A. P. III (1992) Investigating the Archaeological Consequences of Short-Duration Occupations. *American Antiquity* 57:1, p. 99-115.

TIMPSON, M. E.; FOSS, J. E. (1992) The Use of Particle-size Analysis as a Tool in Pedological Investigations of Archaeological Sites. In FOSS, J. E.; TIMPSON, M. E.; MORRIS, M. W., eds. - *Proceeding of the First International Conference on Pedo-Archaeology*. Knoxville: University of Tennessee Agricultural Experiment Station, p. 69-80.

TURNBAUGH, W. H. (1975) *Man, Land and Time*. Evansville: Unigraphic, Inc. 277 p.

WANDSNIDER, L. (1996) Describing and Comparing Archaeological Spatial Structures. *Journal of Archaeological Method and Theory* 3:4, p. 319-384